SOIL CONTAMINATION RESEARCH TRENDS

SOIL CONTAMINATION RESEARCH TRENDS

JAVIER B. DOMÍNGUEZ
EDITOR

Nova Science Publishers, Inc.
New York

For permission to use material from this book please contact us:
Telephone 631-231-7269; Fax 631-231-8175
Web Site: http://www.novapublishers.com

LIBRARY OF CONGRESS CATALOGING-IN-PUBLICATION DATA

Soil contamination research trends / Javier B. Domínguez, Editor.
p. cm.
ISBN 978-1-60456-319-1 (hardcover)
1. Soil pollution--Research. I. Domínguez, Javier B.
TD878.S6635 2008
628.5'5--dc22

2007052099

Published by Nova Science Publishers, Inc. ≃ New York

CONTENTS

Preface		**vii**
Chapter 1	Complementary Approaches for Using Ecotoxicity Data in Soil Pollution Evaluation: Risk Based versus Direct Toxicity Assessments *M. D. Fernandez and J. V. Tarazona*	**1**
Chapter 2	Validated Approaches to Restoring the Health of Ecosystems Affected by Soil Pollution *Ana J. Hernández and J. Pastor*	**51**
Chapter 3	The Impact of Long Term Heavy Metal Contamination on Bacterial Communities in Metallophytic Grassland Soils *Stephen P. Cummings and Lisa Bamborough*	**73**
Chapter 4	Contamination of Vineyard Soils with Fungicide-derived Copper *Michael Komárek and Vladislav Chrastný*	**109**
Chapter 5	Soil Pollution and Response of Microorganisms to Oxidative Stress *Bystrik Polek*	**119**
Chapter 6	Construction of an Efficient Bioremediation System for Petroleum Hydrocarbon-Contaminated Soils Using Specific Hydrocarbon-degrading Bacteria and Bacterial Biomass Monitoring *Kouta Hatayama, Yuri Sakihama, Yoshiki Matsumiya and Motoki Kubo*	**143**
Chapter 7	Heavy Metals in Mediterranean Soils *J. Navarro-Pedreño, I. Gómez, M. B. Almendro Candel and I. Meléndez-Pastor*	**161**
Chapter 8	The Role of Natural Self-purification Processes in In-stream on Site Clean-up Activities *Anatoliy I. Fisenko*	**177**

Chapter 9	Use of Plant Bioassays for the Detection of Genotoxic Compounds in Soils *Miroslav Mišík, Armen Nersesyan, Katarína Mišíková and Siegfried Knasmüller*	**203**
Chapter 10	Statistical Approaches to Establish Background Values of Potentially Toxic Elements *C. Micó, L. Recatalá and J. Sánchez*	**217**
Index		**235**

PREFACE

Soil contamination is the presence of man-made chemicals or other alteration in the natural soil environment. This type of contamination typically arises from the rupture of underground storage tanks, application of pesticides, percolation of contaminated surface water to subsurface strata, leaching of wastes from landfills or direct discharge of industrial wastes to the soil. The most common chemicals involved are petroleum hydrocarbons, solvents, pesticides, lead and other heavy metals. This occurrence of this phenomenon is correlated with the degree of industrialization and intensity of chemical usage. The concern over soil contamination stems primarily from health risks, both of direct contact and from secondary contamination of water supplies. Mapping of contaminated soil sites and the resulting cleanup are time consuming and expensive tasks, requiring extensive amounts of geology, hydrology, chemistry and computer modeling skills. This book presents the latest research from around the world in this field.

Chapter 1 - The assessment of soil quality has usually focused on human health protection as the main objective. Recently, criteria for the protection of ecosystems have been incorporated and ecotoxicological analyses are recommended to estimate the risk to ecological receptors associated with contaminants in soils. The ecotoxicological assessment of soils is mostly based on the toxicity test with selected organisms. Two complementary approaches are available. The first approach consists in the identification of toxicity thresholds for each relevant pollutant, thresholds that are based on the evaluation of effects of chemical substances on selected organisms representing relevant ecological receptors. The results of these assays are used for setting soil quality standards for each pollutant or pollutant class. Risk assessment tools can be used for this purpose, pre-establishing acceptable levels of risk. The contamination level is based on the comparison of the concentration of contaminants measured in the soil with the standards established from the thresholds. Although field and semi-field information can be incorporated in the higher tier steps, the thresholds are mostly developed from standardised toxicity assays conducted under laboratory conditions following international (e.g. OECD, ISO) or national (e.g. USEPA, ASTM) guidelines. In the second approach, toxicity assays are performed directly with the contaminated media (soil, water, sediment). This alternative, performing the assays with environmental samples, constitutes the method called direct (eco)toxicity assessment (DTA), and is based on modified bioassays.

Chapter 2 - The diagnosis of soil pollution in relation to the most commonly used remediation methods is an innovative way of tackling this problem. To this end, the theoretical framework of Ecosystems Health has been instrumental, since this area is presently emerging as a new language for general discourse on pollution. For example, the use of the term stressor for a pollutant implies the study of behavior responses in living organisms (at different levels of organization, from the cell to ecosystem) both in terms of impacts (toxicity) and tolerance towards the pollutant (their adaptation).

Two very common scenarios in the central Iberian Peninsula have allowed us to approach the understanding of this area of study: (i) old abandoned mines and, (ii) old solid waste (urban and industrial) landfills.

The soils of these systems are mainly polluted with heavy metals (Zn, Cu, Pb, Cd), they overlie different lithological substrates (granites, limestones and especially arkoses), and are highly representative of soils of the Mediterranean region (mainly regosols, luvisols and cambisols). The landfills of the mines or waste tips have slopes exceeding 40%, such that any leachates produced transfer pollutants to adjacent ecosystems at lower altitudes (stream bed pastures, marshes and surface water courses). Moreover, since the topsoil layer is usually sandy with low proportions of clay and organic matter, deep leachates also transport pollutants to ground water systems.

We focused on the autotrophic component of the affected ecosystems for several reasons: toxicity tests could be used to examine the physiological and behavioral responses of organisms (mortality, injury, metabolic changes) as well as population (population density, risk of extinction) or community (structure, diversity, biomass, nutrient flow changes) variables. The build-up of a heavy metal in the above-ground part of a plant (phytoaccumulation) consumed by herbivores is also detrimental for health, due to transfer to the trophic network. Root systems may play a role in phytostabilizing heavy metals and in preventing them from passing to deeper soil layers. Finally, given the erosion problems of fine materials in landfills and waste tip slopes, vegetation helps avoid the movement of topsoil layer pollutants to other ecosystems.

The methodological approaches validated by results obtained over the last twenty years can be summarized as: studies of polluted sites based on phytoecological sampling, analysis of soil chemical and physical properties, georeferencing of the heavy metals they contain for further sampling in areas showing the highest levels, collecting and chemically analyzing plants at these sites, and the use of soils with their seed banks to perform experiments in microcosms in controlled conditions. The idea is to use a combination of field and laboratory methods that simulate real scenarios in which soil pollution occurs.

Chapter 3 - Heavy metals are a significant source of pollution in soils that have been demonstrated to exert significant toxic effects on soil microbial communities. Here, the authors investigate the impact of the heavy metals lead and zinc on the bacterial community of an upland metallophytic grassland the UK, contaminated by historical lead mining activity. They focus on the total microbial diversity and that of the Actinobacteria, a significant component of the aerobic soil community that demonstrate a broad metabolic activity and represent useful indicators of soil health.

Two experimental approaches were adopted, the first employed microcosms containing soil from the site of an abandoned lead mine spoil heap, bacterial diversity was investigated using denaturing gradient gel electrophoresis (DGGE) of PCR amplified 16S rRNA DNA with universal primers to amplify the total community and more specific Actinobacteria primers. The Actinobacteria specific PCR primers showed that this taxon represented a diverse and metabolically active element of the microbial community of lead contaminated upland soil. Subsequently, the authors investigated how the stability of the bacterial communities responded to an increase in stress by adding additional Pb and Zn to the soil. There is significant interest in determining whether communities under stress are more or less robust in the face of additional environmental stress. The authors data was equivocal, the total community diversity responded negatively to additional heavy metal input into the soil, whereas, the Actinobacteria community did not show any substantial change in its diversity.

A second experiment characterised the bacterial soil community along a transect of a metallophytic grassland site that had significant differences in the Pb, Zn, organic C and N content across it. In this study the metabolically active bacterial population was studied using reverse transcription of mRNA from16S rRNA gene. Universal primers showed a low diversity of metabolically active bacteria, and both organic C and heavy metal concentrations in the soil seemed to impact on the community diversity. In contrast Actinobacteria showed higher diversities and seemed to be able to maintain this diversity even in soils with low organic C and high heavy metal concentrations suggesting that they are competitive in this stressed environment with other soil flora. The data presented here may prove useful in determining the effect of heavy metal concentrations on soil microbial diversity and community stability, particularly in managing rare metallophytic grassland sites and more widely in remediating contaminated soils.

Chapter 4 - The repeated use of fungicides (e.g., Bordeaux mixture) resulted in the accumulation of Cu in vineyard soils. Due to the fact that Cu is strongly immobilized by soil organic matter and Fe, Mn, Al-(hydr)oxides, elevated Cu concentrations originating from fungicide applications were observed in superficial horizons of vineyard soils. The concentrations in these horizons often exceed 200 mg Cu kg^{-1} (in some cases > 1000 mg Cu kg^{-1}), which surpasses all regulatory limits set for agricultural soils. The highest Cu concentrations were found in the upper ~ 15 cm of the soil profile. Copper concentrations decrease with depth and distance from the treated plants. Abandoning vitriculture and changing to other agronomical practices (e.g., tillage, shift to other agricultural plants) can lead to the mobilization of Cu from such contaminated soils and thus present an important environmental concern. Furthermore, increased Cu concentrations (200–400 mg kg^{-1}) in vineyard soils proved to negatively influence growth of several plant species, especially in acidic soils with a low cation exchange capacity. This review presents an overview on the range and degree of Cu contamination of vineyard soils, together with the biogeochemistry of Cu in these soils and the sorption mechanisms involved.

Chapter 5 - Many pollutants in the soils are subjected to oxidation processes, resulting in the formation of reactive intermediates, mainly radicals that react with oxygen producing the reactive oxygen species (ROS). In general, microorganisms are protected from ROS damage by several enzymes, including catalases. Model microorganisms, bacterial isolates, *Comamonas terrigena* N3H and *Comamonas terrigena* N1C selected from soils contaminated with crude oil, *Comamonas testosteroni* strain isolated from sludge of waste water and isolates of micromycetes *Aspergillus niger* from mine surroundings (As 1651 mg/kg, Sb 362 mg/kg respectively As 400 mg/kg) during the growth exhibited much higher level in total catalase activity than the same control species. Electrophoretic resolution of cell free extract of *C. terrigena* N3H exhibited two isozymes of catalase, catalase-1 and catalase-2 unlike to the control isolate, where only one band of heme catalase was detected. The isolate and the collection culture of *C. testosteroni*, grown in nutrient meat peptone medium, revealed only one band, the constitutively expressed heme catalase. *A. niger* isolates equally as the control culture, in the growth phase of developing hyphae and spore formation three catalase isozymes in the nitrate medium produced. The H_2O_2 – adapted variant of *C. terrigena* N3H, obtained by subculturings with gradually increasing concentration of oxidative stressor exhibited a much higher level of total catalase activity represented mainly by catalase-1, then the original isolate. Phenol peroxidation in *C. terrigena* N3H cells can be also largely attributed to the action of this isozyme, that is essential not only for the protection against oxidative stress, but its peroxidic activity can also play an important role in the degradation. The sequence analysis of *catalase-1* gene confirmed the presence of highly conserved catalase sequence motifs that in other strains (not exposed to pollution) could not be found. This fact can indicate a selective pressure of environment. Catalase expression in microbial isolates as a response to exogenously added stressors was dependent on individual strain, its growth phase and kind of the stressor. The isolates of *C. terrigena* exhibited much higher resistance to exogenous H_2O_2 (20, 40 mM) than collection cultures, mainly in the middle and late exponential growth phases. The results refer to complicated relations between biotic and abiotic effects that may directly affect the processes *in situ* environment.

Chapter 6 - Petroleum hydrocarbons are one of the major energy sources used by humans. Environmental contamination by petroleum hydrocarbons has, however, become a serious problem all over the world, and laws about contaminated soils are enforced in the United States, Germany, the Netherlands and Japan. Incineration and bioremediation systems for the purification of soils contaminated by petroleum hydrocarbons have been investigated and constructed. However, the incineration systems require large-scale equipment and petroleum, and the bioremediation systems need a long time for purification.

To construct an efficient bioremediation system, microorganisms having the ability to degrade long-chain hydrocarbons, which are difficult to degrade in nature, were isolated and characterized (36 isolates belonging to six genera plus one family). The isolates belonging to the genera *Acinetobacter*, *Rhodococcus*, and *Gordonia* degraded long-chain normal alkanes (*n*-alkanes) efficiently. Long-chain cyclic alkanes (*c*-alkanes) were degraded by the isolates from the genera *Rhodococcus* and *Gordonia* without co-oxidation, but the degradation pathway of *c*-alkanes differed between these two genera.

The bacterial biomass in soil influences biodegradation. To enhance the efficiency of the biodegradation, a bacterial biomass measurement method using environmental DNA (eDNA) was constructed. Moreover, a detection method for the hydrocarbon-degrading genus *Rhodococcus* using alkane hydroxylase gene (*alkB*) was developed. The biodegradation of hydrocarbons in soil was accelerated by controlling and maintaining the hydrocarbon-degrading microorganisms using these methods.

Bioremediation of soil historically contaminated by petroleum hydrocarbons (total petroleum hydrocarbons (TPH): about 4,000 mg/kg-soil, main component: long-chain *c*-alkanes) was carried out. *Rhodococcus* sp. NDKK6 was inoculated into the soil (about 5×10^8 cells/g-soil), and the bacterial biomass was maintained above 1×10^8 cells/g-soil by measurement of eDNA. Moreover, *Gordonia* sp. NDKY76A, having a different *c*-alkane degradation pathway from *Rhodococcus* sp. NDKK6, was added to the soil to enhance the bioremediation. Long-chain *c*-alkanes were efficiently degraded, and about 66% of TPH was removed after 35 days.

To establish a large-scale bioremediation system, a 500 kg scale mixing machine (500 kg soil per batch) was developed. The machine homogenously mixed microorganisms and nutrients into soil (range of bacterial numbers in the mixed soil: 2.44 to 2.80×10^8 CFU/g-soil) in a short time compared with a backhoe method (range of bacterial numbers in the mixed soil: 0.22 to 1.84×10^8 CFU/g-soil). In a large-scale bioremediation project (about 2.5 ton, TPH: about 22,000 mg/kg-soil, main component: aromatic hydrocarbons), enhancement of biodegradation (about 20%) by using this bioremediation system was confirmed.

Chapter 7 - The European Environment Agency (EEA) and the United Nations Environment Program (UNEP) stimulate the study about the role of soil in relation to ecological and economical problems derived from the climatic change, industrial development and trade. Both organisms alert about the pollution problems due to mine and industrial activities and mark several areas affected in the Mediterranean, especially in Spain, Italy and Greece (EEA, 2002). The number of environmental compartments that are related to soil and its quality is great due to the interstitial position of soil among water, atmosphere, rocks and life.

The protection of soils is and should be a principal objective of environmental policy. Soils are finite, increasingly scarce and non-renewable resource with varying biological, chemical and physical properties (European Soil Bureau, 1999). That is why the properties that must be determined in a soil are diverse and soil should be studied from an interdisciplinary point of view.

Heavy metals are considered as contaminants capable of causing environmental problems in soils, water, atmosphere and living organism. The meaning of "heavy metals" is closely related to pollution and the potential toxicity of these elements. The presence of metals in soil is, therefore, not indicative of contamination. However, it is so important the need to measure heavy metal concentrations in soils to assess contamination risks and to identify background values as a basis to establish soil quality standards

Chapter 8 - An in-stream on site clean-up approach for stream polluted sites, including the benthic soil and the entire water is proposed. The given approach is based on the proper complex of natural self-purification processes in a stream, leading to formation of froth on a surface with a large amount of all kind of polluting particles - organic and inorganic. Once formed, the natural froth is collected using a boom system and can then be removed and treated appropriately. As a result, we can intervene directly on site in the clean-up of the stream sites that are heavily polluted from various polluting sources, including the non-point ones - runoffs from agricultural and mining operations as well as discharges from sewage treatment plants and others. One important self-purification process in entire water and in/on benthic soil is the decomposition of total organics, including dead fauna and flora and also industrial organic pollution, by fungi, anaerobic and aerobic bacteria and other microorganisms. During the decomposition, the entire water body as well as the benthic soil is periodically enriched with biological surfactants, such as amino, pyruvic, fatty and other acids and generated dissolved biogases – oxygen, ammonia, carbon dioxide, methane, nitrogen and others. Further, stream polluted sites also contain dissolved air and various polluting particles, including man-made surfactants. Therefore, the next step in river self-purification – the generation of biogas and air bubbles, and the attachment of the polluting particles to the bubbles in the presence of surfactants should be considered. For the latter, the proper level of turbulence must exist. Water cascading over weirs, waterfalls and other obstacles, creating the shallow-turbulent character of water current, is the suitable condition. The particle-bubble aggregates as well as decomposed masses from the benthic soil enriched with biogas bubbles, will rise to the water surface and concentrate in the froth and the surrounding thin top layer of surface water upstream from the installed boom system. The resulting froth as well as the thin top layer of surface water comprises a high concentration of all kinds of polluting particles.

Chapter 9 - Plant bioassays (PBA) are widely used for the detection of genotoxins in soils. It is known that the release of DNA damaging agents into the environment may affect the stability of ecosystems and may cause cancer, age related diseases and infertility in exposed humans (e.g. via consumption of plants grown in contaminated soils or via ground waters). The most widely used PBAs are micronucleus (MN) assays with pollen tetrads of Tradescantia and MN and chromosome aberration analyses with root tip cells of *Vicia faba* and *Allium cepa*. In addition, also other endpoints have been developed for plant experiments (e.g. gene mutation assays with *Zea mays*, *Arabidopsis thaliana*, somatic mutation assays as well as the single cell gel electrophoresis test). One of the advantages of the aforementioned assays is that they are sufficiently sensitive to detect the effects of contaminations in native soils in *in situ* experiments whereas for other genotoxicity tests with soils extraction and concentration procedures are required. The evaluation of the currently available data shows that PBAs are quite insensitive towards polycyclic aromatic hydrocarbons (PAHs) and aromatic amines which are often found in soils and sediments but are highly sensitive to certain pesticides, herbicides, heavy metals such as As^{3+}, Cr^{6+}, Cd^{2+}, Ni^{2+} and radionuclides which can not be easily detected with other bioassays (e.g. with bacterial mutagenicity tests which are among the most abundant genotoxic carcinogens found in soils and sediments. These characteristics make PBAs important integral components of test batteries aimed for detection of DNA damaging contamination in soils and sediments.

Chapter 10 - The assessment of potentially toxic element (PTE) contents in soils and the establishment of background values are priority objectives within the European Union and throughout the world. This is a necessary task in order to improve or even maintain the quality of soil resources and for environmental and health protection of livestock and humans. Different methods have been used to derive background values of PTEs in soils. Statistical approaches used elsewhere include descriptive statistics, probability plots, the Bootstrap method, and the recent publication of the International Organization for Standardization (ISO/DIS 19258). Most of these methods do not consider soil properties and characteristics, despite soil components such as organic matter, clay, hydrous oxides and carbonates play a relevant role on the content and dynamics of most contaminants in soils due to their binding properties. Methods based on regression equations take into account some selected soil properties, such as organic matter and clay content, and have been applied in different works. This chapter tries to present an overview of statistical approaches, modifications of existing methods and recent new approaches that have been used to establish background values of PTEs in soils. The application of these various approaches is explored in a representative area of the Mediterranean region, and the limitations and usefulness are also outlined. Moreover, it is discussed to what extent these methods have, up to now, satisfied the expectations or have been satisfactory in particular applications. The correct choice of the most useful statistical approach is finally dependent on the objectives of the background study and on the characteristics of the study area.

In: Soil Contamination Research Trends
Editor: Javier B. Dominguez, pp. 1-50
ISBN: 978-1-60456-319-1

Chapter 1

COMPLEMENTARY APPROACHES FOR USING ECOTOXICITY DATA IN SOIL POLLUTION EVALUATION: RISK BASED VERSUS DIRECT TOXICITY ASSESSMENTS

M. D. Fernandez and J. V. Tarazona
Laboratory for Ecotoxicology,
Department of the Environment,
INIA, Madrid, Spain

1. INTRODUCTION

The assessment of soil quality has usually focused on human health protection as the main objective. Recently, criteria for the protection of ecosystems have been incorporated and ecotoxicological analyses are recommended to estimate the risk to ecological receptors associated with contaminants in soils (Calow, 1993; Stephenson et al., 2002; Loibner et al., 2003; Robidoux et al., 2004b). The ecotoxicological assessment of soils is mostly based on the toxicity test with selected organisms. Two complementary approaches are available. The first approach consists in the identification of toxicity thresholds for each relevant pollutant, thresholds that are based on the evaluation of effects of chemical substances on selected organisms representing relevant ecological receptors. The results of these assays are used for setting soil quality standards for each pollutant or pollutant class. Risk assessment tools can be used for this purpose, pre-establishing acceptable levels of risk. The contamination level is based on the comparison of the concentration of contaminants measured in the soil with the standards established from the thresholds. Although field and semi-field information can be incorporated in the higher tier steps, the thresholds are mostly developed from standardised toxicity assays conducted under laboratory conditions following international (e.g. OECD, ISO) or national (e.g. USEPA, ASTM) guidelines. In the second approach, toxicity assays are performed directly with the contaminated media (soil, water, sediment). This alternative, performing the assays with environmental samples, constitutes the method called direct (eco)toxicity assessment (DTA), and is based on modified bioassays.

Most regulations have developed soil quality standards based on toxicity assays. However, due to the limitations in the lab to field extrapolation, trends were directed towards the combination of chemical analysis and DTA (Peterson et al., 1990; Torstensson, 1993; Tørsløv et al., 1997). In this chapter, both alternatives will be compared. The main difference between both approaches is that in the first case, a reference "uncontaminated" soil sample is spiked with one or a few chemicals at different concentrations, while in the DTA approach real soil samples are collected at the contaminated site, therefore containing a realistic combination of the different pollutants present in the area, the field sample can be then tested and/or "diluted" with "uncontaminated" soil to create a pollution gradient. The toxicity of the spiked or collected/diluted samples is measured and concentration/response relationships obtained in both cases. To understand better this comparison, in this chapter the term "toxicity test" will be used for the first approach: toxicity tests with samples spiked at the lab; while the term "bioassay" will be used for the DTA approach: samples collected at the field.

2. Toxicity Test for the Assessment of Contaminated Soils

Toxicity testing consists in the evaluation of the degree to which an individual substance or a mixture of substances adversely affects living organisms. Tests may be performed at different levels of biological organisation, from molecular level (biochemical changes) to ecosystem. Reproducibility increases from molecular to ecosystem level, whereas the relevance increases in the opposite direction. Therefore, compromise decisions in terms of sensitivity, ecological relevance and predictive potential must be taken and consequently, most protocols are performed at the organism level, constituting the "single-species toxicity tests".

During the test, key test organisms are exposed to toxic agents and the response is observed during and/or at the end of the test. Different endpoints are used to measure the toxic effect. These endpoints are related to a measurable characteristic of the test organism, which can be affected by toxicants. They may be lethal responses (mortality) or sublethal effects (e.g. inhibition of growth or reproduction). The response is also associated to the exposure time; adverse effects may occur after short or long-term exposures. Times should be related to the lifetime of the test organisms. Short-terms exposures are useful to observe acute effects; while extended periods covering a significant part of the life cycle are required to address chronic effects. Taking into account that a pollution situation mostly involves long term exposures to low concentrations, the chronic assessment usually represents more realistic situations.

In toxicology, it is fundamental to establish the relation between the amount of chemical to which an organism is exposed and the magnitude of the response. Thus, a common method for measuring toxicity is to expose the organisms to a concentration gradient. The results may be expressed in two ways: the concentration which will cause a toxic response at a determined level or the highest concentration which will not cause an effect. In the first case toxicity may be expressed as median lethal concentration (LC50) or median effect concentration (EC50), which represents the concentration that produces/causes mortality or the measured effect in 50% of the test organisms, respectively. Other effect levels such as L(E)C10 are also used. The effects can be represented by levels that estimate the concentration at which a specified percentage of impairment occurs. The term IC50 is sometimes employed, instead of the EC50, to express the concentration that inhibits/reduces the response observed in the control by a 50%. In this case, the measure is not a quantal but a quantitative one. For example, in the algae growth inhibition test, the measure is expressed as IC value because it focuses in changes in the growth and not in the number of organisms that show effect. These parameters are usually used in acute assays. Values are estimated by a linear regression performing a transformation of response data, for instance using a logit-probit (Litchfield and Wilcoxon, 1949) or Weibull units (Christensen and Nyholm, 1984). The highest concentration which causes no toxicity is usually expressed as No Observed Effect Concentration (NOEC) and it is used in chronic assays. In addition, LOEC is defined as the lowest concentration at which effects are observed. These endpoints are based on statistical techniques; the effects determined for a particular concentration are compared with the control response using analysis of variance (ANOVA). The soil means must be compared using an appropriate multiple comparison test method. Dunnett´s method is perhaps the most widely used method in ecotoxicology. These methods take account of the multiple statistical tests by adjusting the statistical threshold for declaring significance. The NOEC represents the highest tested concentration at which no statistically significant differences with the control are observed.

The toxicity assays used in soil characterization protocols can be grouped, according to their complexity, in: laboratory assays with single species, multispecies tests and soil microcosms. Larger systems, equivalent to the aquatic mesocosms, and (semi)field assays are often used to assess fate properties but they are scarcely employed for the effect assessment. Assays have been developed to evaluate impacts of chemicals on the diverse communities of microorganisms, plants and invertebrate and vertebrate animals that comprise the terrestrial ecosystem (Stephenson et al., 2002; Loibner et al., 2003). Many single species assays have been standardized to harmonize their use with the aim of reducing variability and comparing results. In contrast to aquatic toxicity, testing protocols for soils are less developed. However, as already mentioned, notable efforts are being made by different organizations (ISO; OCDE; USEPA, ASTM, etc) to standardize soil toxicity assays. These assays can be used to test lab spiked or field contaminated soils and may also be used to monitor remediation processes. In Table 1 the tests to soil and aquatic organisms standardized by OCDE are shown. These tests have been widely used to aid the process of risk assessment.

Table 1. OCDE standardized tests for the determination of the toxicity of chemical substances to soil and aquatic organisms

Taxonomic group	Title	Endpoint	Measurement variables	Assay time (days)	Test Number
Terrestrial invertebrates	Earthworm, acute toxicity test	Survival	Number of living worms	14	207
Terrestrial invertebrates	Enchytraeid, reproduction test	Reproduction	Number of juvenile worms	42	220

Table 1. Continued.

Taxonomic group	Title	Endpoint	Measurement variables	Assay time (days)	Test Number
Terrestrial invertebrates	Earthworm reproduction test	Reproduction	Number of living offspring and cocoon numbers	56	222
Plants	Terrestrial plant test: seedling emergence and seedling growth test	Emergence of seedlings and Inhibition of growth	Emergence, dry shoot weight (fresh weight), shoot weight and assessment of visible detrimental effects	14-21	208
Plants	Terrestrial plants test: vegetative vigor test	Vegetative vigor and growth	Biomass (dry shoot weight) and visible detrimental effects	21-28	227
Microorganisms	Soil microorganisms: Nitrogen transformation tests	Nitrogen transformation	Rate of nitrate production	28	216
Microorganisms	Soil microorganisms: Carbon transformation tests	Carbon transformation	Glucose-induced respiration rates	28	217
Algae Cyanobacteria	Freshwater Alga and Cyanobacteria, Growth inhibition test	Inhibition of growth	Algal biomass: cell counts, cell volume, fluorescence, optical density, etc.	4	201
Aquatic Invertebrates	Daphnia s.p., Acute immobilization test	Survival	Immobilization	1	202
Aquatic Invertebrates	Daphnia magna, reproduction test	Reproduction	Number of living offspring	21	211
Fish	Fish, Acute toxicity test	Survival	Number of living fish	4	203
Fish	Fish, Prolonged toxicity test: 14-day study	Survival, appearance and behavior, and growth	Survival, abnormalities (appearance and behavior), length and weight	14	204
Fish	Fish, early-life stage toxicity test	Survival, appearance and behavior, and growth	Hatching and survival (at different stages), abnormalities (appearance and behavior), length and weight	30-60	210

Taxonomic group	Title	Endpoint	Measurement variables	Assay time (days)	Test Number
Fish	Fish, short-term toxicity test on embryo and sac-fry stages	Survival, appearance and behavior, and growth	Hatching and survival (at different stages), abnormalities (appearance and behavior), length and weight	8-55	212
Fish	Fish, juvenile growth test	Inhibition of growth	Weight	≥ 28	215
Aquatic plants	Lemna, sp. Growth inhibition test	Inhibition of growth	Frond number, total frond area, dry weight or fresh weight	7-10	221

3. Soil Assessment Based on Toxicity Threshold

The assessment of soil quality is generally based on the determination of the concentration of pollutants in soils. The estimated concentration is compared with specific threshold values and the degree of contamination is evaluated. Thus, many regulatory bodies all over the world have developed soil quality values. These values can be developed for three main applications:

- Screening values: representing soil concentration levels that may cause potential ecological dysfunction, and therefore, if exceeded, will require a site-specific assessment.
- Cleanup targets: representing the objectives to be achieved in restoration processes. In some cases, these values represent a similar level of protection as the screening values, while in other regulations the decision has to balance the restoration cost against the ecological benefit.
- Intervention values: representing concentrations which are indicative of seriously contaminated sites that require immediate clean-up or control actions.

Some national guides, i.e. Netherlands Guide (The Ministry of Housing, Spatial Planning and the Environment, 2001), Swiss Guide (Hämmann et al., 1998) or Federal Soil Protection Act in German (BbodSchG, 1998) define different standard values covering several of the above mentioned categories. For a long time, SQSs values have been used as a tool to decide the necessity to undertake actions. Nowadays, in many countries the preferred use of SQSs is that of trigger values for site specific investigations, which allow deciding the remedial actions.

In addition, site specific values can be defined taking into account site specific characteristics such as soil properties, climate, etc. Values obtained in this way are more realistic and usually higher than generic soil quality values. Site-specific values in combination with generic values are used by different regulations (e.g. Canada, The Netherlands) as a valuable tool to determine the risk of contaminated soils.

The SQSs are established on risk basis, and they represent the soil concentration that does not exceed a pre-established ecological risk level. Most developed countries have produced technical guidance documents and protocols to conduct environmental risk assessments (ERAs). In Europe, the so-called Technical Guidance Document (TGD) (EC, 2003) developed by experts from the EU countries under the coordination of the European Chemicals Bureau, represents the basic technical guidance for the risk assessment of industrial chemicals. Specific guidelines for several chemical groups, such as pesticides, feed additives, pharmaceuticals and so on, are also available. A review on the scientific bases of the different European ERA protocols was produced in 2003 by a Task Force of the EU Scientific Steering Committee (SSC-European Commission, 2003).

Risk assessment allows predicting the likelihood of an adverse effect to occur; the assessment of chemicals requires the estimation of the expected level of exposure to potentially toxic chemical substances, and the expected consequences for exposed organisms. Thus, environmental risk assessment includes the following aspects: description of potentially adverse effects to the environment (hazard identification) and estimation of the magnitude of these effects (dose-response assessment); determination of receptors potentially exposed and the conditions of exposure (exposure assessment) and finally, estimation of the probability of an adverse effect to occur based on the comparison between effects and exposure (NRC, 1983; Suter et al., 2000) (Figure 1). An essential part of any proper risk assessment is the identification of the uncertainties in the analysis.

3.1. Determination of Soil Quality Values

To establish soil quality standards, risk assessment is applied in a backtrack approach, which allows setting the level of one or more chemicals in the soil associated to a certain risk level. Initially, the acceptable risk level is established. Secondly, the effects on the environment are determined based on toxicity assays. Finally, the exposure levels that satisfy the predetermined risk levels are selected.

Accordantly, the methodology used to determine SQSs involves the following steps: identification of potential pathways for exposure and relevant receptors for soil contaminants (exposure assessment), identification of effects for the selected receptors in the developed scenario (effect assessment) and determination of SQS for the different receptors/compartments assuming a predetermined risk level (risk assessment). These steps can be associated into risk lines, which define the processes involved in the transference of the chemical from soil to the receptor.

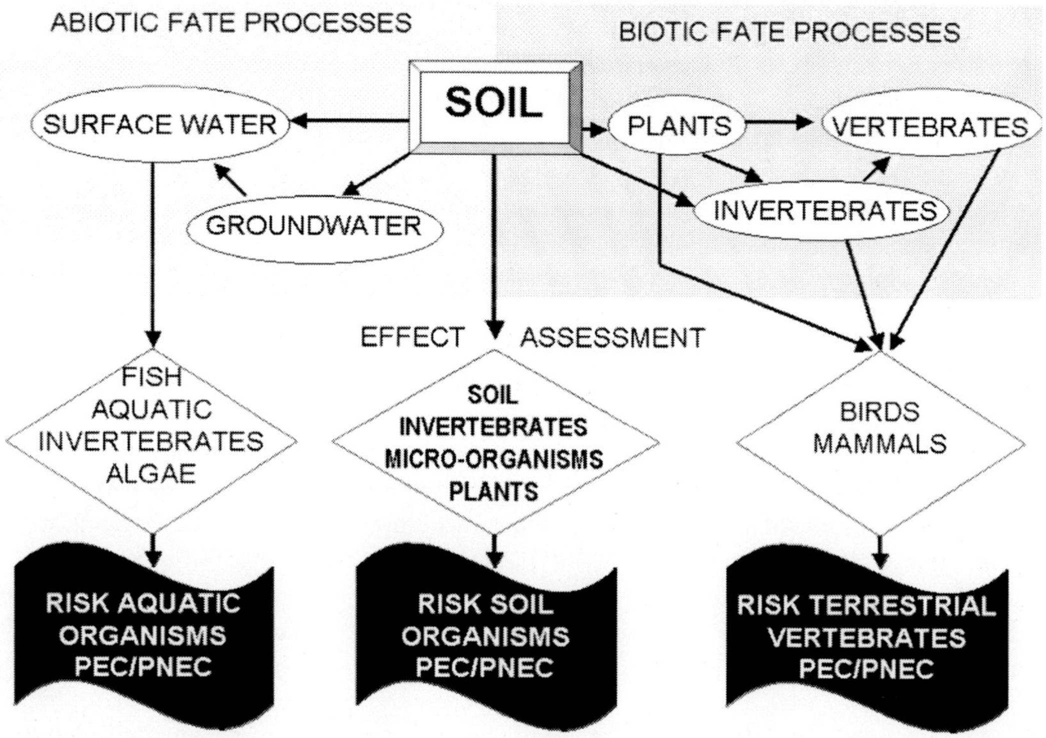

Figure 1. Graphic representation of the generic conceptual model. Symbols represent sources (double square), exposure estimations (ellipses), effect assessments (rhombus) and risk characterization processes (flags).

CONTAMINANT IN SOIL → EXPOSURE PATTERN → RECEPTOR

Different values can be developed depending on the land use patterns. Differences in SQSs for each land use result from the consideration of different levels of protection, different receptors and different exposure pathways. According to some European (Denmark or Switzerland) and non-European authorities (Canadian Council of Ministers of the Environment (CCME) and the Australian EPA) different land uses may not need to support the same diversity of ecological services and functions, and they allow the use of less conservative numbers for commercial, industrial, or residential lands. The Netherlands also defines different land uses, not to modify the values, but to determine if the value for ecological protection should or should not be applied. The approach adopted in Spain considers land uses for human health protection but a single level for ecosystem protection, with regional authorities being responsible for the decision whether to apply or not the ecosystem protection level in each facility depending on the ecological relevance of the area instead of its use pattern. The rationale for this decision is the presence of industrial facilities in areas with ecological values.

3.1.1. Exposure Assessment

Exposure assessment defines the amount of substances available to a receptor by the different exposure routes. Based in these environmental routes, different scenarios must be developed to describe the circumstances that determine the level of exposure. For each scenario relevant ecological receptors and exposure routes are selected; then relationships between the soil concentration and the exposure to the receptor are established. Usually, the scenarios are selected representing different land uses.

Model simulations are accepted for exposure analysis, as environmental measurement data are difficult to obtain. The exposure models relate the concentration of a substance in soil to the potential for exposure or uptake by an ecological receptor. Calculated exposure levels are typically called Predicted Environmental Concentration (PEC). The key elements to determine the exposure to a chemical substance are: magnitude, extension and frequency of exposure, and the fate and transport of contaminants from soil. In any exposure model, three key elements must be defined: partitioning of the substance between the soil phases (air, water, soil), transport of the substance to other mediums (e.g. plant or soil organism, air, surface and groundwater) and direct and indirect exposure routes. With this aim, it is necessary to have information of physiochemical properties (vapor pressure, solubility, etc) and ambient behavior (degradation, bioaccumulation, etc) of soil contaminants. However, for many substances transfer and fate data are lacking.

The scenarios are designed to be protective where conditions may lead to highest exposure i.e. worst possible case. In site specific studies, the specific characteristic of the site should be considered to obtain more realistic exposure data.

3.1.1.1. Selection of Ecological Receptors

Considering the ecological risk assessment principles (e.g. USEPA, 1998) this aspect is an essential part of the problem and requires a combination of regulatory needs and scientifically based assessment possibilities. Receptors can be selected attending to the land use but it is advisable to define a base set of organisms that would insure a minimum function for any land use. These, usually include plants, soil invertebrates and microorganisms.

Ecological receptors, such as aquatic organisms and terrestrial vertebrates, can be affected by soil pollution through indirect routes. Aquatic communities of water bodies located in the vicinity of the polluted site may be affected by the potential contamination of surface and ground waters due to drainage, leaching and run-off. Terrestrial vertebrates may be affected through the consumption of contaminated plants and invertebrates from the area (secondary poisoning). For a higher tier assessment, the complexity of terrestrial ecosystems should be considered (CSTEE, 2000), and specific conceptual models covering this complexity (Tarazona and Vega, 2002; Tarazona et al., 2002) should be applied.

References to water regulations are common in soil protection policies. However, most of the guides do not integrate protection of aquatic organisms in the determination of soil quality values. In Europe, groundwater protection is considered in ca. 50% of the countries in most of the cases related to human health by the consumption of drinking water. Conversely, surface water is only included in Spain and Sweden. In some cases, surface and ground water are considered at site-specific level (Carlon et al., 2007).

Protection of terrestrial vertebrates is considered in most of the guides. However, in general, critical values for these receptors are limited to herbivorous animals or livestock as in Canadian (http://ec.gc.ca/CEQG-RCQE)or British Columbia (http://www.env.gov.bc.

ca/epd/epdpa/contam_sites/guidance) jurisdictions and considered exclusively in the agricultural land use or focused on a species-specific basis as the Australian approach (http://www.epa.gov/ecotox/ecossl/pdf/ecossl_attachment_1-1.pdf). Toxicological relationships are expressed in terms of dietary exposures followed by a variety of methods to estimate a corresponding soil concentration. The bioconcentration factor is used to determine the substance concentration in food. The combination of persistence and potential for bioaccumulation produces and additional hazard: biomagnification through the trophic chain. Biomagnification is considered only in some cases, usually related to persistent substances. Top predators, potentially the most impacted organisms, are the focus of this assessment (Fernández et al., 2006).

3.1.2. Effects Assessment

3.1.2.1. Toxicity Data

Ecosystem integrity was defined by Cairns (1977) as "the maintenance of both structural and functional characteristic of an ecosystem to ensure that ecological services will be available on a sustained basis". Accordingly, the development of SQSs is intended for the protection of the structure and function of ecosystems. Consequently, the most relevant endpoints to set Soil Quality Values are growth, mortality and reproduction. Regarding the preservation of the ecosystem function, microbial endpoints related to the C- and N-cycle are included. Other parameters such as biomarkers and endpoints related to the mode of action, such as endocrine disruption, can be used to trigger further studies on the effects on population.

Toxicity is by definition the measurement of the adverse effects that a substance provokes on a biological system. Ideally, hazard information would be available for all species. However, this can never be achieved given that thousands of species may be potentially exposed. Therefore, tests are performed on a range of representative species and extrapolations are made to account for differences in species sensitivity. In general, chronic NOEC values for ecologically relevant endpoints (growth and reproduction) are preferred, and acute L(E)50s are employed as surrogates if valid chronic data are not found. Nevertheless, other considerations are possible. Thus, different toxicity data (EC10, EC50, etc) are used according to the intended protection level and hence, different Soil Quality Values are obtained (e.g. Germany).

Data on the effects of substances to organisms can be mainly obtained from data bases (ECOTOX, RAIS, HSDB, RAR, etc.) and studies from research institutes. In addition, industry generates data on the effects on organisms for the notification and registration of substances. All these data must be evaluated regarding to their reliability and relevance. Toxicity data are generated from soils with different properties (e.g. pH, percentage of organic matter, etc.). This means that the results of tests conducted in different soils, with different characteristics, cannot be compared or used for risk assessment as such, but should be normalized to standard conditions. For non-ionic organic compounds the data are normalized on the basis of the soil organic matter content, because it is assumed that the bioavailability for non-ionic substances is determined by the organic matter content only (EC, 2003). For metals and other inorganic substances, the organic matter content is not the only factor influencing the bioavailability, factors such as clay content and the pH of the soil are also important. To set environmental quality objectives for metals, different normalization methods are proposed. For example, in The Netherlands, terrestrial toxicity data for metals are normalized on the basis of clay and/or organic matter content. However, the relationship between soil characteristics and bioavailability of metals is very complex and normalization methods lack sufficient scientific validity to be used for metals.

3.1.2.2. Extrapolation

Ecotoxicity data are gathered, screened for their quality and finally, data are extrapolated to cover uncertainties, such as inter- and intra- species variation, extrapolation from subchronic to chronic exposure duration and extrapolation from laboratory to field effects. The commonly used extrapolation procedures are: deterministic methods based on assessment factors and probabilistic approaches based on species sensitivity distributions.

3.1.2.2.1. The Deterministic Method

The deterministic method uses the lowest toxicity value plus an assessment factor to cover the remaining uncertainty. The validated toxicity data for the most sensitive taxonomic group in each compartment is divided by the selected assessment factor. In this way, an ecotoxicological threshold-like concentration, called Predicted No Effect Concentration (PNEC) is estimated for the environmental compartment. This is the methodology described in the TGD (EC, 2003) for the chemical assessment. However, others ERA protocols, in Europe or other parts of the world, do not include this step and the risk quotients are directly calculated as ratios between expected concentrations and toxicity endpoints (SSC-European Commission, 2003).

Assessment factors, used in the deterministic method, are frequently used as a generally accepted practice in soil quality assessment (van Straalen and van Gestel, 1993; Jager et al., 1994; Van de Plassche, et al., 1999). The selection of factors is basically a regulatory decision and the European legislation offers examples of largely different factors applied to the same ecotoxicity data (a revision is presented in SSC-European Commission, 2003). The Assessments Factors depend on the availability of data for the taxonomic groups considered for each compartment. They decrease orders of magnitude if chronic data on all relevant taxonomic groups are available. Generally, the assessment factors for PNEC derivation based on laboratory studies ranges from 1000, when only acute values are available, to 10 if chronic NOECs can be found for all relevant taxonomic groups. The assessment factors can be adjusted if new data become available, and must be selected case-by-case if multispecies tests are used. If the availability of data is scarce, high assessment factors must be used which can result in overprotective Soil Quality Standards. This is especially important in the case of a naturally occurring substance where values of SQS may be very near to or within the background range.

3.1.2.2.2. The Probabilistic Method

The probabilistic method is based on the species sensitivity distributions (SSD) methodology that uses the whole toxicity data set to elaborate a reliable distribution of the sensitivity of the species to the assessed chemical (Postuma, 2002). The method uses all available data, instead of the most sensitive species, to estimate the distribution curve and the selected percentile. Protection levels are defined in terms of percentage of species to be protected. Consequently, a selected percentile of the species sensitivity distribution is estimated and it is used as the final criterion value (van Straalen and Denneman, 1989; Wagner and Løkke, 1991). The main assumptions of the SSD approach are: the protection of a large part of the structure is enough to protect the ecosystem functions, the distribution of species sensitivities follows a theoretical distribution function, and the group of species tested in the laboratory is a random sample of this distribution. An advantage of this method is that it uses the whole sensitivity distribution of species in an ecosystem to derive a Soil Quality Standard, instead of taking always the lowest effect endpoint. Nevertheless, to apply this methodology a sufficiently large reliable data set for different taxonomic groups is necessary, which in occasions is difficult to obtain. In general, at least 10 NOEC values including different taxonomic groups are required.

The species sensitivity distribution approach has been criticized mainly because it assumes that the protection of the selected percentage of species confers an appropriate level of protection on ecosystem structure and function and because of its lack of validation (OECD, 1992a; Forbes and Forbes, 1993; Smith and Cairns, 1993). Moreover, this approach may not be applicable as such for substances with a specific mode of action.

The SSD approach may be also used to determine PNEC, in a combination of deterministic and probabilistic methods. In this case, a toxicity value protective to a selected percentage of species is considered rather than the lowest toxicity value and an assessment factor, usually between 1 and 5, depending on expert judgment, may be applied. Regardless the theoretical discussions, the real comparisons should consider the overall approaches as a whole, certainly, the deterministic approach selects the most sensitive tested species but from a very limited number of species, while the probabilistic approach may appear as less protective as it is based on a percentile, but from a larger data set, representing a significant reduction in the overall uncertainty. The experience tends to demonstrate that in general the deterministic approach is useful for an initial screening assessment and the SSD approach represents a proper refinement alternative.

3.1.3. Risk Assessment

Risk is defined as the estimation of the incidence and severity of the adverse effects likely to occur due to actual or predicted exposure to a substance (Pugh and Tarazona, 1998). SQSs are established on risk basis, and represent the soil concentration that does not exceed a pre-established ecological risk level. A basic problem in the SQSs derivation is to select which levels of risk can be accepted. In countries where different values are defined for different regulatory applications, the level of risk is usually related to the intended use of SQSs. Thus, the need of action (remediation, restrictions in land use, etc) is related to a warning risk; screening values are based on potentially unacceptable risks and target values are based on neglected risk.

In all cases, the exposure is assessed against the effects, according to the maximum permissible risk. In this way, a critical soil concentration associated with the exposure scenario is calculated. Two approaches are possible. In the former (deterministic method), the environmental risk assessment is established for each environmental compartment through the risk characterization ratios (RCRs). These are defined as the ratio between the Predicted Environmental Concentration (PEC) and the Predicted No Effect Concentration (PNEC) for each environmental compartment (water column, soil, wildlife food items). Thus, the development of the soil quality standards followed two steps, first the assessment of the PNEC (or alternative effect assessment) for each compartment; and second, the estimation of the soil concentration which would give an exposure/PNEC ratio of 1. This is the approach followed in the Spanish or Australian regulations.

In the second method, the probabilistic risk assessment "use probabilities or probability distributions to quantify one o more sources of variability and/or uncertainty in exposure and/or effects and the resulting risk" (EUFRAM, 2006). Probabilistic approaches enable variation and uncertainty to be quantified, mainly by using distributions instead of fixed values in risk assessment. A distribution describes the range of possible values of both exposure and toxicity, and shows which values within the range are most likely. The result of risk assessment can also be shown as a distribution, showing the range of environmental impacts that are possible, and which impacts within that range are most likely. This should provide a better basis for decision-making, because the full range of possible outcomes can be taken into account. The SQSs may then represent a fix probability level assumed to be acceptable from a cost/benefit or risk/benefit perspective. In fact this is equivalent to the method used for the human health assessment on non-threshold genotoxic carcinogens.

Up to now, the probabilistic approaches in environmental risk assessment have gain only limited acceptance partly due to a lack of enough knowledge for their use in decision making. Usually, probabilistic tools are used in refinement of exposure and effects at higher tier level. In a single level, the most frequent probabilistic element used is the SSD which are used to represent variation in the sensitivity of different species to a single chemical (Aldenberg et al., 2002). Thus, standards are derived using the percentiles of the distribution, i.e. the concentration that exceeds the toxicity endpoint for a given proportion of species. This value is defined as HCx and represents the concentration of a substance at which x percent of the species in an ecosystem may experience effects. This percentage is a political decision and allows obtaining different values associated to neglected, warning or relevant risk. Thus in The Netherlands the intervention values for ecosystems lie at the serious risk level, i.e. HC50, whereas HC5 is selected for target values (The Ministry of Housing, Spatial Planning and the Environment, 2001). Different values in the range of 5 to 50 percentile are used by different European countries (Carlon et al., 2007).

3.2. Environmental Risk Assessment for Metals

Application of risk assessment methodologies to metal contaminated soils presents a number of problems, due to the especial characteristics of metals that must be accounted:

- Metals are natural occurring substances. Consequently, background concentrations and the exposure due to these background concentrations must be considered in the risk assessment.
- The availability of metals in the soil is highly dependent on soil properties. It is of the utmost importance for the study of exposure and effects assessment.
- Metals can be present in different chemical forms with different chemical and toxicological properties that can be transformed one in other due to environmental changes.
- Some metals are essential elements for metabolism and their levels must be maintained above the deficiency range.

3.2.1. Background Levels

Some substances like metals and some chemicals (e.g. PAHs) occur naturally with background concentrations in soils that vary widely. Consideration of background concentrations is essential in SQSs setting and it is the main concern in metals and metalloids. However, characterizing natural background is not easy and must be site-specific.

Countries considering background concentrations in derivation of standard values use different approaches. In some countries it is taken as a reference value, in other countries it is added to the estimated negligible risk concentrations. For example, the reference values for selected metals in soils in the United States are based on the 95th percentile from the frequency distribution (Holmgren et al., 1993; Ryan y Chaney, 1997). In Spain, the mean plus two times the standard deviation of background values, measured in a surrounding clean area can be used as soil quality values if values based in risk assessment have not been described.

Another possibility for the derivation of soil quality values of metals is the application of the "added risk approach" (Struijs et al., 1997; Crommentuijn et al., 1997; Crommentuijn et al., 2000). The use of the added risk approach only considers the anthropogenic amount of substance for the effect assessment of that substance. Thus, a possible contribution of the natural background concentration to toxic effects is ignored. In this way, the risk is determined according to the added metal, based in "added Predicted Environmental Concentration" (PEC_{add}) and "added Predicted No Effect Concentration" ($PNEC_{add}$).

In countries where risk assessment is applied to metals, assessment factors used in derivation of SQSs can result in concentrations very near to or within the background range of that naturally occurring substance. In these cases, in most of the countries the average natural background levels are adopted as soil quality standards instead of risk based concentrations.

3.2.2. Bioavailability in Metal Risk Assessment

Bioavailability can be broadly defined as the portion of a chemical in the environment that is available for biological action, such as uptake by an organism (Rand, 1995). This fraction is different of the available fraction which can be defined as the fraction of the metal that is extractable from the substrate by chemical (e.g.: neutral salt, water extraction) or physical means (e.g. pore water collection). This fraction is generally considered to be a better estimation as the fraction that is potentially available for organisms, than the total concentration.

The consideration of the bioavailable concentration in the derivation of quality standards would assure a better relation between exposure and effects than the use of total soil concentrations (McLaughlin et al., 2000a, 2000b). Biological availability is often thought to be comparable to chemical availability. However, the bioavailability differs per biological species since it depends on physiological characteristics of the specific organism (exposure route, uptake, etc). Different methods have been used to determine the available fraction: free metal ion activity (Parker and Pedler, 1997), pore-water concentration (Peijnenburg et al. 1997) or the amount of metals extracted using different extraction methods (Adriano, 2001). The bioaccumulation of contaminants in organisms can serve also as a measure of the bioavailability of metals in soil (Van Gestel et al., 2002). A review of the assessment of the bioavailability of metals in different organisms is presented by Hund-Rinke and Kördel (2003).

It has been generally accepted that the phytoavailability of a metal is related to the free ion activity of that metal in solution and that soluble metal-ligand complexes are not readily taken up (Parker and Pedler, 1997). However, in some cases recent findings indicate that activity of the free ion does not always correlate well with availability. Complexation of metals with inorganic or organic ligands in soils can increase plant uptake of metals (McLaughlin et al., 2000b). Moreover, metal free ion activity in an aqueous phase or extract is not a good measure of bioavailability when the exposure is not through the aqueous phase (Janssen et al., 1997).

The difficulty of using soil pore water concentration to determine the bioavailable concentration is that it is difficult to measure and highly variable in temporal and spatial dimensions. Moreover, its use should be limited to those species for which the pore water is the dominant uptake route, as microorganisms and some plant species Other soil dwelling organisms may be also exposed via a combination of uptake routes, such as pore water, food, and direct ingestion of soil (Lock and Janssen, 2001, 2003a).

The chemical extraction technique is the most commonly used method to estimate the fraction of a metal that is bioavailable in the short term. Numerous studies have been performed to estimate bioavailable pool in soils, being the most aimed to relate metal soil speciation and plant uptake (Adriano, 2001). Good correlations between plant uptake and chemical fractions have been obtained using dilute acids (HCl, H_2SO_4, NH_4OAc, etc) and chelating agents such as EDTA o DTPA. However, there is no universal soil extractant that can be used to estimate metal bioavailability. This is because of the complexity of metal ion dynamics in the soil system, environmental factors and mainly, differences between organisms.

A recent advance in the application of bioavailability in the metal risk assessment is the development of Terrestrial Biotic Ligand Model (TBLM). The Biotic Ligand Model (BLM) has been used to develop water quality criteria and perform aquatic risk assessment for metals (Di Toro et al., 2001). This method allows evaluating quantitatively the manner in which water chemistry affects the speciation and biological availability of metals in aquatic systems. Consequently, it allows reducing the toxicity variation due to differences in metal bioavailability. Likewise, the TBLM predict soil toxicity according with metal speciation and soil properties. This method is able to achieve a normalization of the wide variation in toxicological endpoints among soils of different properties and can provide a framework for modeling metals ecotoxicity in soils (Koster, et al., 2006; Thakali, 2006a, 2006b) .

Within the EU, the risk assessment strategies for metals such as zinc or copper are developing alternatives to consider bioavailability in soil risk assessment, following equivalent methods to the Biotic Ligand Model (BLM), which has been used to develop water quality criteria and perform aquatic risk assessment for metals. The combination of these models with the effect background concentrations on the organism responses may produce significant advances in future years.

In the mean time, another approach to the problem is to use the weakly adsorbed and easy extractable fraction rather than the bioavailable fraction in the risk assessment. Soil risk assessment of metals is based on laboratory toxicity tests which are performed in soils freshly spiked with metal salts. However, in the field, metal availability decreases with time, mainly due to aging processes (Ma et al., 2005; Song et al., 2006). Thus, the use of toxicity data obtained with spiked soils can result in an over estimation of the hazard of metal contaminated soils (Lock and Janssen, 2003b: Oorts et al. 2007). An approach to this problem is to define an extraction solution, which is able to extract the weakly adsorbed and easy extractable fraction. This approach could be an alternative to the use of the bioavailable fraction as this chemical fraction would be independent from biological availability which in fact is species-specific.

3.3. Main problems in Derivation of Soil Quality Standards

The aim of risk assessment is the protection of the structure and function of the ecosystems. However, there is not enough information of which endpoints are critical and relevant for the ecosystem. As an alternative this protection is based on toxicity to some selected organisms, determined using toxicity assays. In order to relate effects to the observed concentrations, a large amount of ecotoxicological information is necessary, which is restricted actually, to a few priority pollutants. The lack of enough toxicity data for many substances and the necessity of applying the assessment factor lead to an overestimation of risk.

Soil quality assessment based on standards does not allow an integration of combined effects of contaminants in soil. Thus, synergic and antagonic effects between substances that can modify the toxicity of pollutants in soil are not considered, with some exceptions as some family of compounds with a similar mechanism of toxicity (e.g. chlorophenols) which have SQSs based on the sum of concentrations.

The determination of exposure also presents uncertainties related to the identification of possible routes of exposure and the parameters used to predict the exposure concentration by the indirect routes. Exposure is usually obtained using conservative assumptions that may overestimate the risk. A problem of special concern is addressing the bioavailability in the definition of SQSs. As indicated above, efforts are being made in order to account for the bioavailability of a substance in the prediction of risk.

An added problem is the differences in Soil Quality Standards among countries. In spite of using common methodologies, different soil quality values are defined by the different countries. Differences are mainly due to: regulatory application of the value or values suggested as triggers, land use dependence, selection of ecological receptors (mainly inclusion of aquatic organisms and terrestrial vertebrates, and particularly wild species), consideration of fate processes such as mobilization or biomagnification and the data extrapolation method. International harmonization of Soil Quality Standards was discussed in the concerted actions CARACAS (Ferguson et al., 1998) and CLARINET 2 (Vegter et al., 2003). Presently, the EU is making efforts in order to achieve similar values for the implementation of the Framework Soil Directive.

4. Tier Level Metodologies for Risk Assessment

Risk assessment is usually conducted as a tiered process performed in several steps. The initial steps, covering screening and lower tier risk assessments are based on the precautionary principle, worst-cases are assumed and the spatial and temporal variability is basically not considered. In higher risk assessments, all the available information is used and the specific characteristics of the site are considered to determine the actual risk, if possible in a quantized form and including the evaluation of the remaining uncertainties. The tier approach offers decision making at each level, when the risk is low, no additional efforts are required, when the risk is potentially high, the information required for the subsequent levels must be produced and analyzed. Thus, the degree of uncertainty decreases, while reliability increases. Derivation of generic SQSs is based on simplified risk assessments and the values obtained tend to be overprotective in most of the cases. Usually, these values are also considered as a first step within a tiered approach for decision-making. The following steps consist in data refinement producing a more realistic assessment, in order to allow exonerating sites considered potentially contaminated according to conservative values, or the confirmation of the potential risk.

Tiered protocols have been developed for environmental risk assessment of chemicals (SSC- European Commission, 2003) and pesticides (Campbell et al., 1999; Giddings et al., 2002), mainly aimed at aquatic organisms. There are different methodologies for risk refining, being all of them based in two approaches: i) the use of probabilistic assumptions to cover variability and ii) the use of more complex exposure and/or effect tools to obtain more realistic situations. According to SSC-European Commission (2003), higher tier protocols can be grouped in five levels to refine exposure and five levels to refine effects. The combination of both is also possible. These methodologies can be applied to refine SQSs, obtaining less overprotective and more realistic values. Moreover, in some regulations (e.g. Canada, Spain), screening values may be modified attending to the information collected in a particular site. This information is mainly referred to those physicochemical characteristics affecting chemicals transfer from soil to the aquatic compartment and the terrestrial vertebrates and a selection of relevant receptors in the site (Figure 2). Values which are obtained considering the specific characteristics of the assessed site are referred to as Specific Soil Quality Values (SSQSs) and may be used for the characterization of soils in the site for which they were developed, instead of SQSs.

The refinement process itself is iterative, and based on cost/efficiency estimations. The refinement may focus on the exposure or on the effect sides, or directly in a combination of both approaches, depending on the cost estimations and the likelihood for excluding/confirming risks for each applicable tool. Although this chapter describes the refinement process in terms of subsequent steps, form lower to higher complexity; in reality, the process must be decided case by case. Thus, it is possible to move directly from the screening to the top level if required.

4.1. High Tier Exposure Assessment

The exposure refinement may be conducted following different steps, which have not necessarily to be applied sequentially. As explained above, the site characteristics allow deciding which refinement tool is more adequate.

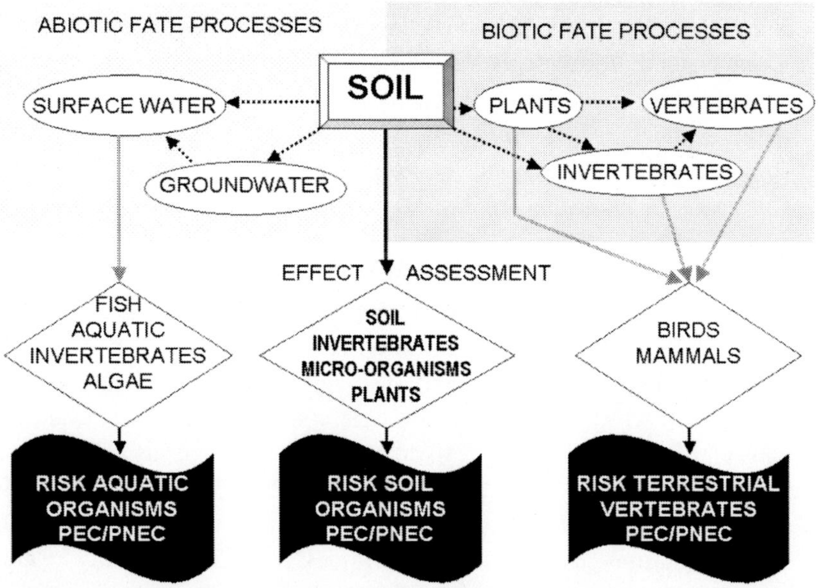

Figure 2. Adaptation of the generic model for site-specific assessments based on site conditions (non continuous lines) and ecological relevance (open arrows).

4.1.1. Level 1. Generic Screening Scenarios

This level constitutes an initial screening evaluation using the least possible information. Threshold values are estimated based on very conservative exposure assumptions. For example, 100% bioavailability is assumed, wildlife is estimated to consume 100% of their diet from the site, and invertebrates are assumed to be stationary in the contaminated portion of the soil. The highest reasonably possible exposure is calculated considering it is constant in time and homogenous in space.

4.1.2. Level 2. Probabilistic Generic Scenarios

At this level, the parameters represented by single data are replaced in the models by the probabilistic estimations. That is, individual data for the variables are substituted by their distribution function, which can be obtained using statistical analysis. The use of distributions as input parameters allows producing the likelihood for the exposure levels as probabilities, as well as, variation and uncertainty in exposure; using tools such as the Monte Carlo analyses. A key step is identifying the main factors and the mechanism that influence the exposure and how they interact.

Distributions may be used to refine some aspects of exposure assessment for aquatic and terrestrial organisms (EC, 2002a; EC. 2002b), to determine the proportion of exposures exceeding some given concentration (e.g. a toxicity endpoint) or to determine the variability and uncertainty within a route of exposure (e.g. concentration of contaminant in the diet of birds of mammals).

4.1.3. Level 3. Realistic Scenarios Considering Spatial and Temporal Variability

At this level, more realistic scenarios are defined, considering relevant receptors and site-specific characteristics affecting the exposure level. Thus, the relevant ecological receptors in the site and compartments that can be affected by soil pollution are selected. Three types of ecological receptors may be affected by soil contamination: soil dwelling organisms, aquatic communities from water bodies located in the vicinity of the polluted site and terrestrial vertebrates. All of these compartments are usually considered in a generic scenario. However, specific site characteristics might be unnecessary considering some of these receptors. On the other hand, some species may require additional protection due to socio/economic or ecotoxicological reasons (e.g. endangered species where individual protection is required to maintain biodiversity).

Regarding exposure level, three factors should be considered, the application of models using specific-site characteristics, the bioavailability of contaminants and the spatial and temporal variability in the exposure level. Initially, the exposure levels coming from indirect routes are determined using models and equations. Depending on the type of assessment, the estimation may focus on worst- and/or realistic-case conditions and use measured or default values. The transfer of chemicals from soil to water and food is usually calculated using partitioning coefficients between different compartments (soil, air, water, food) and standard soil characteristics (organic matter content, bulk density, etc). Default values can be substituted in the equations by site specific values, and Soil Quality Values can be recalculated to obtain Specific Soil Quality Values. In many cases, one limitation is that, equations to obtain SQS are not presented in a transparent way, hence it may be difficult to modify the default conditions to include site-specific characteristics.

Exposure from water can be refined using different models. Many models have been developed to calculate the movement of contaminants through soils. Parkhurst and Appelo, 1999; PELMO (Klein, 1991), PEARL (Leistra et al., 2000) ;etc.). Models simulate the process of transfer from soil to water that can occur in the soil. Its reliability depends on the suitability of the modeled process to actual situation, the equations used to determine the effects and the value of the parameter selected for the model. The application of models requires a sufficient physicochemical characterization of the soil, which is not possible in a generic scenario. At this level, the dilution factors occurring in the site may be considered and models may be applied to estimate the transfer of contaminants from soil to water. An alternative source of information are leaching tests (OCDE, 2004d) performed in the laboratory using standard methods. These tests are relatively short, reproducible and cost-effective methods to estimate the mobility of contaminants in the soil.

Exposure due to the consumption of contaminated food can be also corrected. For example, the determination of secondary poisoning based in worst-case estimations assumes that wildlife consumes 100% of their diet from the site. This is an unrealistic case. Exposure may be corrected to include the real consumption of food. This consumption may be estimated taking into account the time that the organisms stay in the site, which depends on the size of the habitat for this specie. These data can be obtained by radio tracking or visual observations of birds and mammals (EU, 2002). Data from radio tracking studies is obviously the most useful, but it is also the most difficult and expensive to obtain. On the other hand, the time that the organisms stay in the site depends on the size of the habitat for this specie. The quotient between habitat size and surface of the contaminated site is an appropriate factor for an initial correction of expected exposure through the contaminated food. Additional refinement can be obtained through the methodology applied by the European Food Safety Authority (EFSA) for the higher tier assessment of pesticides. The method considers the selection of key species for each site and the identification of food consumption items on the basis of the species metabolic requirements.

The consideration of the bioavailable concentration provides a more appropriate measure of exposure compared to the use of total soil concentration, and should be included in the exposure refinement. Different approaches described in section 3.2.2. to determine the available fraction, such as the use of the extractant solution, the soil pore water concentration, etc. can be applied. For very persistent and bioaccumulation substances, studies must be designed to properly address the potential biomagnification and bioconcentration risk. Studies must cover the potential risk associated with continued or repeated exposures at different trophic levels. The biomagnification risk is considered a key part of the assessment and all potential exposure routes should be considered.

In this level, variations in the exposure level due to time and space changes are considered in the estimation of exposure concentration. In soils, unlike waters, variations in space are more important than variations in time. The spatial consideration may be presented through non-specific or through geo-referenced methods. The first approach requires a lower level of information but the gradients are presented in generic terms not associated to specific locations.

4.1.4. Level 4. Semi Field and Field Studies

Models for exposure scenarios are based in the results obtained in the laboratory assays. These assays do not allow considering some factors occurring in the environment. Semi field and field studies allow more realistic estimations of exposure. The main advantage of the semi-field and field studies is that they are based on a realistic level of complexity. The studies cover specific aspects for refining model estimations. Obviously, the best approach is to conduct the studies within the assessed facility; however, the extrapolation of study outcomes between related field conditions may also offer a proper refinement. Almost any parameter or variable can be studied in the field on soil collected at the facility. These include those factors affecting bioavailability and aging processes.

4.1.5. Level 5. Exposure Assessment Based on Real Measurement

Almost any investigation aimed to characterize soil contamination is based on direct measurements of pollutants in soil. However, the monitoring efforts can also be applied to other samples. Direct exposure measurement instead of model predictions may be used to determine exposure to soil pollutants by indirect routes: in particular for the aquatic compartments (surface and ground waters) and terrestrial vertebrates feeding in the site. For example, for terrestrial vertebrates protection, the concentration in the food item is estimated from the soil-earthworm and the soil-plant bioconcentration factor (BCF) (Romijn et al., 1994). The estimation of the concentration of the contaminant in plants and terrestrial invertebrates using BCFs may be substituted by the direct measurement of the concentration of the substance in plants and invertebrates. As the uncertainties associated to those models are relatively high, this alternative may offer a quite significant level of refinement and uncertainty reduction.

4.2. High Tier Effects Assessment

The objective is to achieve toxicity values close to actual toxicity thresholds, as well as to determine those endpoints that are relevant for the ecosystem. Effects refinement consists in getting more information in order to obtain more realistic SQSs, mainly due to a reduction of uncertainty. Moreover, the specific characteristics of the site influencing the effects may be included in the performance of assays, which allows obtaining more consistent SSQSs. In the schema 1, higher tier protocols to refine effects on ecosystems, grouped in five levels according to SSC-European Commission (2003) are shown.

4.2.1. Level 1. Effects on Single-Species. Deterministic Approach

In this level, risk assessment is based in the deterministic approach carrying out the single-species tests on species selected to cover a minimum of taxonomic groups and assessment factors to account for several forms of uncertainty. In this way, PNECs which are expected to be protective for the concerned compartment, are obtained. More realistic PNEC can be obtained by improving the set of toxicity data. The testing of mores species reduces the uncertainty attributable to inter-species differences in sensitivity and a lower assessment factor may be applied. Likewise, the increase of chronic against acute data to soil and aquatic organisms, and the increase of the exposure time in the assays for terrestrial vertebrates allow decreasing assessment factors applied for short-term to long-term/chronic toxicity extrapolation.

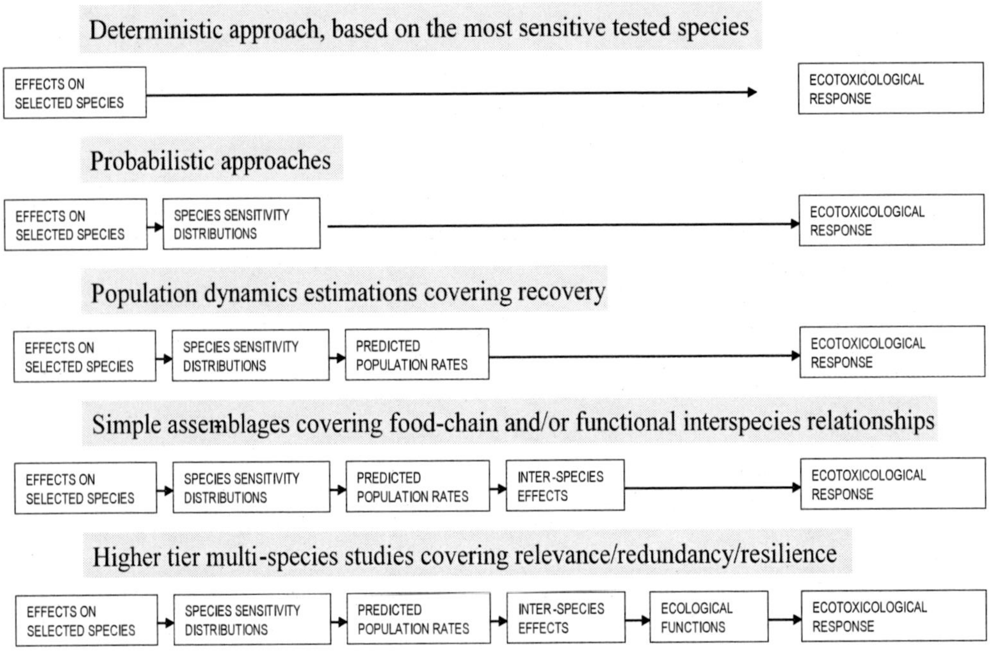

Scheme 1. Higher tier protocols to refine effects on ecosystems, grouped in five levels according to SSC (2003).

In most of the regulations, ecotoxicity data are normalized to standard soils taking into account the organic matter content in the case of organic substances, and the organic matter content and clay in the case of metals. Toxicity values may be modified to account actual soil properties of the site.

4.2.2. Level 2. Effects on Single-species. Probabilistic Approach

The probabilistic method most frequently used to refine effects is the SSD approach, although distributions may be also used to determine other types of variability and uncertainty in the effects assessment. In SSDs, the whole data set is used to elaborate a reliable distribution of the sensitivity of the species to the assessed chemical, unlike the deterministic method, which uses the lowest toxicity value. Two approaches are possible to obtain the species sensitivity distribution. The first consists in estimating SSD separately for each taxonomic group, which can be combined if similarities are observed. The second (TGD) consists in determining SSD using at least ten different species covering at least eight taxonomic groups. Efforts are in progress to validate these models, especially to extrapolate SSD values for the protection of the community and the ecosystem. This method requires a larger amount of data representing a wide spectrum of genera. It is necessary to obtain information for a higher number of species performing additional single-species tests in order to apply the model. Some criticisms of probabilistic approaches were made by Forbes and Forbes, 1993.

It should be noted that the whole concept of the use of SSDs for setting ecotoxicological thresholds has been derived for the aquatic compartment, where all assays are structural endpoints and therefore requires some adaptations when extrapolated to the soil/terrestrial environment (e.g. EC, 2003). The soil assessment includes functional parameters in order to the effects on soil microbial populations. Some guidelines, including the European TGD, recommend the derivation of two complementary distributions, one for structural points (plants and invertebrates) and the other for functional endpoints (soil microorganisms). However, when the actual data used for setting the microbial function distribution are considered, it is observed that actually, the assessment does not describe the distribution of the effects of the chemical on microbial processes, but the distribution observed for the same processes among different soil microbial communities. In fact, the data usually cover just three different microbial-mediated processes, each representing the effects on a different sub-group of soil microbial communities:

- Respiration, or C-mineralization, offers a generic assessment covering potential effects on almost the whole community.
- N-mineralization refers exclusively to a very specific group of species which are involved in the different elements of the N mineralization process such as nitrification or ammonification.
- The enzymatic activities cover potential effects on subgroups of species within the microbial community, with significant taxonomic differences depending on the soil and the measured enzyme.

Thus, the distribution offers the responses of different soil communities to the same endpoints, representing in reality a distribution of "community sensitivities" which is therefore comparable to the SSD concept.

4.2.3. Level 3. Populations

This refinement level is also based in single-species assays, but toxicity tests are performed in more realistic conditions. The assays intend to predict the ecological relevance of the effects. Thus, more realistic conditions of exposure are included in the assays (e.g. assays that include different routes of exposure). In addition, studies on population effects, biomarkers and endpoints related to the mode of action such as endocrine disruption, may be used as toxicity parameters.

An aspect which is usually evaluated at this level is the study of recovery after the exposure. Time assay is extended to allow species recovering. Effects may be considered of low ecological significance if recovery takes place in a given time period. Recovery may be very different according to the reproduction characteristics of the species. Consequently, a main problem of recovery studies is extrapolating the potential for recovery to other species with different ecological strategies. This is particularly relevant when assessing soil pollution, as typically the contamination is not homogeneously distributed within the facility but presents severe spatial differences. When the pollution is concentrated in hot-spots or presents gradients from certain point sources the overall effect should be assessed at the population level. Models to study the effects on populations, such as Predictive Population Dynamic (Ducrot et al., 2007; Lewis and Law, 2007), may be applied.

4.2.4. Level 4. Effects on Communities

Toxicity effects occurring on a species can affect other related species. In this level, species interactions are considered to predict effects at the community level. Single-species assays used in previous levels are substituted by multispecies assays that account for species interactions unlike single-species assays. Multispecie assays are laboratory tests performed in systems denominated microcosms. These systems reproduce part of a terrestrial or aquatic ecosystem and the effects on different organisms are simultaneously studied. Moreover, the exposure conditions may be selected to reproduce the actual conditions. The most used aquatic microcosms reproduces a simplified trophic chain (alga, daphnia, fish). In soils, microcosms include different taxonomic groups (e.g. microorganisms, terrestrial invertebrates and plants).

Aquatic multi-species test systems have been successfully used to assess effects of substances, specially plant protection products, under realistic exposure conditions. Comparable terrestrial use is less developed. However, efforts are directed to prove their application in higher tier assessments. (Fernández et al., 2004; Weyers et al., 2004; Boleas, 2005a; 2005b).

4.2.5. Level 5. Effects on Ecosystems

The aim of risk assessment is to protect the structure and function of the ecosystems. The most direct means of studying the effects on ecosystems are mesocosms and field assays. They allow identifying relevant effects, under simulated field conditions. Mesocosms offer the same advantages as microcosms, but the usually include a wider range of species and generally offer a greater potential to assess the response at the highest level of organization. Furthermore natural fluctuations in climatic conditions enhance the level of field realisms. In particular, they enhance the probability of recovery of some species through e.g. colonization. Mesocosms have been developed to study substance effects in the aquatic medium (Campbell

Bioassays may be used in a screening phase for a fast identification of polluted sites or for a complete ecotoxicological evaluation. Moreover, they represent valuable monitoring tools to evaluate the effectiveness of remediation processes. The main problem of using bioassays is the sensitivity of the selected methods. Assays performed with environmental samples allow the identification of polluted soils but may be inadequate to decide when a soil has a low risk as the additional assessment factor employed to set the threshold cannot be incorporated in the assessment, and the fact that acute (short-term) bioassays can be easily incorporated, while the inclusion of long-term chronic tests is controversial in terms of viability and cost. Moreover, bioassays do not identify substances causing toxicity. It can be a limitation in the remediation processes since the selected technique depends on the characteristics of the pollutants in the mixture.

An important step in soil quality assessment should be correlating chemical concentrations to effects measured with bioassays. If toxicity cannot be explained attending to the concentration of chemical substances in the soil determined by chemical analysis, further investigations into causal agents of toxicity should be performed. However, in occasions, to relate toxic response from bioassays to the concentration measured through chemical analyses may be difficult, mainly if combined toxicity occurs.

5.1. Methods of Ecotoxicity Assay

There are different possibilities to study toxicity with environmental samples. The most widely used method consists in laboratory tests, where organisms are exposed to potentially polluted soils that are collected in the study site and transferred to the laboratory. However, other possibilities to assess soil quality by means of toxicity analysis of environmental samples are possible. Intact soil cores may be transferred to the laboratory and toxic effects may be tested to indigenous biota under controlled laboratory conditions. Another way consists in taking organisms from the site and studying them in the laboratory to determine the bioconcentration, biochemical and physiological effects, etc. Finally, in situ biological monitoring analyzes various parameters of natural populations studying how they are affected by exposure to contaminants and determining the ecological quality of a site (Hopkin, 1993; Beck et al., 2005; Römbke et al., 2005). Although the last studies may be considered more realistic, their implementation presents additional complexities and have a main weakness, as pollution represents previous historical releases, the biological community found at the facility has been adapted to or selected by the pollutants, and therefore may be unrepresentative of the natural population expected for equivalent unpolluted sites. In this chapter, the term bioassay is used for toxicity assays performed in the laboratory with selected organisms.

Bioassays on soil samples may be performed using two approaches. In the former, samples are diluted with a non-contaminated control or reference soil and different soil concentrations are tested. In this way, a dose-response similar to that obtained when assaying pure chemical substances is obtained. In the second approach, assays are performed with non-diluted samples. The approaches can be combined or selected on the basis of the expected level of pollution. The results of the toxicity assessment are compared to either a non-polluted control or a clean reference sample and expressed using conventional methods such as percent of inhibition respect to the control, Toxic Units, or the equivalent to the L(E)Cx or

NOEC expressed in terms of dilution. The first approach is fast and cheap and allows obtaining directly the toxicity results as a percentage of inhibition between 0 and 100% for each sample. In the dose-response method various dilutions must be tested to obtain the L(E)C50 equivalent value, requiring a higher number of replicates and being more expensive. Moreover, the L(E)C50 values may not be obtained if the sample does not present enough ecotoxicity. However, the use of different concentrations allows obtaining additional information about the risk of contaminated soil. Bioavailable pollutants may be released in the environment due to significant changes in soil properties. This fact is particularly important during soil remediation. In the dose-response approach, the dilution processes modify the bioavailability of chemicals in the soil covering risk due the potential bioavailability changes. In addition, assays at low exposure levels allow determining the risk of spreading the contamination to other areas due to surface movement of contaminated soil caused by erosion, flooding, etc. Regarding data analyses, correctly applied, regression models provide estimations of the parameters of interest with higher sensitivity and precision than analyses at a single dose. Moreover, the use of different concentrations allows detecting cross effects such as hormesis or effects due to different physicochemical properties of the test and control soil.

In both approaches, the effects observed for contaminated samples are usually compared with data obtained from control or reference soils, and significant differences are estimated using statistical analyses. A key point in the soil assessment is the selection of a control or reference soil. Organism toxicity can vary with soil characteristics and selecting an unsuitable reference soil may result in erroneous data. In the ideal case, the reference soil should differ from the contaminated soil only by the absence of pollution. Thus, reference soils are sampled at a non-polluted localization close to the contaminated site. When it is not possible, a reference soil with pH, soil organic matter content, soil texture and nutrients comparable to the test soil should be selected. The selection of a proper reference soil is particularly critical for sublethal endpoints such as reproduction or growth. Texture, nutrients availability, etc may be so influential in the control response that the differences related to the presence of toxic chemicals in the sample can be masked or overestimated.

In addition to soil testing using soil organisms, aquatic organisms have been often used to assess soil toxicity based on the assumption that the chemicals affect the soil organisms via the aqueous phase (Hammel et al., 1998; Bispo et al., 1999). However, this method does not seem correct mainly because it assumes that the sensitivity of terrestrial organisms is similar to that of aquatic organisms. Soil quality tests should focus on soil dwelling organisms (Maxam et al., 2000; Van Gestel et al., 2001; Eom, et al., 2007; Leitgib et al., 2007), while aquatic tests on soil leachates should be used for the risk assessment of contaminated soils to ground and surface waters.

Leachates preparation is a crucial step in the assessment of soil toxicity. Leaching tests may be divided into three kinds: batch tests, column tests and field tests. The leachant used in most of the tests is distilled or de-ionized water, however, the nature of the leachant may vary depending on the purpose for which the test was developed. The batch test consists in shaking a sample of the soil to be assessed during a certain time period, defined shaking conditions and a specified volume of leachant. The main variables in this test are the liquid to solid ratio, the time and potency of the shaking method, and the leachant used in the assay. Different standard leaching methods have been developed for soil assessment (e.g. DIN 38414). The soil to liquid ratio ranges between 1:2 and 1:10. In most of methods, water is the solvent selected and a 10:1 is the most frequently proposed liquid to soil ratio. The column test consists in passing the leachant through a column filled with the soil. This method allows selecting assay conditions which are representative of the environmental situation being assessed. The variables which affect the assay are the column size, and the flow and characteristics of the leachant. The leaching test may be also performed in lysimeters under field conditions. The leachates obtained may be used for chemical analyses or ecotoxicological tests, providing information on transfer of contaminants from soil to water. Leachates tests are performed by comparing test data to a blank. However, effects due to nutrients in the soil (e.g. growth-stimulating effect on algae) can hide the toxic effects of soil contaminants. This phenomenon may be avoided by relating test data to the leachate from an appropriate reference soil rather than standard blanks normally used in these tests.

5.2. Test Organisms

The application of a set of assays with different organisms is necessary in order to provide an accurate assessment of environmental quality (Greene et al., 1989; Keddy, 1995, Maxam, et al. 2000); due to the different sensitivity of the organism to the various contaminants (Bierkens et al., 1998). Soil protection is usually not directed to protect the structure (composition of the soil biocenosis) but to protect the soil functions. The main functions of soils are the ones associated with organic matter decomposition, nutrients cycle and synthesis of humic compound. The vegetation coverage and the contribution of soil organisms to the soil texture are also relevant. Every function may be performed by different species (functional redundancy). Therefore, the protection of soil functions is guaranteed if the ecological structure is maintained. However, changes in ecological structure do not necessarily mean loss of function.

Bacteria and fungi play an essential role in the decomposition of organic residues. Other organisms such as mites, springtails (Collembolan), isopods, mollusks, millipedes and earthworms, which disturb soil by ingestion or burrowing, influence soil core structure and texture. Earthworms are the best-known group with an important role in maintaining structure and soil fertility. Soil fauna also contributes to the decomposition process directly or indirectly through soil aeration and by breaking down leaf material into smaller particles, which are more easily decomposed by microbes etc. Finally, plants are the most prominent primary producers in all ecosystems. Moreover, they contribute to soil fertility and structure, regulate the potential for erosion, run-off and water behavior, and define habitats for wildlife.

Ideally, the toxicity of all ecological or commercially important organisms living in or associated to the soil should be tested. However, this is not possible and a small number of assays are selected to cover as much as possible the ecosystem risk. The test organisms selected should be representative of the functional roles played by the organisms encountered in the field and sensitive to the contaminants. The organisms are selected attending to the relevance of the species, the sensitivity of the tests and the simplicity and the costs of the assays. Different toxicity tests have been developed for the assessment of chemical substances and have been standardized by several associations (OECD; ISO, ASTM, DIN, AFNOR, EPA, etc) and may be applied to the assessment of soil samples. There are many different ways to measure toxicity. Biochemical, physiological, reproductive and behavioral effects can provide toxicity data. However, in direct toxicity assays, the usually measured parameters are lethality, reproduction and growth. These parameters are considered to be relevant for the estimation of field effects. A review of the toxicity tests available for ecotoxicity assessment of contaminated sites was compiled by Stephenson et al. (2002).

The most widely used tests with soil invertebrates include earthworms, springtails (Collembolan), enchytraeids and isopods (Callow, 1993). The assays with earthworm are the best established and are used as legislative requirement for the risk assessment of chemical substances. Two earthworm tests, which measure effects on mortality and reproduction, using E*isenia sp.* have been standardized by the OECD (2004a). Other sublethal effects such as avoidance, loss of weight, etc may be measured. The bioconcentration of chemical in earthworms can also be determined and used to predict the exposure of predators through the food.

Collembolans are used in many laboratories to assess the effect of chemicals, although these assays have not got such an advanced legislative stage as earthworms. Collembolans are among the most abundant arthropods. Most species consume fungi in soil and leaf litter. The most widely used specie of Collembolan is a parthenogenetic strain of *Folsomia candida* (ISO, 1999). The effects on reproduction are used as endpoint. However, future testing methods involve the use of more sensitive endpoints such as locomotor activities, behavior, avoidance of contaminated food and effects on population growth rates (Fountain and Hopkin, 2005).

Enchytraeids are ecologically relevant species for soil toxicity. They can be used in laboratory tests as well as in field assays. Many enchytraeids are easy to handle and breed and the reproduction time (4-6 weeks) is less than for earthworms (8 weeks). Soil dwelling annelids of the genus *Enchytraeus* are used in toxicity tests. A reproduction test with *Enchytraeus albidus* has been standardized by OECD (2004b). This species has a worldwide distribution and it is found in marine freshwater and terrestrial habitats, mainly in decaying organic matter.

Tests with isopods are less developed. The species employed are *Porcellio scaber* and *Oniscus asellus* (Drobne and Hopkin, 1994; Van Brummelen et al., 1996). The main problem associated to the use of these species is their low growth rate and long reproductive cycle. They are widespread and act breaking down leaf material into smaller particles, which are more easily decomposed by microorganisms. The isopods tests measure the effects in feeding rates, mainly through feeding repellence.

Plants are an important taxonomic group to test soil quality. Different methods have been developed to study the toxicity of chemicals to plants. Phytotoxicity endpoints are measures of the survival or growth than can be reported as the change in height/length or biomass. Specific metabolic enzymes, total respiration and dark respiration have been also used as measures of endpoints. Sensitivity is highly dependent on plant species. Consequently, the use of several test species, widely distributed within plants taxonomy, is advisable to provide better information on effects in the environment. A list of recommended species for testing may be viewed in (OECD, 2006a). The chemical bioconcentration is usually measured in plants because they act as an exposure path to animals (including humans) through the food chain.

The microbial test consists in measuring effects on their functions, while structural (species oriented) testing is considered of lower relevance. Microorganisms play a fundamental role in the soil ecosystem processes, such as decomposition of soil organic matter and the cycling of nutrient. However, due to functional redundancy, sensitive species can be replaced and the diversity can be altered without a significant effect on the functions. A great number of methods to determine the microbial function and activity have been developed. Two soil microorganism tests have been standardized by OECD (2000a, b): nitrogen and carbon transformation. These tests measure soil respiration and N mineralization which are important processes in the carbon and nitrogen cycles, respectively. In addition, assays are available for a large number of soil enzymes, including dehydrogenases, β-glucosidases, ureases, amidases, phosphatases, arylsulfatases, cellulases and phenol oxidases (Dick et al., 1996).

Aquatic toxicity tests have achieved a higher development than testing protocols for soil organisms. The main exposure route is direct uptake from water (waterborne exposure) although uptake can also occur from food. One limitation to perform aquatic toxicity tests with leachates is the necessity of using static systems. The concentration of the chemical in the water can change due to adsorption and metabolism by the test organism and volatilization or degradation processes and sorption to particles. However, this allows determining effects due to the actual exposure and consequently, assays, if well designed, can be more representative for the expected exposure in the field. Aquatic toxicity is mainly evaluated using fish, daphnids, green algae and vascular plants. Fish, daphnids and algae are required in the risk assessment of chemical substances.

Fish are key organisms in the assessment of waters due to their diversity, their importance in natural ecosystems and their economical value. Moreover, they play an important role as they are secondary consumers. Consequently, fish bioaccumulation is frequently measured in relation to human exposure trough food. Different fish tests have been standardized by OECD (1984b;1992b,c;1998b;2000c). In these assays different species and different stages of the life cycle (eggs, embryo or adults) are employed. Lethal and sublethal effects (hatching, abnormal behaviour, malformations, etc) are assessed at different exposure times. Fish toxicity can be also assessed at the cellular level. The main advantage of *in vitro* assays using cell fish is that they reduce the use of vertebrates in toxicity studies and the volumes of leachates necessary to perform the tests. Moreover, they show a fast response and are usually simple to perform and analyze. However, a bigger effort is necessary to relate effects at the sub-organism level to effects at higher organizational levels and hence to ecosystem effects.

Invertebrates occupy a key position as primary consumers in the aquatic food chain. The toxicity to this taxonomic group has mostly been based on the cladoceran crustacean, *Daphnia*. The selection of daphnids is due to their broad distribution in fresh water bodies, their role in the food chain and the ease of handling in the laboratory. Most of the protocols recommend *Daphnia magna* o *Daphnia pulex* as test substance. An acute and a chronic test have been standardized by OECD (1998a; 2004c), using immobilization and reproductively as endpoints, respectively. A related genus, *Ceriodaphnia*, with a shorter life-span, is frequently used in North America for cost/effective chronic testing.

In algae tests, the effects on growth are determined on a green algal population growing exponentially in an enriched medium (e.g. ASTM, 1990; OECD, 2006b). The algal growth test is a chronic test since growth rate is rapid and effects can be assessed over several generations. Cell concentrations are measured using an electronic particle counter, a microscope with a counter chamber, a fluorimeter, a spectrophotometer, etc. The most frequent species used are unicellular green algae of rapid growth such as *Pseudokirchneriella subcapitata, Desmodesmus subspicatus* and *Chlorella vulgaris*. Ease of culture, visibility and availability are some of the criteria used to select these algal species.

Vascular plants and macrophytes are important in the functioning of freshwater ecosystems. The duckweeds (Family Lemnaceae) are widely distributed throughout the world. *Lemma gibba* and *Lemna minor* have been the most extensive studied species. They are easy to culture and have a rapid reproductive rate. Standardize methods have been published by different associations (e.g. ASTM, 2003; OECD, 2006c). The test endpoint is inhibition of growth and the measurement variable is frond number. Although results of *Lemna* are used in regulatory processes, algae continue being the predominant test species in aquatic phytotoxicity tests.

Finally, two bacterial toxicity tests have been often used to analyze the toxicity of waste and soil leachates. The bioluminescence assay with *Vibrio fischeri* (ISO, 1998) is probably the most widely used. This luminescent marine bacterium is sensible to a wide range of heavy metal chlorides and narcotics (Johnson et al., 1942). Because of the fast metabolism of this bacterial cell, this assay is much faster than other comparable assays and allows a rapid measure of the leachate toxicity. In this test, the addition of contaminants modifies the light emission, which is measured in a photometer after a given interval. The test is sensible, reproducible and precise. However, its relevance has been largely discussed mainly due to the use of a marine bacterial. Soil leachates have been also tested to mutagenicity using *Salmonella typhimurium* strain (Ames et al., 1975). However, the ecological relevance of mutagenicity test results for soils is difficult to evaluate.

5.3. Battery of Single Species Assays

Differences in species sensitivity to different contaminants, stress that a valid assessment of the environmental hazard of contaminated soils can only be obtained on the basis of test covering different key organisms. Soil quality assessment can be performed using a battery of single-species assays on soil organisms and aquatic organisms conducted on the leachates (Van Gestel et al., 2001; Robidoux et al., 2004b). Tests developed for the toxicity analyses of chemicals may be used to test environmental samples without excessive protocol changes. A selection of tests that may be employed for the investigation of soil pollution is given in Table 1. The selection of the appropriate test organisms to measure soil toxicity is complicated. The approach usually consists in defining a simplified hypothetical ecosystem including organisms which belong to different trophic levels, with different feeding and different phylogenetic characteristics. In leachate assessments, the selected organisms usually include algae as primary producers, aquatic invertebrates as primary consumers and fish as secondary consumers. In soil assessments, the organisms belonging to three taxonomic groups are included, plants as producers, terrestrial invertebrates as consumers and microorganisms as decomposers.

The tests can also be selected depending on the use of the soil, the soil functions and the entities to be protected. Dott et al. (1995) proposed a stepwise approach for the assessment of the ecotoxic potential of soils with different land uses and soil functions.

5.4. Microcosms

An alternative to a battery of independent toxicity tests is the use of multispecies systems in microcosms. This method allows assessing at the same time toxicity to several interacting organisms being exposed simultaneously. The method assumes that interactions between species can make a system more sensitive and/or realistic when assessing the effects of contaminants than that expected according to single species toxicity. Moreover, microcosm tests may incorporate and contribute to determine the environmental fate and behavior of contaminants in the soil.

Usually the system consists in a column of soil with a device to collect the leachate from below. Water is applied at regular times and it is collected using a funnel. Water addition simulates rainwater percolating through the soil. Two types of microcosms have been often used: intact soil cores with autochthonous soil organisms (US EPA, 1996; Sheppard, 1997; Knacker et al., 2004) and artificial assemblages adding soil organisms on sieved soil columns (Salminen and Haimi, 1997; Chen and Edwards, 2001; Fernández et al. 2004). Both systems have been developed to test potentially harmful substances.

The application of the first approach to the assessment of contaminated soils is difficult. Intact soil cores with naturally occurring organisms allow analyzing effects to the indigenous biota under defined laboratory conditions. However, these organisms have been exposed to, and selected by the toxicants for long periods and consequently the development of resistance to exposed contaminants and the selection of the less sensitive species is expected (Loibner et al., 2003). In this case, the effects studied are limited to the ecosystem functions such as soil respiration or nitrification capacity. For ecotoxicological testing of soil quality, the latter approach seems to be more appropriate. It allows determining toxicity to a set of selected organisms not previously exposed to contaminants. Moreover, experimental variations between replicates units are minimized (van Straalen and van Gestel, 1993). It is also possible to adapt these types of soil microcosms to the conditions required to test contaminated soils, the MS·3 approach developed at INIA and presented later on in this chapter represents a good example.

5.5. Data Interpretation

5.5.1. Statistical Analyses

Two different types of statistical analyses are possible depending on the experimental design. In the first, the experimental conditions comprise a control plus a number of concentrations replicated a number of times. The aim is to obtain a quantitative concentration-response relationship by regression analyses. In the second a non-diluted sample is assessed, and the effects on contaminated soil are compared to a reference or control soil.

In the first case, the data may be fit to a linear regression by performing a transformation of response data for instance using a logit-probit (Litchfield and Wilcoxon) or Weibull units (Christensen and Nyholm, 1984). Through a process known as inverse estimation, a concentration corresponding to a specified percentage effect relative to the control is estimated. As a difference with the analyses of chemical substances, data are referred as dilution of contaminated soil instead of concentration (i.e. the dilution of tested soil corresponding to a given percentage of effect is estimated). Confidence intervals can be calculated and hence, the precision of L(E)Cx obtained may be determined. These methods are appropriate to use on quantal data (e.g. mortality). For continuous data (e.g. growth) other methods are more appropriate (OECD, 2005).

In the second method, the test soil and the control soil are compared using analysis of variance (ANOVA). A disadvantage of testing only a soil concentration is that if experimental variability is relatively high, the sensitivity of the analyses to detect differences from control will be relatively low. This implies that only large differences from the control can be detected. The sensitivity of statistical analyses can be improved, increasing the number of replicates. The detection of no effect for toxic samples (false negative results) can be also due to insufficient sensitivity of the test method. This is especially important in the assessment of remediation processes where the levels of the contaminants have been reduced. It is a limitation of the use of bioassays since they allow detecting toxicity effects in polluted soils but not always in remediated soils.

5.6. Hazard Soil Classification

Bioassays may be applied for screening purposes to identify toxic samples. In this case, a simple differentiation of test results using toxic or not toxic is sufficient. However, this qualitative approach is not sufficient to quantify the risk potential of contaminated soils (Maxam et al., 2000). To incorporate direct toxicity assays in regulatory frameworks, toxicity classification systems which allow attributing a hazard score to polluted samples in accordance with increasing levels of toxicity are desirable. Thus, the evaluation of the results should be made with the help of threshold values, similar to the threshold values used in the fields of the chemical analyses.

Different methods have been proposed for ranging toxicity in assays performed with complex samples. However, none of the classification methods so far has fond general acceptance at the international level. Most of them are based on batteries of short-term toxicity tests, performed on dilution series of original samples. Some methods are based in the value of Toxic Units (TU), described as the inverse of the L(E)C50 expressed in % (Sprague and Ramsay, 1965). This expression is the dilution factor, which must be applied to the sample to obtain a 50% effect. Different categories of soil toxicity are established according to the TU value determined. These categories range from non toxic (TU<1) until extremely toxic (TU>100) (Bulich, 1982). This method has been widely applied to waste waters and waste leachates classification (Clément et al. 1996; Persoone et al., 2003).

Another approach attributes soil to classes of hazard on the basis of the lowest ineffective dilution (LIDx), which is defined as the dilution at which the effect is below an established percentage (x). This method has been applied to aqueous soil extracts (Maxam et al., 2000; Hund-Rinke et al., 2002) which are considered toxic or not toxic using a different percentage depending on the test specie, ranging from 10 % (e.g. algae test) to 20 % (e.g. luminescence test). A classification of hazard for the sample is given on the basis of the LID obtained for the most sensitive test species by Selivanovskaya and Latypova (2003). Accordingly, threshold values of LID_{10} are established allowing designed samples to four classes of hazard (hazardous to non-hazardous)

The "Chemical equivalent method" expresses L(E)C50 values as chemical equivalent, which mean the chemical concentration that would cause the same toxicity as the pollution in the soil. Soil toxicity is related to the content of potent reference compound. These compounds are selected according to soil contamination (Leitgib et al. 2007).

Complex index including toxicity responses, consideration of routes of exposure, level of response (acute lethal or chronic sublethal) and sensitivity of the assays can also ser defined (e.g. Bombardier and Berminghan, 1999, developed for sediments).

In the Spanish regulation, dilution values are defined, which, if exceeded, indicate a high ecotoxicological risk. The proposed criteria consider a high risk if the short-term L(E)C50 value for the most sensitive soil taxonomic group is lower than 10 mg sample soil /g total soil or if short term L(E)C50 values of soil leachates to the most sensitive specie of aquatic organisms is lower than 10 ml leachate /l solution. This is equivalent to a limit of 100 TUs described above. On the other hand, potential risk exists if toxicity in one test performed with non-diluted samples indicates toxicity.

Threshold values can also be based on percentages of toxicity. For example, Phillips et al. (2001) established a threshold based in the difference in organism response between the sample and control. This threshold is established as a percentile of minimum significant difference value. Value of percentile is specific to each toxicity test.

In all the approaches mentioned above to classify soils the last step consists in deciding which level of soil toxicity is acceptable. This decision should be based in the potential risk to the ecosystem. With this aim, the probability that the effects measured in the laboratory will occur in the field, has to be quantified. Consequently, the development of models which allow predicting ecological effects from laboratory assays, is necessary. In summary, bioassays endpoints are quantitative measures of toxicity, which can be used to characterize contaminated soils and priories soils to remediation. However, before the extensive apply of bioassays to regulations, more information must be obtained as how the results of bioassays are related to the hazard classification of soil.

6. Soil Characterization Using MS-3. A Case Study

Terrestrial microcosms are a powerful tool to study soil quality. These systems can be used to evaluate qualitatively the fate of chemicals, their effects on structural and functional aspects of ecosystems and to develop conceptual models of chemical transport, fate and effects (Gillett and Witt, 1979). In this section the application of a microcosm system denominated "Multispecies Soil System" (MS-3) for soil characterization is presented. This soil microcosm system was initially designed in the Laboratory for Ecotoxicology (INIA) to test pure chemicals and complex mixtures (Fernández et al., 2004; Boleas et al., 2005a, 2005b). However the usefulness of MS-3 for the characterization of polluted soils has been proved in different assays (Fernández et al., 2005).

The Multispecies Soil System (MS-3) consists of soil columns coupled to a leachate collector system. Representative soil macro-organisms (plants and terrestrial invertebrates) are introduced into these soil columns. These organisms, as well as the soil microbial community, are exposed to chemicals under simulated realistic environmental conditions. During the exposure period, the system is watered to simulate rainfall events, and artificial sunlight is provided for plant growth and to provide a photoperiod and temperature gradient approaching natural conditions. Leachates are collected and tested for toxicity using representative aquatic organisms (i.e. algae, daphnia and fish cell-lines) for which test procedures have been standardized.

In order to determine soil toxicity to soil organisms, after 21 days MS-3 columns are opened and the number of adult earthworms surviving as well as emergence of seedlings and aerial biomass production are recorded. The effects to microorganisms may be determined using the soil respiration test and nitrogen transformation test following the principles of standardized methods (OECD, 2000a; 200b) or by measurement of enzymatic activities: dehydrogenases and phosphatases (Boleas et al., 2005b). Toxicity of soil to aquatic organisms is determined for the leachate samples obtained from the soils in the microcosms using aquatic tests that included an algal test (OECD, 2006b; Ramos et al., 1996), the daphnid test (OECD, 2004c) and the RTG-2 fish cell line test (Babín et al., 2001), to measure growth, immobilization, and cytotoxicity of leachate respectively. Cytotoxicity is determined to RTG-2 fish cell line (ATCC, CL N° 55) using four different endpoints parameters: enzymatic activities as 7-ethoxyresorufin O-deethylase (EROD) (Bols et al., 1999) and β-Galactosidase (Pablos et al., 1998), uptake of neutral red stain (NR) to evaluate cell viability (Borenfreund and Puerner, 1985) and FRAME KB protein assay to evaluate cell detachment from a substrate (Knox et al., 1986).

The goal of the system was to develop a cost-effective system which provided useful information on both, adverse effects of pollutants on soil dwelling organisms and the capability of the pollutants to migrate through the soil matrix, providing an indication of their mobility and the possible risk to reach the aquatic compartment and produce adverse effects on aquatic organisms. Next, a practical example of the application of this approach for the assessment of polluted sites is presented. Five soils taken from different sites contaminated with organic and/or inorganic substances were assessed using the MS-3 microcosm system. The main physicochemical characteristic of these soils were shown in Table 2. Chemical characterization of soils was based on the historical use of the site. All soils were contaminated with metals at different concentrations. Soil 1 showed the highest metal concentrations. Soil 2 and soil 5 were also contaminated with organic substances. Soil 2 presented levels of polychlorinated byphenyls (PCB) and hexachlorocyclohexanes (HCH). Soil 5 contained also concentrations of petroleum hydrocarbons. Procedures for soil toxicity assessment using MS-3 were accomplished as described by Fernández et al. (2005).

Two approaches were used to direct toxicity assessment of contaminated soils. In the former one, soils were tested in non diluted samples and toxicity data were obtained as percentage of inhibition compared to the control soil. Toxicity responses obtained in test soils were compared with those in the control soil by one-way analysis of variance (ANOVA), with Fisher's least significant difference procedure (LSD, $P< 0.05$). In the second approach soils were tested at four dilutions of contaminated soil: 6.25, 12.5, 25 and 50 % (w/w) for subsequent calculation of L(E)C50 using probit analyses.

Table 2. Main physicochemical characteristics and soil chemical concentrations in the control and test soils

	Texture					Metals mg/kg					PCB	HCH	TPH
	Clay %	Silt %	Sand %	pH w (1-2.5)	Organic matter %	As	Cd	Cu	Pb	Zn	µg/kg	µg/kg	mg/kg
Control Soil	7.8	18.8	73.4	7.27	1.9	nm	0.2	27	31.3	62	nm	nm	nd
Soil 1	17.4	19.7	62.9	5.56	17.5	2492	209	13547	24127	7458	nm	nm	nm
Soil 2	12.3	29.4	58.3	6.9	10.3	484	44	372	4084	1056	189	43	nm
Soil 3	5.15	nm	nm	7.9	1.1	nm	9.6	67	480	5900	nm	nm	nm
Soil 4	25.6	nm	nm	6.3	4.6	nm	9.2	nm	431	1254	nm	nm	nm
Soil 5	15.5	43.8	40.7	7.82	3.8	68	1.6	600	986	1476	98	199	974

Note. nm: not measured; nd: not detected.

Toxicity data to soil organisms obtained in the assessment of non diluted samples of soils using the MS-3 microcosm system are shown in Table 3. Soils were assessed at 100 % soil concentration except in the carbon mineralization test, where soils were tested at 10 % (w/w) concentration of contaminated soil, to minimize interferences in the test due to differences in microbial biomass between control and test soils. In contaminated soils, an active microbial population may exist as a consequence of the adaptation of soil microbial populations to the contaminants (Bruins et al., 2000). Assessment of diluted soils allows testing the effects of contaminated soils to non-adapted populations of the control soil.

As expected, the toxic effects varied substantially depending on soil and test species. All soils were toxic to plants. However, effects varied with the plant specie tested and the endpoint measured (emergence of seedling or growth). Effects to earthworms were observed only for soil 1 and soil 3. Inhibition of carbon mineralization was observed with soil 3, soil 4 and soil 5. The lowest percentages of inhibition were observed in the carbon mineralization test, however, these do not indicate a lower sensitivity of microorganisms to these soils since in this test, soils were assessed at a lower concentration (10% soil concentration).

Soil 1 presented high effect on earthworm survival with 100 % of mortality. In the plant test statistical analyses revealed significant differences on wet mass of shoots between treatment and control soil to three species tested, but not on seedling emergence. Wet weight was significantly reduced with respect to control soil in *T. aestivum* (52 ± 9 %), *B. napus* (81 ± 20%) and *T. pratense* (51 ± 11 %). No inhibition of carbon mineralization was observed for soil 1, indicating that the microbial communities were not adversely affected in this soil.

Soil 2 was the least toxic presenting toxicity only for one among the eight endpoints measured. A decrease on *T. pratense* weight equal to 47±9 % was the only adverse effect observed. Earthworms and microorganisms were not affected.

Soil 3 produced toxic effect to all taxonomic groups tested. This soil caused a 25±1 % of earthworm mortality. In the plant assay, soil 3 adversely affected seedling emergence and plant growth for *T. aestivum*, with values equal to 72±10 % and 48±5 %, respectively. Microorganisms were also adversely affected with 32±1 % inhibition of carbon mineralization.

Soil 4 was toxic to plants and microorganisms. In the plant test, wet weight was significantly reduced with respect to control soil only in *T. aestivum* growth (44 ± 7 %)

whereas a decrease of seedling emergence was observed in B. *napus* (68 ± 6 %). Soil 4 produced a slight (9.9 ± 0.2%) but significant inhibition of carbon mineralization.

Soil 5 presented toxicity to plants and microorganisms. This soil did not adversely affect earthworm survival. A significant decrease of seedling emergence compared to the control soil was observed in *T. aestivum* (85±8 %) and T. *pratense* (65±11 %), whereas growth was not affected in any of plant species tested. An significant inhibition of carbon mineralization of 14 ± 1 % was observed.

A high toxicity should be expected for these soils according to soil metal contents. However, toxicity was lower than the expected from total metal concentrations, suggesting a low bioavailability of metals in this soil. In all soils, the soil lead level (431-24127 mg kg^{-1} soil dry wt) was higher than the data of L(E)C50 values described in the literature (ECOTOX Database) for plants (EC50=50-100 mg kg^{-1} soil dry wt) and for earthworm (EC50=3.7-10 mg kg^{-1} soil dry wt). However only adverse effects to earthworm survival were observed with soil 1 and soil 3. Moreover, effects in soil 3 were 25 ± 1 % of mortality despite that the lead level in this soil was 480 mg kg^{-1} soil dry wt. Likewise, soil 2 with high contain in lead and other metals did not present toxicity to any soil organisms tested except a decrease of *T. pratense* weight (47 ± 9 %). This seems to indicate that the contaminants in this soil are not biologically available. Total concentrations can overestimate the real risk, as aging processes can strongly reduce the bioavailability, and, subsequently, the toxicity of pollutants (Song et al., 2006; Oorts et al., 2007). Consequently, it is difficult to estimate the potential risk of a soil contaminated by a complex mixture of substances attending exclusively to chemical characterization. Direct toxicity assays complement soil characterization based in chemical analyses since they allow direct assessment of the bioavailable fraction of the soil contaminants as well as synergistic and antagonistic interactions.

Table 3. Toxicity data to soil organisms exposed to non diluted samples of soils 1, 2, 3, 4, 5 and 6, respectively, in the microcosm system MS-3

	Percentage of inhibition compared with control soil (%)							
	Earthworms[*]	Plants						**Microorganisms**[**]
		Triticum aestivum		**Brassica napus**		***Trifolium pratense***		
	Mortality	Seedling Emergence	Wet weight plant	Seedling Emergence	Wet weight plant	Seedling Emergence	Wet weight plant	C. Transformation
Soil 1	100±0[a]	ns	52±9[a]	ns	81±20	ns	51±11[a]	ns
Soil 2	ns	ns	ns	ns	ns	ns	47±9[a]	ns
Soil 3	25±1[b]	72±10[a]	48±5[a]	ns	ns	ns	ns	32±1[a]
Soil 4	ns	ns	44±7[a]	68±6	ns	ns	ns	9.9±0.2[b]
Soil 5	ns	85±8[a]	ns	ns	ns	65±11	ns	14±1[b]

Note. Shown data were significantly different from control (P<0.05) by LSD procedure; ns: not significantly different from control.

Different letters indicate significantly different values using one-way ANOVA (LSD; P<0.05);

[*] In earthworm, the percentage of mortality was determined in relation to the number of organisms in the test (10 organisms in each microcosm).

[**] Microorganisms toxicity was assessed at a concentration of 1.0 g test soil /10 g soil.

Table 4. Toxicity data to soil organisms and aquatic organisms exposed to diluted samples of soils 3 and 5, in the microcosm system MS-3

Soil	Test organism	Endpoint	L(E)C50 % (w/w) soil concentration	TU	Description
Soil 3	Plants *(B. napus)*	Wet weight plant	48 (36-87)	2.1 (2.8-1.2)	Acute Toxicity (1<TU<10)
	Microorganisms	C. transformation	50 (42-62)	2.0 (2.4-1.6)	Acute Toxicity (1<TU<10)
Soil 5	Earthworms	Mortality	10 (2-15)	10.0 (50.0-6.7)	High Acute Toxicity (10<TU< 50)
	Plants *(T. aestivum)*	Seedling emergence	21 (15-28)	4.8 (6.7-3.6)	Acute Toxicity (1<TU<10)
	Plants *(T. pratense)*	Seedling emergence	18 (14-23)	5.6 (7.1-4.3)	Acute Toxicity (1<TU<10)
	Microorganisms	C. transformation	30 (26-35)	3.3 (3.8-2.9)	Acute Toxicity (1<TU<10)
	Algae	Growth rate	30 (23-40)	3.3 (4.3-2.5)	Acute Toxicity (1<TU<10)

Data are expressed as L(E)C50 values.

Bioassays were also performed on dilution series of the original samples of soil 3 and soil 5. A dose response relation was observed for some of the endpoints measured and L(E)C50 was determined (Table 4). Ecotoxicity of aqueous leachates obtained from soil 3 and soil 5 was tested to algae, daphnia and fish. Toxicity of leachates to aquatic organism indicates transfer of contaminants from soil to leachates and the potential risk of soil to surface and ground waters. In soil 3, a dose-response relation with percentages of effects higher than 50 % was observed only for growth of *B. napus* and carbon mineralization. For soil 5, L(E)C50 could be estimated for soil organisms including the three taxonomic groups and for one aquatic organism (algae) exposed to the leachates of this soil.

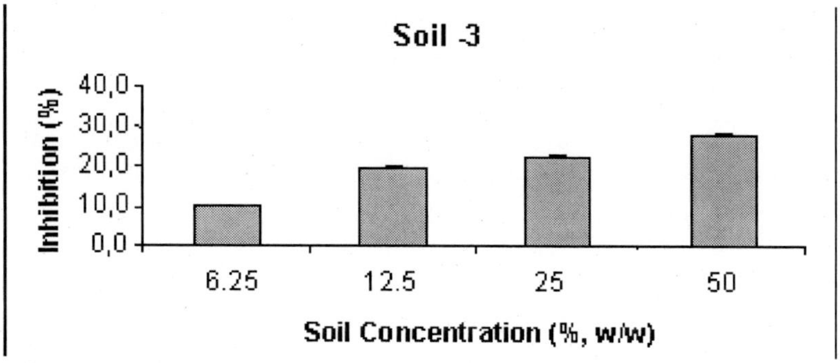

Figure 4. Inhibition growth rate of algae exposed to leachates from soil 3 determined according to OECD (2006b).

To soil organisms, soil 3 showed a concentration-dependent effect on *B. napus* growth and carbon mineralization with an EC50 = 48 (36-87) % and EC50 = 50 (42-62) % (w/w) soil concentration, respectively. To aquatic organisms, soil 3 produced a dose-response inhibition of algae growth rate, which was lower than 50% of effect at the highest concentration tested (Figure 4). Effects were also observed to daphnia and fish, which cannot be explained from the soil concentration. In both organisms, the least effect was observed when these organisms were exposed to leachates obtained from soil at the highest concentration tested (50 %). In the daphnids test, significant differences were observed for *Daphnia magna* exposed to the leachates from control soil and soil 3 at 6.25, 12.5 and 25 % soil concentration (mean value = 51 ± 10 %), whereas leachates obtained from soil 50% did not cause any adverse effect. Likewise, EROD activity decreased at 6.25, 12.5 and 25 % (mean value = 54 ± 3 %) whereas at the highest concentration tested (50%, w/w, soil concentration) the lowest effect was measured (27 ± 3 %). Toxicity of leachates to aquatic organism depends, in part, on the nature of the contaminants as well as on the transfer of pollutants from soil to leachates. Changes in the physicochemical soil properties can affect the behaviour of contaminants in the soil. Dilution processes may release pollutants and this effect increases as high dilution factor increases. Differences observed at 50% soil concentration may be due to a lower availability of metals at this concentration.

Soil 5 produced a toxic effect greater than 50% on all soil taxonomic groups. A concentration-dependent effect on earthworm survival, carbon mineralization and seedling emergence of *T. aestivum* and *T. pratense* was observed. To aquatic organisms, leachates obtained from soil 5 adversely affected algae growth rate. The assay with RTG-2 cells indicated a toxic response only when EROD and β-galactosidase activity were measured. A decrease of EROD activity was observed at all concentrations tested, which a percentage of effect equal to 93 ± 12 % at the highest concentration tested (50 % soil concentration). However, EC50 cannot be estimated. Leachates from soil 5 caused inhibition of β-galactosidase activity (90±7 %) only at the highest dose tested (50% soil concentration).

It is interesting to compare toxicity data obtained in the assessment of soil 3 and 5 in both approaches: non diluted samples and dose-response assay. Differences of toxicity to earthworm obtained with the soil 5 in both assays are especially remarkable. Survival of earthworm was highly affect in the assay performed with diluted samples with a value of LC50=10 (2-15) (w/w) soil concentration. Conversely, the assay performed with non diluted samples of soil 5 did not show effect in earthworm survival. In addition, inhibition of *B. napus* growth was observed with soil 3 in the dose-response assay whereas the assay with non diluted soils did not affected this plant specie.

Differences observed between both approaches may be due to the differences in the method used to statistical analyses of data in both or to changes in the contaminant bioavailability in the soil affecting toxicity. The statistical analysis is different in both approaches. In the assay with non diluted samples the data obtained in the test soil are compared to the control using analysis of variance (ANOVA). If experimental variability is relatively high, then the sensitivity of the analyses to detect differences from control will be relatively low. This implies that only larger differences from the control can be detected and false negatives can be obtained. On the other hand, toxicity of contaminants in the soil may decrease with time due to adsorption and aging processes. However, some of these processes may be reversible by changes in physicochemical soil properties (Alexander, 2000).

Contaminants may be released and become bioavailable. Moreover, these changes in the bioavailable fraction may be different for each specie and exposure route. In the bioassays with diluted samples, dilution processes can modify the bioavailability of chemicals in the soil, and hence their toxicity. In the earthworm assay with non diluted soil 5, the variation coefficient for the control and the test soil was 0%, which indicates that the lack of toxicity observed is not due to the analysis method. Consequently, toxicity observed in diluted samples seems due to pollutants released from soil as a consequence of dilution processes. A disadvantage of using diluted samples is that if the sample is not enough toxic, L(E)C50 cannot be determined. For example, in the assay with non diluted samples (100 % soil concentration), soil 3 affected *T. aestivum* growth (48±5 %). However, the EC50 value cannot be determined with the concentrations assessed since 50% of inhibition was not got in the dose response test.

Toxicity values may be used for ranking soils considering possible risk for the environment. With this end, toxicity data obtained for soil 3 and soil 5 were compared. In the assay with non diluted samples, soil 3 was toxic to all taxonomic groups tested whereas soil 5 did not affected earthworm survival. Levels of toxicity may be also compared in the different assays. Inhibition of carbon mineralization was significantly higher for soil 3 (32±1 %) than soil 5 (14±1 %). Regarding the plant test, the effects were different depending on the plant specie and the measured parameter. Both soils affect emergence of *T. aestivum* but no significant differences were observed between them. Soil 3 also affected growth of *T. aestivum* (48±5 %) whereas soil 5 showed significant effects on emergence of *T. pratense* (65±11 %). Comparison of results obtained in this assay for soil 3 and soil 5 seems to indicate increasing toxicity of soil 3 since this soil showed toxicity for a higher number of taxonomic groups and the higher toxicity observed in microorganisms.

In the second approach, soils are compared using a classification system, developed for wastes leachates (Persoone et al., 2003), based in the L(E)C50 values obtained for the different taxonomic groups. The L(E)C50 values are recalculated as Toxic Units, and cut-off values based in TU are established (Table 4). According to this classification, the toxicity of soil 5 is higher than the toxicity of soil 3. Thus, soil 3 would be considered as toxic (1<TU<10) for all endpoint measured and soil 5 as highly toxic since L(E)C50 is reached in the 10-fold dilution (10<TU<100) for at least one test (earthworm mortality).

The Spanish legislation (Ministerio de Presidencia, 2005) includes assays with diluted and non diluted samples. Thus, a threshold to classify a soil as contaminated, based on the L(E)C50 values to soil and aquatic organisms is established. According to this criteria, if the short-term L(E)C50 value for the most sensitive soil taxonomic group is lower than 10 mg sample soil /g total soil or if short term L(E)C50 values of soil leachates to the most sensitive specie of aquatic organisms is lower than 10 ml leachate /l solution, the soil is declared contaminated. In addition, the lack of short-term toxicity in non diluted soil samples and leachates is essential to classify a soil as non-contaminated. Soil 3 and soil 5 toxic effects are observed below this threshold and consequently they cannot be declared as contaminated. However, as toxicity was observed with both soils for some taxonomic groups, the characterization of these soils requires further site-specific risk assessment.

The assay of soil at different concentrations to obtain a dose-response similar to chemical substances requires the analysis of a higher number of samples than the assay with non diluted samples. However, the information obtained in this work seems to indicate that the assessment of different concentrations provides additional information on the potential of soils to release contaminants to the environment due to physicochemical changes in the soil and the soil risk for spreading the contamination to other areas due to surface movement of contaminated soil caused by erosion, flooding, etc.

In summary, biological test and conventional chemical analyses of soils complement each other in the assessment of soil pollution. As showed in this study, risk analysis based in total soil concentration of chemicals can overestimate the real risk since the toxicity of metals in soils was less than that expected to correspond to metal soil concentration. The use of mutispecies laboratory terrestrial models to soil characterization allows making a more realistic approach for risk assessment. In the MS-3 system, the organisms were selected from different trophic levels and included taxonomic groups that cover essential ecological roles for the sustainability of soil use. The mobility, and possible risks for the aquatic compartment (groundwater and superficial water), can also be determined through a watering and subsequent leaching processes.

7. Conclusion

The assessment of soil quality regarding chemical pollution is complex, but a significant process has been achieved in recent years. Part of the complexity is related to the intricacy and variability of the relationships between the soil, as compartment, and its associated terrestrial community. Most species are not linked to the soil itself, but to the ground/above-ground interface. The main consequence is that the assessment must cover a multitude of exposure routes for receptors from different taxonomic groups. In addition, soil, as a complex matrix, has relevant physical-chemical interactions with the chemical pollutants, obviously, issues such as bioavailability and ageing are not exclusively observed for soil, but achieve particular relevance when assessing soil pollution. The main consequence is that is possible to set generic thresholds of potential concern, but they must be conservative for covering the differences in toxicity among soils for the same chemical. Thus, potential risk can be identified, but the confirmation of a real likelihood for ecological consequences requires additional considerations. The combination of chemical analysis and direct toxicity testing on the same soil sample offers a proper and cost/effective solution. Both, single-species tests and soil microcosms are nowadays available, offering the basis for a sound assessment of soil quality in relation to chemical pollution.

8. References

Adriano, DC. Trace elements in terrestrial environments. *Biogeochemistry, bioavailability and risk of metals*. 2nd Ed. New York: Springer-Verlag; 2001.

Aldenberg, T; Jaworska, JS; Traas, TP. Normal species sensitivity distributions and probabilistic ecological risk assessment. In: Posthuma L, Suter GW, Traas TP, editors. *The use of Species Sensitivity Distributions in ecotoxicology*. Boca Raton, FL, USA: Lewis Publishers; 2002; 49-102.

Alexander, M. Aging, bioavailability, and overestimation of risk from environmental pollutants. *Environmental Science and Technology,* 2000, 34, 4259-4265.

ASTM (American Society for Testing and Materials). Standard guide for conducting static 96h toxicity test with microalgae, E1218-90. Philadelphia: ASTM; 1990

ASTM (American Society for Testing and Materials). Standard guide for conducting static toxicity test with *Lemna gibba*, G3, E1415-91. Philadelphia: ASTM; 2003.

Ames , BN; McCann, J; Yamasaki, E. Methods for detecting carcinogens and mutagens with the Salmonella/mammalian-microsome mutagenicity test. *Mutation Research,* 1975, 31, 347–364.

Babín, MM; García, P; Fernández, C; Alonso, C; Carbonell, G; Tarazona JV. Toxicological characterization of sludge from sewage treatment plants using identification evaluation protocols based on in vitro toxicity tests. *Toxicology in vitro,* 2001, 15, 519-524.

Bols, NC; Schirmer, K; Joyce, ME; Dixon, DG; Greenberg, BM; Whyte, JJ. Ability of polycyclic aromatic hydrocarbons to induce 7-ethoxyrresorufin-o-deethylase activity in a trout liver cell line. *Ecotoxicology and Environmental Safety,* 1999, 44, 118-128.

BbodSchG (Bundes-Bodenschutzgesetz). *Gesetz zum Schutz des Bodens in der Fassung der Bekanntmachung vom* 17.03.1998. Bundesgesetzblatt I. 1998.

Beck, L; Römbke, J; Breure, AM; Mulder, C. Considerations for the use of soil ecological classification and assessment concepts in soil protection. *Ecotoxicology and Environmental Safety,* 2005, 62, 189-200

Bierkens, J; Klein, G; Corbisier, P; Van Den Heuvel, R; Verschaeve, L; Weltens, R; Schoeters, G. Comparative sensitivity of 20 bioassays for soil quality. *Chemosphere,* 1998, 37, 2935-2947.

Bispo, A; Jourdain, MJ; Jauzein, M. Toxicity and genotoxicity of industrial soils polluted by polycyclic aromatic hydrocarbons (PAHs). *Organic Geochemistry,* 1999, 30, 947-952.

Boleas, S; Alonso, C; Pro, J; Babín, MM; Fernández, C; Carbonell, G; Tarazona, JV. Effects of sulfachloropyridazine in MS-3 arable land: a multispecies soil system for assessing the environmental fate and effects of veterinary medicines. *Environmental Toxicology and Chemistry,* 2005a, 24, 811-819.

Boleas, S; Alonso, C; Pro, J; Fernández, C; Carbonell, G; Tarazona, JV. Toxicity of the antimicrobial oxytetracycline to soil organisms in a multispecies-soil system (MS-3) and influence of manure co-addition. *Journal of Hazardous Materials,* 2005b, 122, 233-241.

Bombardier, M; Bermingham, N. The SED-TOX index: Toxicity-directed management tool to assess and rank sediments based on their hazard. Concept and application. *Environmental Toxicology and Chemistry,* 1999, 18, 685-698.

Borenfreund, E; Puerner, JA. Toxicity determined in vitro by morphological alterations and neutral red absorption. Toxicology Letters, 1985, 24, 119-124.. Bruins, MR; Kapil, S; Oehme, FW. Microbial resistance to metals in the environment. *Ecotoxicology and Environmental Safety,* 2000, 45, 198-207.

Bulich, AA. A practical and reliable method for monitoring the toxicity of aquatic samples. Process Biochemistry, 1982, 17, 45-47.

Burton, GA Jr.; Batley, GE; Chapman, PM; Forbes, VE; Smith, EP; Reynoldson, T; Schlekat, CE; den Besten, PJ; Bailer, AJ; Green, AS; Dwyer, RL. A weight of evidence framework for assessing sediment (or other) contamination: improving certainty in the decision-making process. *Human and Ecological Risk Assessment*, 2002a, 8, 1675-1696.

Burton, GA Jr.; Chapman, PM; Smith, EP. Weight of evidence approaches for assessing ecosystem impairment. *Human and Ecological Risk Assessment,* 2002b, 8,1657-1673.

Cairns, J Jr. Quantification of biological integrity. In: Ballentine RK, Guarria LJ, editors. *Integrity of water.* Washington DC: USEPA Office of Water and Hazardous Materials; 1977; 171-187.

Calow, P. *Handbook of Ecotoxicology,* Vol 1. 1st Ed. Oxford: Blackwell Scientific Publications; 1993.

Campbell, PJ; Arnold, DJS; Brock, TCM; Grandy, NJ; Heger, W; Heinbach, F; Maund, S; Streloke, M. *Guidance document on higher tier aquatic risk assessment for pesticides* (HARAP). Brussels: SETAC-Europe; 1999.

Carlon, C; D´Alessandro, M; Swartjes, F. Derivation methods of soil screening values in Europe. A review and evaluation of national procedures towards harmonization. Report, EUR 22805 EN, 2007, European Chemical Bureau. *Ispra: Joint Research Centre y el Directorate General Environment de la Commission European;* 2007.

Clément, B; Persoone, G; Janssen, C; Le Dû-Delapierre, A. Estimation of the hazard of landfills through toxicity testing of leachates. *Chemosphere,* 1996, 33, 2303-2320.

Crommentuijn, T; Polder, MD; van de Plassche, EJ. Maximum permissible concentrations and negligible concentrations for metals, tacking background concentrations into account. *RIVM report Nº 601501001;* 1997.

Crommentuijn, T; Polder, M; Sijm, D; de Bruijn, J; van de Plassche, E. Evaluation of the Dutch environmental risk limits for metals by application of the added risk approach. *Environmental Toxicology and Chemistry,* 2000, 19, 1692-1701.

Crossland, NO. Extrapolating from mesocosms to the real world. *Toxicology and Ecotoxicology News,* 1994, 1, 15-22.

CSTEE (Scientific Committee on Toxicity, Ecotoxicity and the Environment). CSTEE opinion on the available scientific approaches to assess the potential effects and risk of chemicals on terrestrial ecosystems. C2/JCD/esteeop/Ter91100/D(0), Brussels: *Reports of the Scientific Committee on Toxicity, Ecotoxicity and the Environment (CSTEE)* ; 2000, 178 pp.

Chen, SK; Edwards, CA. A microcosm approach to assess the effects of fungicides on soil ecological processes and plant growth: comparisons of two soil types. *Soil Biology and Biochemistry,* 2001, 33, 1981-1991.

Christensen, ER; Nyholm, N. Ecotoxicological assays with algae: Weibull dose–response curves. *Environmental Science and Technology,* 1984, 18, 713–718.

Dick, RP; Breakwell, DP; Turco, RF. Soil enzyme activities and biodiversity measurements as integrative microbiological indicators. In: Doran JW, Jones AJ, editors. Methods for assessing soil quality. Madison, WI: *Soil Science Society of America;* 1996; 247-271

DIN 38414, Teil 4: Deutsche einheitsverfaheren zur Wasser-, Abwasser- und Schalammuntersuchung; Schalamm und Sediment (Gruppe S). Bestimmung der Eluierbarkeit mit Wasser (S4), 1984

Di Toro, DM; Allen, HE; Bergman, HL; Paquin, PR; Santore, RC. Biotic ligand model of the acute toxicity of metals. 1. *Technical basis. Environmental Toxicology and Chemistry,* 2001, 20, 2383-2396.

Dott, W; van Afferden, M; Ahlf, W; Crößmann, G; Daei, B; Dorgerloh, M; Eikmann, Th; Fabig, W; Gauglitz, U; van Gestel, CAM; Hartmann, A; Hund, K; Kästner, M; Kalnowski, G; Kanne, R; Müller-Wegener, U; Müller-Markgraf, W; Pfeifer, F; Rippen, G; Schäfer-Treffenfeld, W; Schulz-Berendt, V; Sellner, M; Sprenger, B; Weißenfels, WD; Werner, D; Wiegand-Rosinus, M; Wiesner, J; Wilke; BM. DECHEMA-Ad-hoc-committee Methods for toxicological/ecotoxicological assessment of soils; Frankfurt am Main, Germany: DECHEMA Deutsche Gesellschaft für Chemisches Apparatewesen, *Chemische Technik und Biotechnologie;* 1995.

Drobne, D; Hopkin, SP. Ecotoxicological laboratory test for assessing the effects of chemicals on terrestrial isopods. *Bulletin of Environmental Contamination and Toxicology,* 1994, 53, 390-397.

Ducrot, V; Pery, AR; Mons, R; Queau, H; Charles, S; Garric, J. Dynamic energy budget as a basis to model population-level effects of zinc-spiked sediments in the gastropod *Valvata piscinalis*. *Environmental Toxicology and Chemistry,* 2007, 26,1774-1783.

EC (European Commission). *Guidance document on aquatic ecotoxicology in the context of Directive* 91/414/EEC. SANCO /3268/2001 rev.4 (final). Brussels: European Commission; 2002a.

EC (European Commission). *Guidance document on terrestrial ecotoxicology in the context of Directive* 91/414/EEC. SANCO/10329/2002 rev.2(final). Brussels: European Commission; 2002b.

EC (European Commission). *Technical Guidance Documents in support of the Commission Directive 93/67/EEC on risk assessment for new notified substances and the Commission Regulation (EC) 1488/94 on risk assessment for existing substances.* Ispra: European Chemical Bureau; 2003.

Eom, IC; Rast, C; Veber, AM; Vasseur, P. Ecotoxicity of a polycyclic aromatic hydrocarbon (PAH)-contaminated soil. *Ecotoxicology and Environmental Safety,* 2007, 67, 190-205.

EU (European Union) Guidance Document on risk assessment for birds and mammals under Council Directive 91/414/EEC. SANCO/4145/2000, *European Commission Health & Consumer Protection Directorate-General;* 2002; 44 pp.

EUFRAM. Concerted action to develop a European Framework for probabilistic risk assessment of the environmental impacts of pesticides. Introducing probabilistic methods into the ecological risk assessment of pesticides. *EUFRAM Report,* Vol. 1; 2006. http://www.eufram.com.

Ferguson, CC; Darmendrail, D; Freier K; Jensen, BK; Jensen, J; Kasamas, J; Urzelai, A; Vegter, J. *Risk Assessment For Contaminated Sites In Europe. Scientific Basis.* Vol. 1. Nottingham: LQM. Press; 1998.

Fernández, C; Alonso, C; Babín, MM; Pro, J; Carbonell, G; Tarazona, JV. Ecotoxicological assessment of doxycicline in aged pig manure using multispecies soil system. *Science of the Total Environment,* 2004, 323, 63-69.

Fernández, MD; Cagigal, E; Vega, MM; Urzelai, A; Babín, M; Pro, J; Tarazona, JV. Ecological risk assessment of contaminated soils through direct toxicity assessment. *Ecotoxicology and Environmental Safety,* 2005, 62, 174-184.

Fernández, MD; Vega, MM; Tarazona, JV. Risk-based ecological soil quality criteria for the characterization of contaminated soils. Combination of chemical and biological tools. *Science of the Total Environment,* 2006, 366, 466-484.

Forbes TL, Forbes VE. A critique of the use of distribution-based extrapolation models in ecotoxicology. *Functional Ecology,* 1993, 7, 249-254.

Fountain, MT; Hopkin, SP. Folsomia Candida (Collembola): An "standard" soil arthropod. *Annual Review of Entomology,* 2005, 50, 201-222.

Giddings, J; Brock, TCM; Heger, W; Heimbach, J; Maund, SJ; Norman, S; Ratte, HT; Schäfers, C; Streloke, M. Community level aquatic system studies. Interpretation citeria

(CLASSIC). Pensacola FL, USA: *Society for Environmental Toxicology and Chemistry* (SETAC), 2002.

Greene, JC; Bartels, CL; Warren-Jicks, WJ; Parkhurst, BR; Linder, GL; Peterson, S; Miller, WE. Protocols for shot term toxicity screening of hazardous waste sites. OR, 600/3-88-029. *Corvallis: United States Environmental Protection Agency;* 1989; 102pp.

Hämmann, M; Gupta, SK; Zihler, J. Protection of soils from contamination in Swiss legislation. *Advances in GeoEcology,* 1998, 31, 629-636.

Hammel, W; Steubing, L; Debus, R. Assessment of ecotoxic potential of soil contaminants by using a soil-algae test. *Ecotoxicology and Environmental Safety,* 1998, 40, 173-176.

Holmgren, GGS; Meyer, MW; Chaney, RL; Daniels, RB. Cadmium, lead, zinc, cupper and niquel in agricultural soils of United States of America. *Journal of Environmental Quality,* 1993, 22, 335-348.

Hopkin, SP. *In situ* biological monitoring of pollution in terrestrial and aquatic ecosystems. In Calow P, editor. *Handbook of Ecotoxicology,* Vol 1. 1st Ed. Oxford: Blackwell Scientific Publications; 1993; 397-427.

Hund-Rinke, K; Kördel, W; Hennecke, D; Eisenträger, A; Heiden , S. Bioassays for the ecotoxicological and genotoxicological assessment of contaminated soils (Results of a Round Robin Test. *Journal of Soil and Sediments,* 2002, 2, 43-50.

Hund-Rinke, K; Kördel, W. Underlying issues in bioaccessibility and bioavailability: experimental methods. *Ecotoxicology and Environmental Safety,* 2003, 56, 52-62.

ISO (International Organization for Standardization). Water quality. Determination of the inhibitory effect of water samples on the light emission of Vibrio fischeri (Luminescent bacteria test). Part 1: *Method using freshly prepared bacteria.* No. ISO 11348-1:1998 ; Geneva: Int. Stand. Organ.; 1998; 16 pp.

ISO (International Organization for Standardization). *Soil quality-inhibition of reproduction of Collembolan (Folsomia candida) by soil pollutants.* No. ISO 11267:1999. Geneva: Int. Stand. Organ.; 1999; 16 pp.

Jager, DT; Vermeire, TG; Slooff, W; Roelfzema, H. Uniform system for the evaluation of substances. II Effects assessment. *Chemosphere,* 1994, 29, 319-335.

Janssen, RPT; Peijnenburg, WJGM; Posthuma, L; van den Hoop, MAGT. Equilibrium partitioning of heavy metals in Dutch field soils. II. Prediction of metal accumulation in earthworms. *Environmental Toxicology and Chemistry,* 1997,16, 2479-2488.

Johnson, FH; Caruer, CM; Harryman, WK. Luminous bacterial auxanograms in relation to heavy metals and narcotics, self-photographed in color. *Journal of Bacteriology,* 1942, 44, 703-713.

Keddy, CJ; Greene, JC; Bonnell, MA. Review of whole-organism bioassays: soil, freshwater sediment, and freshwater assessment in Canada. *Ecotoxicology and Environmental Safety,* 1995, 30, 221-251

Klein, M. PELMO: *Pesticide Leaching Model. Fraunhofer: Institut für Umweltchemie und Ökotoxicogie,* D57392 Schmallenberg; 1991.

Knacker, T; van Gestel, CAM; Jones, SE; Soares, AMVM; Schallnaβ, H-J; Förster, B; Edwards, CA. Ring-testing and field-validation of a Terrestrial Model Ecosystem (TME). An instrument for testing potentially harmful substances: conceptual approach and study design. *Ecotoxicology,* 2004, 13, 9-27.

Knox, P; Uphill, PF; Fry, JR; Beuford, J; Balls, M. The FRAME multicentre project on *in vitro* cytotoxicology. *Food Chemical Toxicology,* 1986, 24, 457-463.

Koster, M; de Groot, A; Vijver, M; Peijnenburg, W. Copper in the terrestrial environment: Verification of a laboratory-derived terrestrial biotic ligand model to predict earthworm mortality with toxicity observed in field soils. *Soil Biology and Biochemistry,* 2006, 38, 1788-1796.

Leistra, M; van der Linden, AMA; Bvoesten, JJTI; Tiktak, A; van den Berg, F. PEARL model for pesticide behaviour and emissions in soil-plant systems. *Description of processes.* Alterra report 13, RIVM report 711401009; 2000.

Leitgib, L; Kàlmán, J; Gruiz, K. Comparison of bioassays by testing whole soil and their water extract from contaminated sites. *Chemosphere,* 2007, 66,428-434.

Lewis, HM; Law, R. Effects of dynamics on ecological networks. Journal of Theoretical Biology, 2007, 247, 64-76.

Litchfield, JT; Wilcoxon, F. A simplified method of evaluating dose effect experiments. *Journal of Pharmacology and Experimental Therapeutics,* 1949, 96, 99-113.

Lock, K; Janssen, CR. Ecotoxicity of zinc in spiked artificial soils versus contaminated field soils. *Environmental Science and Technology,* 2001, 35, 4295-4300.

Lock, K; Janssen, CR. Comparative toxicity of a zinc salt, zinc powder and zinc oxide to Eisenia fetida, *Enchytraeids albidus and Folsomia candida. Chemosphere,* 2003a, 53, 851-856.

Lock, K; Janssen, CR. Influence of ageing on zinc bioavailability in soils. *Environmental Pollution,* 2003b, 126, 371-374.

Loibner, AP; Szolar, OHJ; Braun, R; Hirmann, D. Ecological assessment and toxicity screening in contaminated land analysis. In: Thompson KC, Nathanail CP, editors. *Chemical analysis of contaminated land.* Oxford, UK: Blackwell Publishing Ltd; 2003; 229-267.

Long, ER; Chapman, PM. A sediment quality triad: measures of sediment contamination, toxicity and infaunal community composition in Puget Sound. *Marine Pollution Bulletin,* 1985, 16, 405-415.

Ma, Y; Lombi, E; Nolan, AL; McLaughlin, MJ. Short term natural attenuation of copper in soils: effects of time temperature and soil characteristics. *Environmental Toxicology and Chemistry,* 2005, 25, 652-658.

Maxam, G; Rila, JP; Dott, W; Eisentraeger, A. Use of bioassays for assessment of water-extractable ecotoxic potential of soils. *Ecotoxicology and Environmental Safety,* 2000, 45, 240-246.

McLaughlin, MJ; Hamon, RE; McLaren, RG; Speir, TW; Rogers, SL. Review: A bioavailability-based rationale for controlling metal and metalloid contamination of agricultural land in Australia and New Zealand. *Australian Journal of Soil Research,* 2000a, 38, 1037-1086.

McLaughlin, MJ; Zarcinas, BA; Stevens, DP; Cook, N. Soil testing for heavy metals. *Communications in Soil Science and Plant Analysis,* 2000b, 31, 1661-1700

Mendonca, E; Picado, A. Ecotoxicological monitoring of remediation in a coke oven soil. Environmental Toxicology, 2002, 17, 74-79.

Ministerio de Presidencia. REAL DECRETO 9/2005, de 14 de enero, por el que se establece la relación de actividades potencialmente contaminantes del suelo y los criterios y estándares para la declaración de suelos contaminados. *España: Boletín Oficial del Estado* 15/2005; 2005; 1833-1843.

NRC (National Research Council). *Issues in Risk Assessment.* Washington DC: National Academy Press; 1983; 374 pp.

OECD (Organization for Economic Cooperation and Development). *Guidelines for Testing of Chemicals.* Earthworm, acute toxicity tests. Test guideline N° 207. Paris, France; 1984a

OECD (Organization for Economic Cooperation and Development). Guidelines for Testing of Chemicals. Fish, Prolonged toxicity test:14-day study. *Test guideline N° 204.* Paris, France; 1984b.

OECD (Organization for Economic Cooperation and Development). Report of the OECD workshop on the extrapolation of laboratory aquatic toxicity data to the real environment. *OECD Environment Monograph 59,* OECD/GD (92)169. Paris, France; 1992a.

OECD (Organization for Economic Cooperation and Development). Guidelines for Testing of Chemicals. Fish, Acute toxicity test . *Test guideline N° 203.* Paris, France; 1992b.

OECD (Organization for Economic Cooperation and Development). Guidelines for Testing of Chemicals. Fish, early-life stage toxicity test. *Test guideline N° 210.* Paris, France; 1992c

OECD (Organization for Economic Cooperation and Development). Guidelines for Testing of Chemicals. Daphnia magna, reproduction test. *Test guideline N° 211.* Paris, France; 1998a

OECD (Organization for Economic Cooperation and Development). Guidelines for Testing of Chemicals. Fish, short-term toxicity test on embryo and sac-fry stages. *Test guideline N° 212.* Paris, France; 1998b

OECD (Organization for Economic Cooperation and Development). Guidelines for Testing of Chemicals. Soil micro-organisms, nitrogen transformation test. *Test guideline N° 217.* Paris, France; 2000a.

OECD (Organization for Economic Cooperation and Development). Guidelines for Testing of Chemicals. Soil micro-organisms, carbon transformation test. *Test guideline N° 217.* Paris, France; 2000b.

OECD (Organization for Economic Cooperation and Development). Guidelines for Testing of Chemicals. Fish, juvenile growth test. *Test guideline N° 215.* Paris, France; 2000c.

OECD (Organization for Economic Cooperation and Development). Guidelines for Testing of Chemicals. Earthworm reproduction test. *Test guideline N° 222.* Paris, France; 2004a.

OECD (Organization for Economic Cooperation and Development). Guidelines for Testing of Chemicals. Enchytraeid reproduction test. *Test guideline N° 220.* Paris, France. 2004b.

OECD (Organization for Economic Cooperation and Development). Guidelines for Testing of Chemicals. Daphnia s.p., acute immobilisation test. *Test guideline N° 202.* Paris, France; 2004c

OECD (Organization for Economic Cooperation and Development). Guidelines for Testing of Chemicals. Leaching in Soil Columns. *Test guideline N° 312.* Paris, France; 2004d.

OECD (Organization for Economic Cooperation and Development). *Current approaches in the statistical analyses of ecotoxicity data: a guidance to application.* Paris, France; 2005.

OECD (Organization for Economic Cooperation and Development). Guidelines for Testing of Chemicals. Terrestrial plants, growth test. *Test guideline N° 208.* Paris, France; 2006a

OECD (Organization for Economic Cooperation and Development). Guidelines for Testing of Chemicals. Alga, growth inhibition test. *Test guideline N° 201.* Paris, France; 2006b

OECD (Organization for Economic Cooperation and Development). Guidelines for Testing of Chemicals. Lemna s.p. Growth inhibition test. *Test guideline N° 221.* Paris, France; 2006c

Oorts, K; Ghesquiere, U; Smolders, E. Leaching and aging decrease nickel toxicity to soil microbial processes in soils freshly spiked with nickel chloride. *Environmental Toxicology and Chemistry,* 2007, 26, 1130-1138.

Pablos, MV; Boleas, S; García, P; Carbonell, G; Tarazona, JV. Inhibición de la actividad B-GAL como parámetro de toxicidad subletal interespecífico. *Cuadernos de Investigación Biológica,* 1998, 20, 471-474.

Parker, DR; Pedler, JF. Reevaluating the free-ion activity model of trace metal availability to higher plants. *Plant and Soil,* 1997, 196, 223-228.

Parkhurst, DL; Appelo, CAJ. User's guide to PHREEQC. A computer program for speciation, batch-reaction, one-dimensional transport and inverse geochemical calculations. *U.S. Geological survey water-resources investigations Report 99-4259*; 1999.

Peijnenburg, WJGM; Posthuma, L; Eijsackers, HJP; Allen, HE. A conceptual framework for implementation of bioavailability of metals for environmental management purposes. *Ecotoxicology and Environmental Safety,* 1997, 37, 163-172.

Persoone, G; Marsalek, B; Blinova, I; Torokne, A; Zarina, D; Manusadzianas, L; Nalecz-Jawecki, G; Tofan, L; Stepanova, N; Tothova, L; Kolar, B. A practical and user-friendly toxicity classification system with microbiotests for natural waters and wastewaters. *Environmental toxicology,* 2003, 18, 395-402.

Peterson, SA; Greene, JC; Miller, WE. Toxicological assessment of hazardous waste samples extracted with deionised water or sodium acetate (TCLP) leaching media. In: Friedman D, editors. Waste testing and quality assurance. Vol 2 ASTM STP 1062. Philadelphia, USA*: American Society for Testing and Materials;* 1990; 107-129.

Phillips, BM; Hunt, JW; Anderson, BS; Puckett, HM; Fairey, R; Wilson, CJ; Tjeerdema, R. Statistical significance of sediment toxicity test results: Threshold values derived by the detectable significance approach. *Environmental Toxicology and Chemistry,* 2001, 20, 371-373.

Posthuma, L; Suter, GW; Traas, TP. *The use of Species Sensitivity Distributions in ecotoxicology.* Boca Raton, FL, USA: Lewis Publishers; 2002.

Pugh, DM; Tarazona, JV. Environment & Policy. Vol 15. *Regulation for chemical safety in Europe: analysis, comment and criticism.* Dordrecht, The Netherlands: Kluwer Academic Publishers; 1998; 194 pp.

Ramade, F. Précis d'écotoxicologie. *Collection d'écologie 22.* Masson, Paris; 1992; pp 293.

Ramos, C; de la Torre, AI; Tarazona, JV; Muñoz, MJ. Desarrollo de un ensayo de inhibición de Chlorella vulgaris utilizando un test en microplacas. *Revista de Toxicología,* 1996, 13, 97-100.

Rand, GM. *Fundamentals of Aquatic Toxicology*. Washington, DC, USA: Taylor & Francis; 1995.

Robidoux, PY; Dubois C; Hawari J; Sunahara GI. Assessment of soil toxicity from an antitank firing range using Lumbricus terrestris and Eisenia andrei in mesocosms and laboratory studies. *Ecotoxicology,* 2004a, 13, 603-14.

Robidoux, PY; Gong, P; Sarrazin, M; Bardai, G; Paquet, L; Hawari, J; Dubois, C; Sunahara, GI. Toxicity assessment of contaminated soils from an antitank firing range. *Ecotoxicology and Environmental Safety,* 2004b, 58, 300-313.

Robidoux, PY; Svendsen, C; Sarrazin, M; Thiboutot, S; Ampleman, G; Hawari J; Weeks, JM; Sunahara, GI. Assessment of a 2,4,6-trinitrotoluene-contaminated site using Aporrectodea rosea and Eisenia andrei in mesocosms. *Archives of Environmental Contamination and Toxicology,* 2005, 48, 56-67.

Römbke, J; Breure, AM; Mulder, C; Rutgers, M. Legislation and ecological quality assessment of soil: implementation of ecological indication systems in Europe. *Ecotoxicology and Environmental Safety,* 2005, 62, 201-210.

Romijn, CAFM; Luttik, R; Canton, JH. Presentation of a general algorithm to include effect assessment on secondary poisoning in the derivation of environmental quality criteria. Terrestrial food chains. *Ecotoxicology and Environmental Safety,* 1994, 27, 107-127.

Ryan, JA; Chaney, RL. Issues of Risk Assessment and its utility in development of soil standards: the 503 methodology as an example. In: Prost R, editor. Contaminated Soils: Proc. T*hird International Symposium on Biogeochemistry of Trace Elements,* Paris, May 15-19, 1995, Colloque No. 85. Paris, France: INRA Editions; 1997; 393-414.

Salanitro, JP; Dorn, PB; Huesemann, MH; Moore, KO; Rhodes, IA; Jackson, LMR; Vipond, TE; Western, MM; Wisniewski, HL. Crude oil hydrocarbon bioremediation and soil ecotoxicity assessment. *Environmental Science and Technology,* 1997, 31, 1769-1776.

Salminen, J; Haimi, J. Effects of pentachlorophenol on soil organisms and decomposition in forest soil. *Journal of Applied Ecology,* 1997, 34, 101-110.

Selivanovskaya, SY; Latypova, VZ. The use of bioassays for evaluating the toxicity of sewage sludge and sewage sludge-amended soil. *Journal of Soils and Sediments,* 2003, 3, 85-92

Sheppard, SC. Toxicity testing using microcosms. In: Tarradellas J, Bitton G, Rosell D, editors. *Soil Ecotoxicology.* Boca Raton, FL, USA: Lewis Publishers, 1997; 345-373.

Smith, EP; Cairns, J Jr. Extrapolation methods for setting ecological standards for water quality: Statistical and ecological concerns. *Ecotoxicology*, 1993, 2, 203-219.

Song, J; Zhao, F-J; McGrath, SP; Luo, Y-M. Influence of soil properties and aging on arsenic phytotoxicity. *Environmental Toxicology and Chemistry,* 2006, 25, 1663-1670.

Sprague, JB; Ramsay, BA. Lethal levels of mixed copper-zinc solutions for juvenile salmon. *Journal of the Fisheries Research Board of Canada, 1*965, 22, 425-432.

SSC-European Commission. *Harmonisation of the ecological risk assessment of chemicals.* Scientific Steering Committee. Final draft for consultation, 2003.

Stephenson, GL; Kuperman, RG; Linder, GL; Visser, S. Toxicity tests for assessing contaminated soils and ground water. In: Sunahara GI, Renoux AY, Thellen C, Gaudet CL, Pilon, A, editors. *Environmental Analysis of Contaminated Sites.* New York, USA: John Wiley and Sons. Ltd; 2002; 25-43.

Struijs, J; van de Meent, D; Peijnenburg, WJGM; van den Hoop, MAGT; Crommentuijn, Tl. Added risk approach to derive maximum permissible concentrations for heavy metals: *How to take natural background levels into account Ecotoxicology Environmental Safety,* 1997, 37, 112-118.

Suter, GW; Efroymson, RA; Sample, BE; Jones, DS. *Ecological Risk Assessment for Contaminated Soils.* Boca Raton, FL, USA: Lewis Publishers; 2000.

Tarazona, JV; Vega, MM. Hazard and risk assessment of chemicals for terrestrial ecosystems. *Toxicology,* 2002, 181-182, 187-191.

Tarazona, JV; Hund, K; Jager, T; Salonen, MS; Soares, AMVM; Skaare, JU; Vighi, M. Standardizing chemical risk assessment, at last. *Nature,* 2002, 415: 14.

Tarazona, JV; Fernández, MD; Vega, MM. Regulation of contaminated soils in Spain. A new legal instrument. *Journal of Soils and Sediments,* 2005, 5, 121-124.

Thakali, S; Allen, HE; Di Toro, DM; Ponizovsky, AA, Rooney, C P, Zhao, F-J, McGrath, S P. A Terrestrial Biotic Ligand Model. 1. Development and application to Cu and Ni toxicities to barley root elongation in soils. *Environmental Science and Technology,* 2006a, 40, 7085-7093

Thakali, S; Allen, HE; Di Toro, DM; Ponizovsky, AA; Roonev, CP; Zhao, F-J; McGrath, SP; Criel, P; Van Eeckhout, H; Janssen, CR; Oorts, K; Smolders, E. Terrestrial biotic ligand model. 2. Application to Ni and Cu toxicities to plants, invertebrates, and microbes in soil *Environmental Science and Technology,* 2006b, 40, 7094-7100.

The Ministry of Housing, Spatial Planning and the Environment. Environmental quality standards in the Netherlands (1999). *A review of environmental quality standards and their policy framework on the Netherlands.* Netherlands: Kluwer, Alpenaan den Rijn; 2001, 627 pp.

Tørsløv, J; Samsøe-Petersen, L; Rasmussen, JO; Kristensen, P. Use of waste products in agriculture. *Environmental project 366.* Copenhagen, Denmark: Danish Environmental Protection Agency; 1997.

Torstensson, L. Guidelines. Soil biological variables in environmental hazard assessment. Report No 4262. Stockholm, Sweden: *Swedish Environmental Protection Agency;* 1993

USEPA (US Environmental Protection Agency). Ecological Effects Test Guidelines OPPTS 850.2450 Terrestrial (Soil-Core) Microcosm Test. EPA 712-C-96-143. Washington, DC, USA: *US Environmental Protection Agency;* 1996

USEPA (US Environmental Protection Agency). Guidelines for Ecological Risk Assessment. EPA/630/R095/002F. Washington, DC, USA: *US Environmental Protection Agency, Risk Assessment Forum;* 1998.

Van Brummelen, TC; Van Gestel, CAM; Verweij, RA. Long-term toxicity of five polycyclic aromatic hydrocarbons for the terrestrial isopods Oniscus asellus and Porcellio scaber. *Environmental Toxicology and Chemistry,* 1996, 15, 1199-1210.

Van de Plassche, EJ; de Bruijn, JHM; Stephenson, RR; Marshall, SJ; Feitjel, TCJ; Belanger, SE. Predicted no-effect concentration and risk characterization of four surfactants: linear alkyl benzene sulfonate, alcohol ethoxilates, alcohol ethoxylated sulfates and soap. *Environmental Toxicology and Chemistry,* 1999; 18, 2653-2663.

Van Gestel, CAM; van der Waarde, JJ; Derksen, JGM; van der Hoek, EE; Veul, MEX; Bouwens, S; Rusch, B; Kronenburg, R; Stokman, GNM. The use of acute and chronic bioassays to determine the ecological risk and bioremediation efficiency of oil-polluted soils. *Environmental Toxicology and Chemistry,* 2001, 20, 1438-1449.

Van Gestel, CAM; Henzen, L; Dirven-Van Breemen, EM; Kamerman, JW. Influence of soil remediation techiques on the bioavailability of heavy metals. In: Sunahara GI, Renoux AY, Thellen C, Gaudet CL, Pilon A, editors. *Environmental Analysis of Contaminated Sites.* New York, USA: John Wiley and Sons. Ltd; 2002; 361-388.

Van Straalen, NM; Denneman, CAJ. Ecotoxicological evaluation of soil quality criteria. *Ecotoxicology and Environmental Safety,* 1989, 18, 241-251.

Van Straalen, NM; Van Gestel, CAM. Soil invertebrates and micro-organisms. In: Calow P, editor. *Handbook of Ecotoxicology,* Vol 1. 1st Ed. Oxford: Blackwell Scientific Publications; 1993; 251-277.

Vegter J, Lowe, J, Kasamas H. *Sustainable management of contaminated land: an overview.* Austria: Umweltbundesamt GmbH; 2003.

Wagner, C; Løkke, H. Estimation of ecotoxicological protection levels from NOEC toxicity data. *Water Research,* 1991, 25, 1237-1242.

Weyers, A; Sokull-Klüttgen, B; Knacker, T; Martin, S; Van Gestel, CAM. Use of Terrestrial Model Ecosystem data in environmental risk assessment for industrial chemicals, biocides and plant protection products in the EU. *Ecotoxicology,* 2004, 13,163-176.

In: Soil Contamination Research Trends
Editor: Javier B. Dominguez, pp. 51-72
ISBN: 978-1-60456-319-1

Chapter 2

VALIDATED APPROACHES TO RESTORING THE HEALTH OF ECOSYSTEMS AFFECTED BY SOIL POLLUTION

***Ana J. Hernández*[1,*] *and J. Pastor*[2,†]**
[1]Dpto. Ecología, Universidad de Alcalá. Edificio Ciencias, Campus Universitario. Madrid, Spain
[2]Dpto. de Ecología de Sistemas. CCMA, CSIC, c/ Serrano 115, Madrid 28006, Spain

ABSTRACT

The diagnosis of soil pollution in relation to the most commonly used remediation methods is an innovative way of tackling this problem. To this end, the theoretical framework of Ecosystems Health has been instrumental, since this area is presently emerging as a new language for general discourse on pollution. For example, the use of the term stressor for a pollutant implies the study of behavior responses in living organisms (at different levels of organization, from the cell to ecosystem) both in terms of impacts (toxicity) and tolerance towards the pollutant (their adaptation).

Two very common scenarios in the central Iberian Peninsula have allowed us to approach the understanding of this area of study: (i) old abandoned mines and, (ii) old solid waste (urban and industrial) landfills.

The soils of these systems are mainly polluted with heavy metals (Zn, Cu, Pb, Cd), they overlie different lithological substrates (granites, limestones and especially arkoses), and are highly representative of soils of the Mediterranean region (mainly regosols, luvisols and cambisols). The landfills of the mines or waste tips have slopes exceeding 40%, such that any leachates produced transfer pollutants to adjacent ecosystems at lower altitudes (stream bed pastures, marshes and surface water courses). Moreover, since the

* E-mail: anaj.hernandez@uah.es
† E-mail: jpastor@ccma.csic.es

topsoil layer is usually sandy with low proportions of clay and organic matter, deep leachates also transport pollutants to ground water systems.

We focused on the autotrophic component of the affected ecosystems for several reasons: toxicity tests could be used to examine the physiological and behavioral responses of organisms (mortality, injury, metabolic changes) as well as population (population density, risk of extinction) or community (structure, diversity, biomass, nutrient flow changes) variables. The build-up of a heavy metal in the above-ground part of a plant (phytoaccumulation) consumed by herbivores is also detrimental for health, due to transfer to the trophic network. Root systems may play a role in phytostabilizing heavy metals and in preventing them from passing to deeper soil layers. Finally, given the erosion problems of fine materials in landfills and waste tip slopes, vegetation helps avoid the movement of topsoil layer pollutants to other ecosystems.

The methodological approaches validated by results obtained over the last twenty years can be summarized as: studies of polluted sites based on phytoecological sampling, analysis of soil chemical and physical properties, georeferencing of the heavy metals they contain for further sampling in areas showing the highest levels, collecting and chemically analyzing plants at these sites, and the use of soils with their seed banks to perform experiments in microcosms in controlled conditions. The idea is to use a combination of field and laboratory methods that simulate real scenarios in which soil pollution occurs.

Introduction: The Complexity of Degraded Ecosystems within the Framework of Systems Science

With the title "From Biodiversity to Biocomplexity", the monograph of the June 2001 issue of *BioScience* was devoted to the multidisciplinary step we need to take if we are to understand our environment. The main questions addressed as challenges for biologists at this start of a century are perfectly apt to introduce the subject of the work we describe in this chapter. There is no question that the field of biocomplexity will be in the forefront of research endeavors in the years to come, given it is a property of all ecosystems; or in other words, it is the thread that links many complex systems that are structured or influenced by living organisms, their components or biological processes.

Interactions between live beings and all the facets of their external environment are an evident fact and research into these interactions, which involve multiple levels of biological organization and/or several spatial or temporal scales, is of great importance for everything related to environmental impacts. To undertake such research, long term data collection and monitoring are required to examine interactions at different spatial scales of the landscape and assess interplay between air, soil, water and the ecosystems they sustain. These issues will guide decision-making and help understand the functions of ecosystems and the changes they suffer in an ever-evolving world. Notwithstanding, the study of complex systems, among which we can include terrestrial ecosystems at polluted sites, is in large measure at a very early stage. Analyzing the individual components of a system offers no information about the properties of its combined components. The emergence of new properties when components change is a common phenomenon, but deciphering what makes new features emerge is a topic that has only just started to attract the attention of scientists.

With these brief considerations only, we can begin to understand the current impact that the topic of restoring degraded ecosystems is having for biologists because of its complex nature. Biocomplexity is nevertheless difficult to describe and experimentally assess owing to its non-linear nature. Indeed, understanding biological complexity requires the development of new paradigms that cross temporal, spatial and conceptual barriers. Certainly, in the journal mentioned it is recognized that the development of the ecosystemic focus and its use to try to understand and resolve environmental problems, has been one of the major advances of Biology in the past 50 years. For many years now this has been our approach (Hernández, 1989 and 1991) and in the subsequent sections of this chapter we describe some of the concepts we have established and tested to address the subject of restoring degraded ecosystems.

Remediating and recovering degraded ecosystems is really a topic of systemic reality. This dictates a need to juggle the basic principles of systems analysis (complexity, interaction, uncertainty, multicausality) in an episystemic focus on ecological restoration. The applicability of this focus stems from the fact that ecosystems are dynamic systems that evolve and coevolve with human activity, though always evoking their stability, a term frequently used in relation to the response of an ecosystem to perturbation. This in reality is the picture we are faced with: both EU community policies and global warming that affect the dynamics of abandoned cereal fields or grassland systems, along with the environmental impacts of new emerging systems in structured landscapes such as landfills. The remediation of landscapes by reforesting with tree species is met with changes in soil structure, a feature not usually considered in practice that complicates the system to be restored. In the case of solid waste landfills capped with soils taken from surrounding areas, we have to face the complexity that arises from the secondary ecological succession, arising from the seed bank of the covering soil, intermixing with the primary succession of these new ecosystems. Not only does the impact of waste disposal need restoring but the "waste system" itself also needs to be remediated.

ECOSYSTEMS' HEALTH

The health of an ecosystem is one of the pillars that supports sustainable development, and is emerging as a new language for general discourse on pollution. Thus, ecosystems' health has been defined as an incipient science or systems approach to prevent, diagnose, and predict factors useful for managing the system and establish links between the health of an ecosystem and human health (Calow, 1995; Di Giulio and Monosson, 1996; Rapport et al 1995 and 2003). Thirty years ago research into geochemistry and health commenced as one of the UNESCO's lines of investigation in the Man and Biosphere (MAB) program. The hypothesis that heavy metals generated by geochemical actions in some tropical ecosystems, besides being related to the productivity of the system, could have effects on human health prompted us to assess the bioavailability of these metals in the main crops used for human consumption or forage in this region.

This was the basis for an experimental study conducted at several abandoned mine sites of the central Iberian Peninsula (Hernández and Pastor, 2007) and for our recently initiated study on tropical ecosystems (Hernández et al., 2007). The above flow diagram (Figure 1) indicates the general protocol inferred from our long-term results that has best served us to evaluate the relationship between the geochemistry and the health of an ecosystem.

The ecosystems found at these sites are essentially those of grasslands and forage or cereal crops. In each of these settings, we find soils containing more than one heavy metal in their top layer as point-source pollution that affects both the plant populations and their consumers with the risk of transfer of these pollutants to ground and surface water aside from the food chain.

METHODS VALIDATED FOR DIFFERENT MEDITERRANEAN SCENARIOS

According to the issues discussed in the preceding sections, we selected different sites on substrates and soils representative of the central Iberian Peninsula. The former include granite, limestone and gypsum and mainly arkosic substrates (Madrid facies), and the soils are those also found across Spain (regosols, luvisols and cambisols). On the soils of arkosic regions in environments that are almost semiarid we find agrosystems of cereals, vines, fallow land, pastures and shrubs with a compacted topsoil layer poor in organic matter and of slightly acid pH. These three properties makes them good candidates for use as reference systems to address the revegetation of sites with soils impoverished by these or similar soil degradation processes that are to change use. A common example is the change from past industrial use to quality housing developments with green leisure zones.

The arkosic substrate also generates soils that are highly silted (scarce clay and much sand) besides containing slight amounts of carbonates. These two features reflected in agroecosystems and pastures of the area, make the latter ideal references for old soil-capped landfills with steep slopes along which the finer elements of the covers are lost and transported downstream. Fallow systems are perhaps the most analogous to the most recently sealed urban solid residue (USR) landfills if not subsequently intervened with. Moreover, their understanding can also be a reference for the adequate restoration of agrosystems with naked soils.

Many factors of great ecological interest are often not included in environmental impact analysis (EIA) studies, despite the fact they need no vast amount of experimentation, nor do they require long sampling periods nor generate huge costs. Thus, for example, the analysis of real floristic composition and not only the potential composition (inferred from the literature), the diversity, the relative abundance of the main taxa, dominance, succession dynamics or the conservation state of the vegetation, although addressed in some studies and mentioned in the corresponding reports and statements, very few EIA are accompanied by fieldwork. Even fewer of these studies include soil analyses, despite integrated soil and plant variables providing relevant information on nutrient flow and on several functional properties of the ecosystem itself, which will need to be considered if restoration is in mind (Hernández et al., 1998a).

We have also observed that in many areas of central Spain, there are zones corresponding to old mining areas in which rubble landfills exist and soils are polluted to a greater or lesser extent by heavy metals. The plant communities they sustain are grassland and shrub-pasture formations and some wooded meadows used by cattle, sheep and wild animals. Given that most heavy metals and trace elements are components of biogeochemical cycles whose main compartments are the soil and vegetation, their importance in the food chain means that these elements warrant detailed investigation.

Heavy metals and trace elements can be absorbed by plants, be lost by deeper leaching to reach ground water, or can be lost through erosion (surface runoff), affecting surface water channels. The importance of the different transfer routes of these elements to other compartments of the trophic network varies considerably depending on the element in question, the plant species present or the use given to the pasture (whether used by livestock in situ or for fodder or elaborating feeds).

Several studies have shown that animals exhibit toxic element concentrations when grazing on polluted soils (Morcombe et al. 1994; Petersson et al., 1997; FAO, 2000; López Alonso, 2002; Wilkinson et al., 2003). Thus, it is necessary to control metals in terrestrial ecosystems. Although the work reported in this chapter has been undertaken over the last twenty years, we will try to present a main line of progress regarding the solution of the problems arising during our efforts to ecologically restore the environments described above. Given the impossibility of detailing many results, the reader is referred to the corresponding bibliographic references cited. Summary tables of the results obtained are nevertheless provided as a general guide.

Before any effort is made to ecologically restore an area, a detailed description of the zone has to be made along with an estimate of its negative impact on the environment, the landscape and the structure and function of the ecosystems in the affected area. These preliminary evaluations should of course be conducted by mulidisciplinary research teams.

Our present understanding of these issues is, nevertheless scarce, especially in terms of which wild species should be used to restore degraded Mediterranean ecosystems. Neither does the literature provide many ecotoxicolgical analyses of grasslands. The experience described here can be, therefore, viewed as the ground-work for the remediation of soils contaminated with heavy metals.

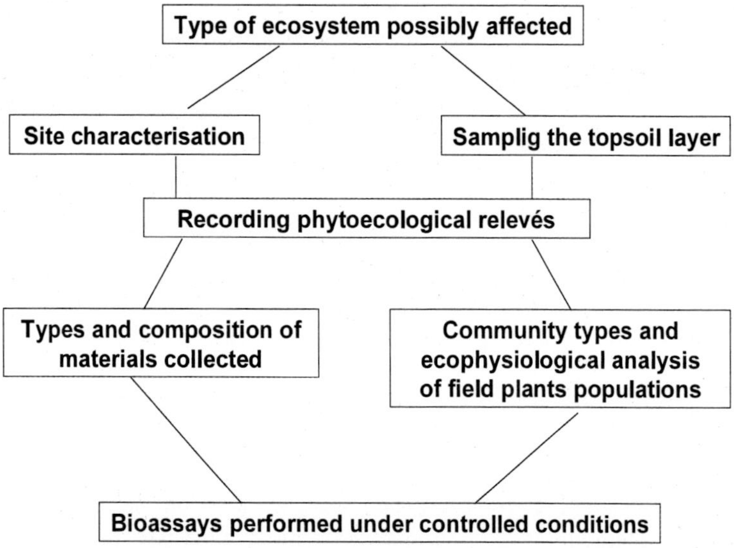

Figure 1. Protocol.

Assessing the Ecotoxicity of Soil Pollutants and Design of Bioassays

It is uncommon to find in the literature on the ecological restoration of terrestrial ecosystems valid protocols for use in studies on this subject. The following section describes certain concepts that may help decide upon the best procedure to use.

A *bioassay* is a method in which organisms (e.g., plants or animals), biological systems (e.g., tissues), or biological processes (e.g., enzyme activities) are used to measure the biological effects of a substance. In the context of managing sites with dangerous chemical residues, bioassays may be defined as the exposure of biological indicators to environmental samples collected in the field aimed at detecting the presence of toxicity and/or identify the toxic effects produced on the resident species. Generally a bioassay on a site containing hazardous residues involves laboratory tests (on the soil, soil leachates, water or sediment samples) using a standard set of test organisms under controlled lab conditions. Test organisms that have been widely used for ecotoxicology studies include *Selenastrum capricornutum* (a freshwater alga); *Daphnia magna* (a macroinvertebrate); *Pimephales promelas* (freshwater fish); *Rattus norvegicus* (rat) *and Lactuca sativa* (lettuce). Nonetheless, any nonstandard organism may be appropriate for a bioassay when: a), the standard organisms used shows no response to known or likely pollutants; b) the response of a particular non-standard organism is more specific for the given pollutant; or c), if the response of a particular organism not included in the list of standards needs to be assessed. It is likely however that tests based on non-standard organisms will be more costly owing to the difficulty in obtaining, culturing or cultivating new organisms and consequently standardizing new bioassays (including quality assurance procedures). Moreover, a large number of preliminary tests using the non-standard organisms will be needed to confirm their validity.

A further feature to consider is the realistic planning of bioassays: (i) the design of bioassay studies involves establishing the field samples to be collected and their lab analysis; (ii) it is important that any project be completely planned in detail, from its objectives to expected results, before embarking on any work. A poorly planned project will waste time and resources.

The steps of the protocol we followed for our bioassays designed for the purpose of restoring polluted soils were:

- Perception of a real-life possible ecotoxicity problem.
- Field and lab studies on the ecosystems possibly affected by the pollutants.
- Selection of species appropriate for evaluating toxicity based on the results of lab tests.
- Knowledge of the ecosystems possibly affected; knowledge of the chemical nature of the pollutants and their normal behavior or behavior in normal growth conditions of the populations and/or communities to be used in bioassays. Literature search on the case examined.
- Programming the different experimental designs necessary to fulfill the study's objectives. Designing an adequate statistical analysis optimizing the numerical treatment of the information with the number of situations or possible cases (to limit

the number of samples analyzed). Establishing the exposure doses, time and the toxicity tests to use.

- Collecting the material needed for the experimental procedure.
- Modifying and validating the necessary techniques for the chemical and biological tests that will be needed, (Landis and Yu, 1999). To this end, the variables to be determined in the different bioassays should be analyzed, and according to the proposed objectives and the information obtained in the first step, tables should be drawn up to record data during biomonitorization. For this, we should consider the different levels of organization of living organisms (figure 3). Thus, the entire set of techniques should be designed to address the following issues: bioaccumulation/ biotransformation/ biodegradation; biochemical monitoring; physiological and behavioral monitoring; population variables; community variables (figure 4).
- Performing bioassays, analyzing the material used in the bioassays (population and community ecology, soil and plant chemical analysis, use of the most appropriate techniques to demonstrate the toxicity of contaminants).
- Numerical treatment of the information, analyzing and discussing the results.

It may be noted from the above description of our approach to scenarios of heavy metal soil pollution along with Figures 2 and 3, that an ecotoxicological analysis involves many complexity levels: spatial (geologic-edaphic), biological (populations), as well as many forms of ecotoxicological characterization. Hence, it is important to integrate different scales through the use of *microcosms.*

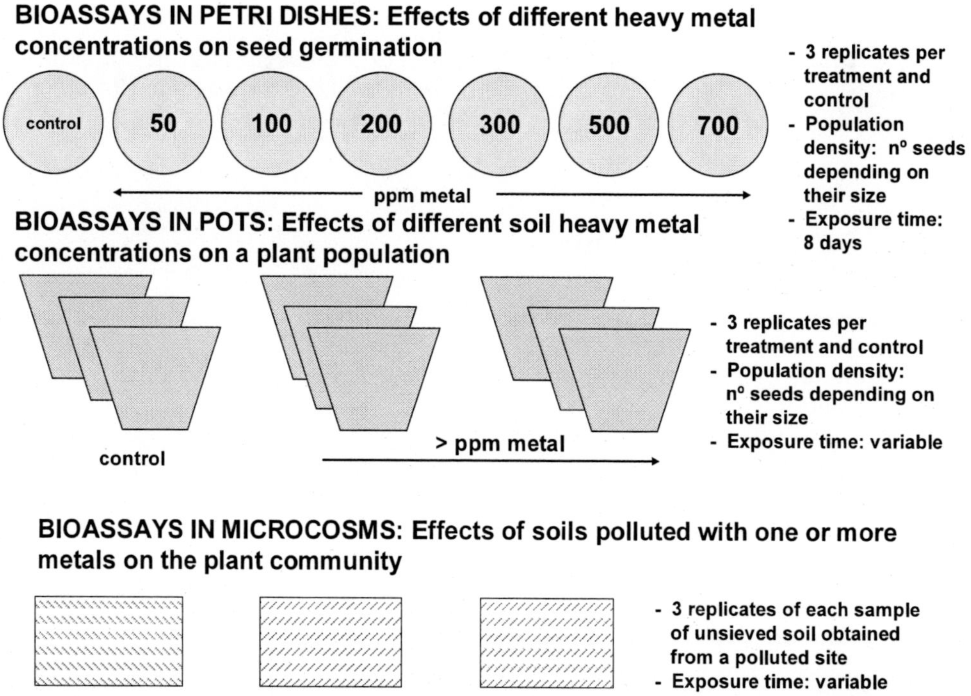

Figure 2.

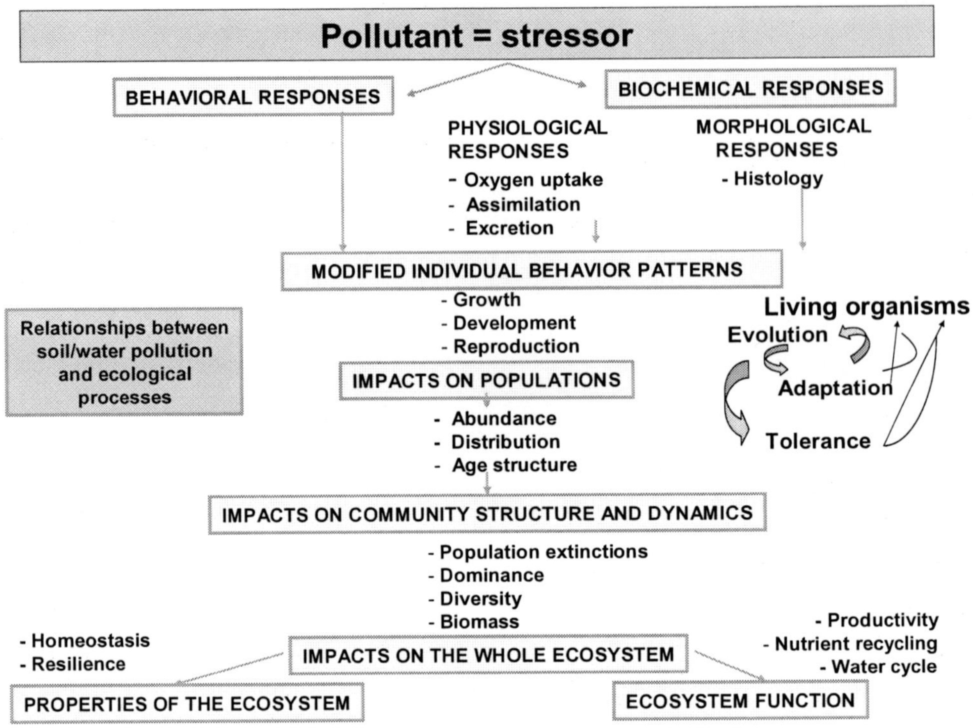

Figure 3. Inspired by Ramade (1995).

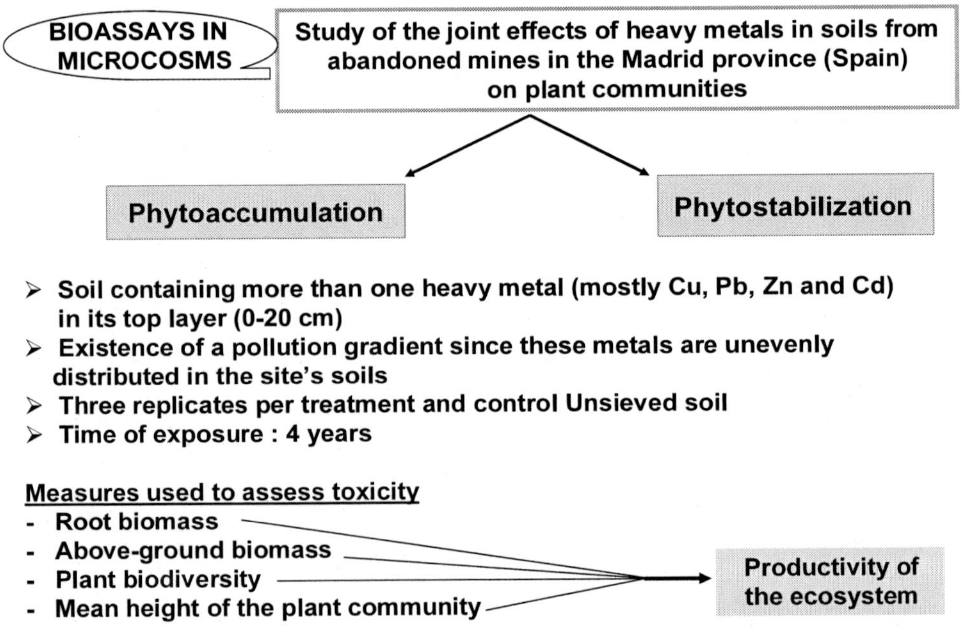

Figure 4.

Table 1. Total heavy metal contents (mg/kg) of polluted soils obtained from several sites of an abandoned silver mine (Guajaraz, Toledo) used to prepare microcosms for phytoremediation trials

Soil sample	Zn	Cu	Pb	Cd	As
Control	250	16	410	0	0
1	5095	85	3850	37	326
2	2865	50	2430	11	284
3	835	20	1420	0	197
4	2940	46	2250	15	239
5	2290	22	1635	0	220
6	820	13	1205	0	193
7	1490	10	180	0	180
8	1005	20	1845	0	0
9	1585	24	1770	0	261
10	855	17	1220	0	0
11	515	9	720	0	294
12	4160	110	910	29	163
13	2410	31	1215	7	239

Case Studies. Abandoned Mine Sites with Heavy Metal-polluted Soils

The soils of landfills and areas adjacent to old disused mines sustain different types of ecosystems and contain several heavy metals (Tables 1 and 2).

Owing to their efficiency, phytoremediation techniques (phytoextractiom and phytostabilization) are considered the most appropriate for restoring these sites. A realistic application of such techniques is dependent on the ecotoxicologic diagnosis of the site in terms of its mineral paragenesis (see Tables 3, 4 and 5), and on knowledge of the response mechanisms of plant species, both at the population and community levels, to the combined action of several heavy metals.

The chronological order of the steps to be taken for the ecological restoration of such sites we found to be most useful is indicated in Figure 5. This methodology may be viewed as an advance in ecotoxicological studies on native plants that play an important role in the trophic networks of ecosystems polluted with heavy metals. Knowledge of the main characteristics of the heavy metal contamination of a given site, along with the ecotoxicological problems they cause enables us to: a) to identify possible metal-accumulating, -tolerant or- excluding species that grow on landfills and areas adjacent to abandoned mines; b) to quantify the effect of different types of grass communities in terms of impairing the transport of the metals present in the polluted landfills and soils in leachates or surface runoff; c) design revegetation programs based on optimizing conditions for the phytoextraction, phytostabilization and physical stabilization of metals; d) evaluate potential "collateral environmental effects" associated with the different types of plant communities of the zones selected; e) establish potential benefits linked to the revegetation of the polluted sites, and f), to develop a protocol for lab experiments to obtain additional data on soil-plant relationships in these settings.

Table 2. Mean total (mg/Kg) and available (mg /100 g) heavy metal contents and organic matter percentages of soil samples taken from several zones of a disused copper mine (Garganta de los Montes, Madrid)

Soil origin	Zn	Cu	Pb	Cr	Cd	OM %
Landfill (rubble)	200 / 1.1	145 / 2.6	45 / 0	< 2	< 2	0.03
Landfill base	175 /1.1	228 / 3.4	40 / 0	< 2	< 2	0.94
Floodplain	145 / 0.8	315 /4.7	105 / 0	< 2	< 2	0.94
Ash woodlands	150 / 3.3	791 / 58.8	95 / 1.3	< 2	< 2	6.70
Temperate grasslands	100 / 1.7	741 / 1.7	47 / 1.1	< 2	4.5 / 1.1	2.69
Wet grasslands	800 / 4.2	1480 /107	3750 / 1	< 2	340 / -	6.45

Ecologically restoring abandoned mine sites whose soils are polluted with heavy metals

1. **Perform an ecotoxicologic diagnosis of the mine site**
2. **Identify any native plant species that could behave as phytoextractors and determine their ecological behavior within the study area**
3. **Evaluate through experiments performed in microcosms** *the extent of metal* **phytostabilization of the soils from the sites** *with and without the use of chelating agents*
4. **Characterize the types of mixed-species vegetative covers that could be used to restore soils polluted with several heavy metals and determine the possible situations in which they may be used**

Figure 5.

Table 3. Species numbers recorded in the plant communities of rubble landfills and ecosystems of the copper mine Garganta de los Montes, Madrid

Plant families	Landfill	Landfill base	Floodplain	Ash woodland	Temperate grassland	Wet grassland
Gramíneae	8	8	6	6	2	2
Leguminosae	2	2	6	7	3	2
Compositae	6	4	4	6	2	0
Others	14	16	9	9	4	4
Total	30	30	25	28	11	7
Plant cover %	60	39	80	95	85	95
Most Abundant Species	*Melilotus alba* (15%)	*A. castellana* (6%) *M. alba* (4%)	*T.campestre* (23%) *A.castellana* (5%) *P.coronopus* (15%)	*A. castellana* (40%) *Adenocarpus* (8%)	*Corrigiola* (30%) *A. castellana* (9%)	*A. castellana* (60%) *H .lanatus* (2%) *T. pratense* (2%)

Table 4. Mean heavy metal contents (mg/Kg) of grassland species (grouped as families) containing at least three metals in their above-ground mass growing in soils of the copper mine Garganta de los Montes (Madrid)

Plant families	Cu	Zn	Cd	Cr	Ni	Pb
GRAMINEAE	566 ± 1568	223 ± 244	6 ± 0.9	7 ± 11	2 ± 4	3 ± 4
LEGUMINOSAE	28 ± 26	87 ± 22	3 ± 5	1 ± 1.4	1.3 ± 1.8	0 ± 0
OTHERS	37 ± 50	153 ± 175	6 ± 8	0.04 ± 0.09	0.9 ± 1.1	0.7 ± 1.5

Table 5. Plant species growing in soils containing one or more heavy metal in the central Iberian Peninsula

Agrostis castellana	Convolvulus arvensis	Lolium rigidum	Spergularia purpurea
Anagallis arvensis	Crepis capillaris	Plantago afra	Stipa lagascae
Andryala integrifolia	Crepis vesicaria	Plantago coronopus	Thymus zygis
Avena barbata	Dactylis glomerata	Plantago lagopus	Trifolium striatum
Bromus hordaceus	Diplotaxis catholica	Plantago lanceolata	Trisetum paniceum
Bromus madritensis	Echium vulgare	Pulicaria paludosa	Vulpia ciliata
Bromus rubens	Hirschfeldia incana	Sanguisorba minor	Vulpia myuros
Bromus tectorum	Jasione montana	Scirpoides holoschoenus	
Carduus pycnocephalus	Leontodon taraxacoides	Sonchus asper	

The following is brief description of the most appropriate protocol used for phytostabilizing heavy metals:

a) Characterizing the plant communities growing at different abandoned mine sites.
b) Analyzing heavy metals in the topsoil layer (0-10 cm) by stratified sampling of the different morphological units of the landscape and plant communities of the sites (landfills, slopes and pastures of affected and non-affected areas). The reference values provided by the EU and the US EPA should be taken into account.
c) Chemical analysis of the root systems of the most abundant species that grow in the most polluted sites.
d) Selecting sites polluted with three or more heavy metals and turfing up the topsoil layer (avoiding excessive handling) to set up microcosms. Experiments are run under controlled conditions for 3-4 years (approximate timeframe for a grass community to become established from the seed bank with the appearance of biannual and perennial species) and watered using deionized water. Yearly species relevés are recorded, species are cut back after flowering-fruiting (to simulate consumption by herbivores or harvesting) and leachates are periodically collected to assess the export of pollutants.
e) Examining the microcosms: determining root biomass and chemical analysis of the species (individual above-ground mass and combined root systems).
f) Examining the effects of heavy metals in tissues of the root systems of individual species by electron microscopy. We use LTSEM, SEM-SE and controlled pressure SEM with an EDX detector to obtain information on the mineral composition of tissues and to locate the heavy metals in the tissue.

ECOSYSTEMS DEGRADED BY THE EROSION-POLLUTION DICHOTOMY

The urban solid residue (USR) landfills of central Spain, which at the end of the 1980s were sealed, or capped, simply by applying a layer of soil taken from the surroundings, have also been the subject of our research efforts. From the perspective of revegetating these systems appearing in our landscapes, these sites are probably the most complex we have come across in terms of their ecologically restoration and the environmental impacts they produce on the surroundings.

The characteristics of old USR tips makes their revegetation particularly difficult. These characteristics include: their height, orientation, slope, type and depth of the soils used to cap them along with their continued unauthorized use for waste disposal or other uses (e.g., for shooting practice); some waste dumps even have platforms with scarce slopes that are sometimes sown with cereals. Indeed, reality has surpassed the difficulties we initially predicted (Hernández, 1994). Accordingly, one of our main study tasks has been to identify the communities and autoecology of the species, mainly grasses, that grow on waste tip slopes, using as reference the study, also autoecological, of communities growing on the banks of rural roads in the same settings as these dumps. These are considered analogous, especially for the purpose of determining the mechanisms of the ecological succession involved in the spontaneous revegetation process (Estalrich et al., 1992 and 1997). They are also useful for identifying the best candidate species for mitigating erosion or the loss of fine soil components. In a second stage, we went on to analyze the behavior of many of these species in response to the salt and heavy metal pollution commonly suffered by these landfills (Adarve et al., 1998; Hernández et al., 1998; Pastor and Hernández, 2002 a, 2002 b and 2004; Pastor et al., 2003 and 2007).

From a scientific standpoint, this type of scenario allows us to gain insight into what we call the erosion-pollution dichotomy. This phenomenon is currently among the most frequent environmental impact problems and is practically undescribed in the literature. We consider the problem to be of special interest when realistically approaching restoration based on ecological principles.

The scheme provided in Figure 6 indicates a need to start by determining the directions of ground water flows and their relationship with the surface water courses of the area (Adarve et al., 1994 a and b; 1996). This feature is linked to "distance contamination" or to the non-point source pollution observed in ecosystems outside the affected area (e.g., aquatic ecosystems in surface waters such as rivers or streams, and terrestrial ecosystems e.g., intoxication effects on animal populations that modify the structure of an ecosystem).

A further issue to consider, is the composition of leachates and directions of their main flows in relation to discharge zones, since this is linked to what we can denote "the spatio-temporal pollution of the ecosystems of the landfill environment". The different level of the water table in discharge zones contributes to the "point-source pollution" of stable ecosystems in the environment (mainly marsh and streambed grasslands).

The nature and depth of the landfill's soil cover is related to different aspects of the soil physics that help mitigate pollution, but that often determine that the slopes of the capped landfill are not ideal for the development of the ecological succession in the landfill system. The loss or retention of compounds and chemical elements by the soil cover are related to what we refer to as "pollution in the landfill system". Passage to the autotrophic component of that contamination is linked to pollution at a distance (of other ecosystems in the landfill's surroundings) (Pastor et al., 1993; Urcelai et al., 1994 and 2000). Hence, it is also important that the study should examine ecotoxicological aspects analogous to the issues considered in the foregoing section (Pastor et al., 1994) within the conceptual framework of "risk analysis".

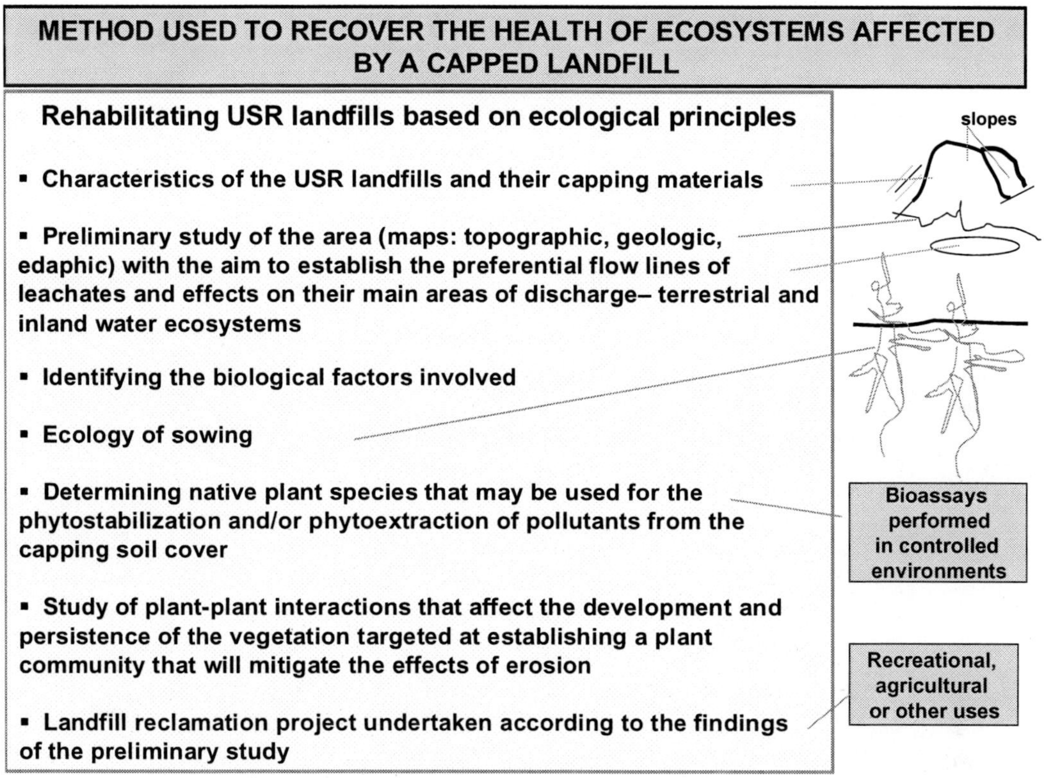

Figure 6.

Table 6. Present characteristics (2006-07) of USR sealed landfills on arkosic, limestone and gypsum substrates in the *Communidad de Madrid* (Spain)

USR landfills Granites and gneiss	1st year of capping	Ecosystem Main discharge	After-uses	No. slopes
Colmenar Viejo	1986	Stream and pastures grazed by cattle	Itinerant shepherding; fencing and reforesting with pines	3
San Lorenzo			Open burning, itinerant shepherding; houses constructed in the discharge area	3
El Escorial		Stream and ash woodland	Unauthorized rubble disposal/cattle grazing	4
Arkosic				
Móstoles	1986	Stream and wetland	Cereal cultivation and itinerant shepherding; cereals and horse-riding paths	3

Table 6. (Continued)

USR landfills Granites and gneiss	1st year of capping	Ecosystem Main discharge	After-uses	No. slopes
Villaviciosa	1987	Slope and wetland	Recreational use	1
Navalcarnero	1989	Sheep pasture	Cereal cultivation; shooting practice; reused for waste disposal; housing	-
El Álamo	1995	Wetland	Restored using covers comprised of native grasses	2
Limestone, loams, clays				
Alcalá de Henares	1986	River	Sown with acacias; other trees; grasses irrigated with river water	1
Torrejón de Ardoz 1	1982	Wetland	Pines; reused for waste disposal and continuously infilled by the wetland	3
Torrejón de Ardoz 2	1991	Wetland	Reused for rubble disposal; newly sealed in 1994; fenced off but unauthorized dumping continues	several
Mejorada del Campo	1986	Slope, streambed and river	Itinerant shepherding; shooting practice; restructured by the high-speed train, or AVE; reforested with pines; sown with alien grass species	3
Getafe	1986	Wetland	Reused for inert waste and rubble disposal	12 overlapping
Pinto 1	1986	Slope	Reused for waste disposal; use currently controlled	4
Pinto 2 a	-	Cereal crops	Reforested with pines	3
Pinto 2 b	-	Cereal crops	Reused for waste disposal	3
La Poveda	-	Stream	Reused to deposit rubble from the AVE works	2
Arganda	1987	Stream and streambed pastures	Uncontrolled - waste and inert materials continue to be deposited	2
Gypsum				
Aranjuez	1990	Stream	Open burning, steep slopes corrected by breaking-up after capping	1

Twenty years after many of these landfills were capped (Table 6), we intend to demonstrate the complexity that both their revegetation (spontaneous colonization of the vegetation from the seed bank of the soil cover) and phytoremediation using species that can adapt to their conditions supposes.

a) Characterizing Capped Landfills: Platforms, Slopes and Discharge Areas

The waste materials deposited in the landfills are of a mixed nature (urban solid, industrial and inert) and are not pretreated in any way. The soil used to cover the landfills is no deeper than 40 cm. Slopes are generally over 15 m in height and occasionally overlap due to subsequent reuse of the landfill by tipping waste on top of the sealing soil cover. Gradients are high, and frequently surpass 40%. These slope features besides affecting the plant colonization of these systems, also affect the extent of the leachate discharge area as surface runoff. Even when a landfill has a single slope, runoff occurs as a fan and thus differentially affects the biodiversity of the discharge zone. Moreover, a single slope often shows considerable variation in soil factors apart from exhibiting obvious differences with respect to the cover soils derived from different substrates.

Our phytoecological and edaphic studies performed in the past years both on the capping soils (Table 7) and soils of the discharge zones have revealed certain features that may be of help when planning to ecologically restore this type of environmental impact setting. The following is a systemized list of the main questions addressed: differences in the values of some edaphic variables and biodiversity between the landfills and the reference ecosystems, as well as differences in edaphic variables among the different landfill slopes.

Table 7. Soil variables recorded for the soil covers (mean values and standard deviation) of landfills overlying substrates representative of the central Iberian Peninsula five years after they were capped

Soil variables	Granites and gneiss	Arkoses	Limestones	Gypsums
pH	7.0±0.2	7.1±0.4	7.6±0.1	7.6±0.3
OM (%)	1.5±0.5	0.6±0.3	1.5±0.03	0.36±0.3
Total N (%)	0.080±0.030	0.033±0.010	0.094±0.056	0.048±0.004
P (mg/100g)	21.1±19.2	13.2±9.0	6.3±2.5	9.0±1.3
Na (mg/100g)	1.8±0.7	6.9±3.7	1.3±0.2	1.9±1.5
K (mg/100g)	17.2±9.5	21.8±3.8	32.3±9.9	13.1±1.1
Ca (mg/100g)	350.0±172.9	335.0±144.5	715.0±65.0	1396.7±1108.2
Mg (mg/100g)	9.6±3.6	37.2±23.3	25.1±2.8	6.60±1.7
Zn (mg/Kg)	125.5±69.8	83.5±146.0	57.5±10.5	33.3±5.7
Cu (mg/Kg)	8.7±19.4	150.9±730.2	13.0±11.0	5.0±5.2
Pb (mg/Kg)	7.7±5.6	72.8±297.0	28.5±3.5	0.0±0.0
Cd (mg/Kg)	0.0±0.0	1.5±3.1	0.0±0.0	0.0±0.0
Cr (mg/Kg)	0.0±0.0	4.4±4.5	0.0±0.0	0.0±0.0
Ni (mg/Kg)	22.2±3.6	15.7±8.1	22.5±3.5	17.7±1.2
Co (mg/Kg)	0.0±0.0	1.5±2.3	0.0±0.0	0.0±0.0

From this analysis, we were able to conclude that the events occurring and/or after-uses given to the landfills after their initial capping, along with the particular characteristics of each landfill pose many difficulties for the phytorestoration of their cover soils. The reasons for this are explained in more detail in the following section.

b) Cover Soil-vegetation Eco-chemical Relationships

Floristic relevés were recorded in twenty capped landfills of the central peninsula found on different substrates (granite and gneiss, arkoses, gypsum, limestone and marl). We also examined specialization of the flora to the different levels of disturbance as well as the adaptive strategies used by the species colonizing these environments, manifested by different biological features. Notwithstanding, based on the observations made (Tables 8 and 9), we focused on certain characteristics of the perturbation effects on the species and communities that most often appear in this type of setting.

Evident differences related to the diversity of both the plant species (all herbaceous) and nematodes in the soil (a soil mesofaunal indicator of nutrient recycling) were noted between the soil cover of the landfills and the corresponding reference ecosystems (Urcelai et al. 2000). An initial response to ecochemical relationships was that associated with the salinity of the landfills. The data shown in Table 10 indicate high levels of anions, especially of chloride ions soils detected in soils devoid of vegetation, and those provided in Table 11 reflect salinity differences among zones. These observations prompted an analysis of the link between salinity and pasture species to understand the ecological behavior of these plants and apply this knowledge to our restoration strategies (Adarve, et al., 1998; Hernández et al., 1998 b; Pastor et al., 2002 a and b).

Table 8. Significantly different results (99.9%) obtained in 36 soil samples taken from landfills overlying arkosic materials (in the fifth year after their initial sealing) and 55 samples collected from reference ecosystems of the same region

Biotic factors	Landfills	Reference ecosystems
Cover. Total vegetation (%)	34.9 ± 17.1	60.6 ± 25.0
Mean vegetation height (cm)	14.9 ± 9.1	22.7 ± 10.2
Plant diversity (no. sp / m^2)	15.5 ± 7.3	29.5 ± 11.4
Density of nematodes (no./100 cm^3)	45.6 ± 38.3	122 ± 50.7

Table 9. Plant biodiversity (species richness) variation in the discharge areas of three landfills of the *Comunidad de Madrid* five years after their initial sealing affected by surface runoff (N: reference ecosystem; A to H: Landfill sites)

Alcalá			Torrejón					Móstoles								
N	A	B	N	A	B	C	D	N	A	B	C	D	E	F	G	H
75	36	46	61	32	45	35	50	73	32	32	35	35	37	40	35	71

Table 7 indicated that Zn is a metal found in greatest quantities in all the landfills. Along with the high chloride contents, this led us to design a set of bioassays conducted under controlled conditions applying different concentrations of Zn chloride to the soil and then sowing with several native species collected from the field (analogous to those growing on the landfills), as well as other commercially available ecotypes: pasture gramineas (*Lolium rigidum, Hordeum murinum, Bromus hordaceus* and *B. rubens*), commercial gramineas (oats, wheat, maize), pasture legumes (*Trifolium subterraneum, T. glomeratum, T tomentosum and Lupinus angustifolium*) and forage plants (alfalfa, vetch and lupin), along with the crucifer *Hirschfeldia incana* (Pastor et al., 2003 and 2004).

Table 10. Levels of anions (mg/Kg) recorded in the soil cover of three landfills found on an arkosic substrate six years after they were initially capped according to the revegetation arising from the seed banks of the cover soil used

USR landfill	Sulfates	Chlorides	Nitrates	Fluorides
Mejorada				
Soil under Gramineae	10.5	14.8	10.0	1.2
Soil under Leguminosae	23.4	20.8	7.3	1.3
Naked soil	47.8	374.4	36.7	3.3
Móstoles				
Soil under Gramineae	11.0	5.6	0.9	1.4
Soil under Leguminosae	15.9	10.6	0.9	0.9
Naked soil	11.0	3.3	0.9	0.9
Navalcarnero				
Soil under Gramineae	11.1	8.3	3.8	0.9
Naked soil	123.4	145.6	43.3	0.9

Table 11. Soil conductivity values recorded ten years after capping in a single landfill (Mejorada) with several slopes and a landfill (Móstoles) whose only slope has very different zones in its low, middle and high parts

Soil cover of the Mejorada landfill				Soil cover of the Móstoles landfill			
		pH	Conductivity (μS/cm)			pH	Conductivity (μS/cm)
SLOPE 1	Soil 1	7.8	297	ZONE 1	Soil 1	7.1	706
	Soil 2	7.8	361		Soil 2	7.1	484
	Soil 3	7.9	460		Soil 3	7.4	452
SLOPE 2	Soil 1	7.6	395	ZONE 2	Soil 1	7.3	450
	Soil 2	7.7	553		Soil 2	7.4	494
	Soil 3	7.7	514		Soil 3	4.2	669
SLOPE 3	Soil 1	7.9	564	ZONE 3	Soil 1	3.4	1032
	Soil 2	7.6	551		Soil 2	3.2	1882
	Soil 3	7.6	810		Soil 3	7.4	394
SLOPE 4	Soil 1	7.9	282	ZONE 4	Soil 1	2.1	2810
	Soil 2	7.7	364		Soil 2	2.7	2690
	Soil 3	7.9	405		Soil 3	2.6	2620

We also examined the effects of soil Zn contamination on individual species (Pastor et al. 2003) and the plant communities that grow in the ecosystems of the natural surroundings of the landfill areas. Only a few community species were able to tolerate high soil Zn levels and may thus be considered appropriate as phytoremediators of soils with similar concentrations of this metal, around 700 ppm (Tables 12 and 13).

Table 12. Plant diversity and cover determined in a bioassay using microcosms prepared with soil containing different concentrations of zinc chloride

	Treatment (ppm Zn)	Mean and SD
Diversity	0	20.0 ± 2.0 a
(total no. vascular sp.)	300	11.7 ± 2.3 a
	500	6.0 ± 1.0 b
	700	4.7 ± 0.6 b
% Cover vascular sp.	0	70.0 ± 8.9 a
	300	53.7 ± 2.1 b
	500	30.0 ± 4.0 c
	700	14.7 ± 4.2 d
% Cover mosses	0	17.0 ± 2.6 a
	300	2.0 ± 2.0 b
	500	0.3 ± 0.6 b
	700	0.3 ± 0.6 b

Table 13. Zn contents (mg/Kg) of the above-ground mass of several species of the community grown in microcosms containing different soil Zn chloride concentrations

Species	Control	300	500	700
Vulpia myurus	74,6	228	765	1157
Polypogon maritimus	75,0	278	-	957
Juncus buffonius	75,3	120,5	149	893
Echium sp.	259	811	-	-
Scirpoides holoschoenus	28	36,5	64	-
Trisetum panicetum	63	373	-	-
Lolium rigidum	-	255	-	-
Gaudinia fragilis	-	251	-	-
Anagallis arvensis	-	86	-	-

As we examined behavior at the species and community level in controlled conditions (microcosms), we also performed experiments in field conditions. These experiments revealed some responses of the plants to physical abiotic factors (slope gradient and orientation) and biotic factors (such as the species growth habit in relation to the horizontal soil cover, rooting types in shallow soils, and possible seed production). These features need to be considered for revegetating landfills besides ecochemical interactions.

The real situation shown by sealed waste landfills, whether solid urban residues, industrial residues or mixed (as most of those in the Comunidad de Madrid), means we are faced with a rather complex restoration process mainly due to the following characteristics:

1) Their many slopes with steep gradients and different orientations are environmental factors that any revegetation project will have to deal with.
2) The scarce layer of capping, or sealing, material (soils of the landfill surroundings) along with the steep slopes leads to the loss of fine materials that affect the implantation of root systems and cause fertility losses of the soils proper to the

landfill system and silting of discharge areas (this often occurs in wetlands where the water table fails to crop out or in surface water channels in which sediments build up).

3) Besides these effects of physical degradation processes, those arising from the contamination that the landfills also exhibit determines that any restoration process will have to jointly deal with both types of effect, which we denote the erosion-pollution dichotomy of soils. In many cases, plant species that could be implanted to prevent erosion are unable to tolerate any of the pollutants found in these systems, or vice-versa.
4) In landfills capped with soils from their respective surroundings, the two processes of ecological succession intermix: on the one hand, there could be a primary succession since a new community starts in these "recently arising" systems, and on the other, a secondary succession could arise through germination of the seed bank present in the soil used to cover the landfill.

The two most important questions that should be highlighted are: a need to arrest erosion and its effects such as silting of water courses, eutrophication and pollution of surface waters; and in a subsequent stage, a need for adequate surveillance to avoid the constant reuse of a sealed landfill for waste disposal, which will hinder any restoration project.

CONCLUSION

After several years of exploring the best way of going about restoration based on ecological principles, we would like to highlight the following conclusions. The first is the highly complex nature of the processes involved in restoring ecosystems degraded through human actions. In our introduction to this subject, we mentioned the basic aspect of the complexity of ecosystems. This complexity is exaggerated by any human land management measures, mainly inappropriate land use, waste disposal and use of fertilizers. Without a good understanding of the "normal" behavior of an ecosystem, we cannot even attempt to identify the components of the ecosystem affected by an environmental impact and will not be able to undertake a realistic restoration. This determines the importance of finding a reference ecosystem that resembles in as much as possible the degraded ecosystem, unless of course, we have information of the previous state and behavior of the system before its perturbation.

Despite this complexity, we feel the different methodologies used to restore the terrestrial ecosystems of the degraded scenarios most frequently arising in the Mediterranean semi-arid and arid environments of Spain could render autochthonous plants with a capacity to restore polluted soils and help detain the erosion of sites with pronounced slopes. Our approach has also provided combinations of plant species that may be used for sites whose soils are polluted with several heavy metals. Although we used low cost techniques requiring a considerable length of time to adapt them for use in polluted environments, they all offer information for the restoration of specific polluted sites of the area. Thus, it seems that no general ecological restoration concept can be applied.

The many levels of complexity involved in an ecotoxicological study: spatial (geologic-edaphic), biologic (populations), and the many possible ways of evaluating the ecotoxicological nature of an environment means we should integrate scales through the use of *microcosms* (see figure 9).

Experiments in microcosms provide a quantitative idea of the optimal field conditions for the phytostabilization (in highly polluted soils) and phytoextraction (in moderately polluted soils) of metal contaminants. Both field and laboratory work is needed to identify the plant species that determine community structure in these environments and to complement previous bioassays by establishing further levels of heavy metal tolerance and toxicity.

ACKNOWLEDGMENTS

The studies mentioned were funded by the EIADES program of the CM and several M° MA and MEC projects (CTM2005-02165/TECNO).

REFERENCES

Adarve, Mª J.; Hernández, A. J.; Sánchez, A. J.; Rebollo, L. F. (1994 a). La contaminación de las aguas subterráneas por lixiviados de un vertedero sellado de residuos sólidos urbanos localizado en Torrejón de Ardoz (Madrid). En: *Agua y Medioambiente* pp. 162-170, TIASA, Madrid.

Adarve, Mª J.; Pastor, J.; Rebollo, L. F.; Hernández, A. J. (1994 b). Incidencia de un vertedero con residuos industriales sobre las aguas subterráneas y el suelo de su entorno. En: *Comunicaciones Técnicas II° Congreso Nacional del Medio Ambiente, 4:* 115-130, Madrid.

Adarve, Mª J.; Rebollo, L.; Hernández, A. J. (1996). Ensayos experimentales del flujo del agua y la distribución de la conductividad eléctrica en perfiles de la zona no saturada próximos a un vertedero de residuos sólidos urbanos. *Geogaceta, 20*: 1311-1314.

Adarve, MªJ; Hernández, A.J.; Gil, A; Pastor, J. (1998). B, Zn, Fe and Mn content in four grassland species exposed to landfill leachates. *J. Environmental Quality, 27:* 1286-1293.

Calow, P. (1995). Ecosystem Health. A critical análisis of concepts. In: *Evaluating and Monitoring the Health of Large-Scale Ecosystems*, pp. 33-41. NATO, ASI Series 138, Springer-Verlag, Berlín.

Crabbé, P., Holland, A. J., Ryszkowski, L., Westra, L. Eds. (2000). *Implementing Ecological Integrity. Restoring Regional and Global Environmental and Human Health.* NATO Science Series IV, Earth and Environmental Sciences Vol. 1., Kluwer Academic Publishers. Dordrecht, The Netherlands, 492 p.

Di Giulio, R. T.; Monosson, E. Eds. (1996). Interconnections between Human and Ecosystems Health. Champman & Hall. Norwell, MA, USA.

Estalrich, E.; Hernández, A. J.; Pastor, J. (1992). Conocimiento de ecotipos de especies fijadoras en taludes de carretera en clima semiárido. En: *III° Simposio Nacional sobre taludes y laderas inestables*, pp. 781-790. Ediciones de la Universidad de La Coruña, La Coruña.

Estalrich, E.; Hernández, A. J.; Aranda, L.; Pastor, J. (1997). Estudio de factores edáficos relacionados con la erosión y escorrentía en taludes de fuerte pendiente; ensayos de revegetación. *Edafología, 2:* 161-167.

FAO (2000). Infections and intoxications of farm livestock associated with feed and forage. *www.fao.org/es/ESN/animal/animapdf7annex-4.pdf*

Hernández, A. J. (1989). *Metodología sistémica en la enseñanza universitaria*. Narcea, Madrid.

Hernández, A. J. (1991). Ecosystems Global Problems. In: De Cleris, M (Ed.). *Handbook of Systems Science*, pp. 253-259.European Systems Union Hellenic Systems Society and UNESCO, Atenas.

Hernández, A. J. (1994). Restauración del impacto ambiental de vertederos. En: *Contribución a la Educación Ambiental. El tratamiento de los residuos urbanos*. Álvarez, J. B., Polo A. (Comps.) Cuadernos del ICE 11, pp. 35-47. Ediciones de la Universidad Autónoma de Madrid. Madrid.

Hernández, A. J.; Adarve, Mª J; Pastor, J. (1998 a). Some impacts of urban waste landfills on Mediterranean soils. *Land Degradation & Development, 9*: 21-33.

Hernández, A. J; Adarve, Mª J; Gil, A; Pastor, J. (1998 b). Soil salination from landfill leachates: effects on the macronutrient content and plant growth of four grassland species. *Chemosphere, 38*: 1693-1711.

Hernández, A. J.; Pastor, J. (2007). Ecosystems health and geochemistry: concepts and methods applied to abandoned mine sites. In: Loredo, J. (Ed.) *Exploring our Environment. 23rd International Applied Geochemistry Symposium,* 10 p. The Association of Applied Geochemistry, Oviedo (Spain).

Hernández, A. J, Pastor, J., Alexis, S., Vizcayno, C. (2007). Geoquímica y salud: una aproximación al estudio en ecosistemas tropicales. In: Preto, E., Alencoao A. M. (Coords.) *Geoquímica, VI Congreso Ibérico Geoquímica,* pp. 406-409. Planetaterra , Vila Real (Portugal).

Landis, W. G.;Yu, M. H. 1999 *Environmental Toxicology. Impacts of Chemical upon Ecology Systems.* Lewis Publishers. Boca Raton 390 p.

López Alonso, M.; Benedito, J.L.; Miranda, M.; Castillo, C.; Hernández, J.; Shore, R.F. (2002). Cattle as Biomonitors of Soil Arsenic, Copper, and Zinc Concentrations in Galicia (NW Spain). *Arch. Environ. Contam. Toxicol. 43*: 103-108.

Morcombe, P.W.; Petersson, D.S.; Masters, H.G.; Ross; P.J. Edwards, J.R. (1994). Cd concentrations in kidney of sheep and cattle in western Australia. I. Regional distribution. *Austr. J. Agric. Res. 45:* 851-862.

Pastor, J.; Urcelay, A.; Oliver, S.; Hernández, A. J. (1993). Impact of Municipal Waste on Mediterranean Dry Environments. *Geomicrobiology Journal, 11*: 247-260.

Pastor, J.; Alía, M.; Hernández, A. J.; Adarve, M. J.; Urcelay, A.; Antón, F.A. (1993). Ecotoxicological studies on effects of landfill leachates on plants and animals in central Spain. *Science of the Total Environment, 140*: 127-134.

Pastor, J; Hernández, A J. (2002 a). Estudio de suelos de vertederos sellados y de sus especies vegetales espontáneas para la fitorrestauración de suelos degradados y contaminados del centro de España. *Anales de Biología, 24*: 159-167.

Pastor, J; Hernández, A.J. (2002 b). Evaluation of the suitability of two grass species for phytorestoration of contaminated soils from landfills under field and experimental conditions. In: Rubio, J.L. et al. (Eds.), *Man and Soil at the Third Millennium,* pp.1687-1701. Geoforma, Valencia (Spain).

Pastor, J.; Hernández, A.J.; Prieto, N; Fernández-Pascual. (2003). Accumulating behaviour of *Lupinus albus* L. growing in a normal and a decalcified calcic luvisol polluted with Zn. *J. Plant Physiology, 160*: 1457-1465.

Pastor, J.; Hernández, A. J. (2004). Evaluating the response of plant community to Zn polluted soil as a basis for the revegetation/remediation of polluted sites. In: Faz, A., Ortiz, R., García, G. (Eds.) *Rehabilitation of Degraded Lands, IV International Conference on Land Degradation,* pp. 399-403. Quaderna Editorial, Cartagena (Spain).

Pastor, J; Martín-Aparicio, A; Gutierrez-Maroto, A; Hernández, A.J. (2007). Effects of two chelatinh agents (EDTA and DTPA) on the autochthonous vegetation of a soil polluted with Cu, Zn and Cd. *Science of the Total Environment, 378*: 114-118.

Petersson, K.P.; Thierfelder, T.; Jorhem, L.; Oskarsson, A. (1997). Cd levels in kidneys from Swedish pigs in relation to environmental factors-temporal and spatial trends. *Science of the Total Environment 208*: 111-122.

Ramade, F, 1995. Qualitative and quantitative criteria defining a "healthy" ecosystem. In: Rapport, D. J. et al. (Eds). *Evaluating and Monitoring the Health of Large-Scale Ecosystems,* pp. 43-61. NATO ASI Series, vol. 128, Springer-Verlag. Berlín.

Rapport, D. J.; Gaudet, C. L. & Calow, P. (1995). Evaluating and Monitoring the Health of large Scale Ecosystems, pp.33-41. NATO, ASI Series, vol. 128, Springer-Verlag. Berlín.

Rapport, D.J.; Lasley, W.L.; Rolston, D.E.; Nielsen, N.O., Qualset, C.O.; Damania A.B. Eds. (2003). Managing For Healthy Ecosystems. Lewis Publishers, Boca Raton, FLA, USA, p. 1184.

Urcelai, A.; Pastor, J.; Oliver, S.; Sánchez, A.; Hernández, A. J. (1993). Use of herbaceous species in land reclamation of dump areas in Central Spain. En: *Aire-Residuos, Proc. III Congreso de Ingenieria Ambiental,* pp. 524-532. Bilbako Nazioarteko Erakustazoka, Bilbao.

Urcelai, A.; Hernandez, A. J.; Pastor, J. (2000). Biotic indices based on soil nematode communities for assessing soil quality in terrestrial ecosystems. *Science of the Total Environment 247*: 253-261.

Wilkinson, J.M.; Hill, J.; Phillips, C.J. (2003). The accumulation of potentially-toxic metals by grazing ruminants. *Proceedings of the Nutrition Society*, *62*: 267-277.

In: Soil Contamination Research Trends
Editor: Javier B. Dominguez, pp. 73-107
ISBN 978-1-60456-319-1

Chapter 3

THE IMPACT OF LONG TERM HEAVY METAL CONTAMINATION ON BACTERIAL COMMUNITIES IN METALLOPHYTIC GRASSLAND SOILS

Stephen P. Cummings [*] ***and Lisa Bamborough***
School of Applied Sciences, University of Northumbria,
Newcastle-upon-Tyne, UK

ABSTRACT

Heavy metals are a significant source of pollution in soils that have been demonstrated to exert significant toxic effects on soil microbial communities. Here, we investigate the impact of the heavy metals lead and zinc on the bacterial community of an upland metallophytic grassland the UK, contaminated by historical lead mining activity. We focus on the total microbial diversity and that of the Actinobacteria, a significant component of the aerobic soil community that demonstrate a broad metabolic activity and represent useful indicators of soil health.

Two experimental approaches were adopted, the first employed microcosms containing soil from the site of an abandoned lead mine spoil heap, bacterial diversity was investigated using denaturing gradient gel electrophoresis (DGGE) of PCR amplified 16S rRNA DNA with universal primers to amplify the total community and more specific Actinobacteria primers. The Actinobacteria specific PCR primers showed that this taxon represented a diverse and metabolically active element of the microbial community of lead contaminated upland soil. Subsequently, we investigated how the stability of the bacterial communities responded to an increase in stress by adding additional Pb and Zn to the soil. There is significant interest in determining whether communities under stress are more or less robust in the face of additional environmental stress. Our data was equivocal, the total community diversity responded negatively to additional heavy metal

[*] Correspondents address: Stephen Cummings, School of Applied Sciences, Ellison Building, University of Northumbria, Newcastle-upon-Tyne. NE1 8ST; Tel/ fax. 0191 227 3176; Email; Stephen.cummings@unn.ac.uk

input into the soil, whereas, the Actinobacteria community did not show any substantial change in its diversity.

A second experiment characterised the bacterial soil community along a transect of a metallophytic grassland site that had significant differences in the Pb, Zn, organic C and N content across it. In this study the metabolically active bacterial population was studied using reverse transcription of mRNA from16S rRNA gene. Universal primers showed a low diversity of metabolically active bacteria, and both organic C and heavy metal concentrations in the soil seemed to impact on the community diversity. In contrast Actinobacteria showed higher diversities and seemed to be able to maintain this diversity even in soils with low organic C and high heavy metal concentrations suggesting that they are competitive in this stressed environment with other soil flora. The data presented here may prove useful in determining the effect of heavy metal concentrations on soil microbial diversity and community stability, particularly in managing rare metallophytic grassland sites and more widely in remediating contaminated soils.

INTRODUCTION

Heavy metals are a poorly defined group of elements that lack a coherent scientific definition, however, the term is commonly applied to a group of toxic metallic elements including As, Cd, Cr, Cu, Pb, Hg, Ni, Se and Zn that exert biological toxicity (Duffus 2002). These metals find their way into the environment through a range of anthropogenic activities and cause in significant problems due to their toxicity and persistence. In soil, high concentrations of heavy metals often arises due to anthropogenic activities such as mining (Bamborough and Cummings, 2006), the smelting of ore (Almås et al. 2004), from vehicle exhaust emissions and agricultural practices such as sewage sludge or fertiliser application to arable land (Kozdrój and van Elsas, 2001). The impact of heavy metals on soil fertility can be significant, affecting crop plants, livestock, and the indigenous microbial community (Giller et al. 1998). Such effects are of concern because in the case of crops and livestock, there is a potential for human exposure through the food chain. While the microbial community is responsible for key nutrient cycling activities in the soil essential for its continuing fertility. As a consequence there is an extensive literature on the impact of heavy metals on soil microbial communities and the processes that they drive.

Here we present a recent study that examines the effect of long term heavy metal contamination on soil, focussing particularly on Actinobacteria. These are a group of soil prokaryotes that have been reported to be the most important sub-population of bacteria in soils with high heavy metal content (Gremion et al. 2003). Our aim was to determine whether these bacteria are indeed significant contributors soil microbial communities exposed to long term heavy metal contamination and how the population responds in such soil to additional heavy metal stress. We contextualise this study within the recent literature describing the interactions between microbial communities and heavy metals in soil. This discussion highlights a number of problems that remain to be solved before we can accurately describe the impact of heavy metal contamination on soil processes. Amongst the most important of these are (i) how we relate the complexity of the microbial community at both a phylogenetic, and functional level in soil to the effects of heavy metal disturbance. (ii) How the toxicity of a particular heavy metal in soil measured?

Many studies on the impact of heavy metals on soils have attempted to measure how the microbial community structure changes in response to their addition (Sandaa et al. 2001; Hery et al 2003; Nakatsu et al. 2005). However, caution must be exercised in describing both the significance of such changes and whether they have consequences for the ecological processes driven by the community under scrutiny. Recent work on mathematically modelling soil bacterial populations to predict their diversity from small sample sizes, typical of those derived using clone libraries, indicated a high probability of diversities of between 10^5 to10^6 taxa in pristine soils (Lunn et al. 2004; Gans et al. 2005). In addition, related work has illustrated gaps in our understanding, in particular, the microbial community structure *in situ* that is being sampled and the appropriate scale to represent and characterise such communities (Sloan et al. 2007). Moreover, inherent problems in the methodology used, particularly PCR biases (Narang and Dunbar, 2004) and the acceptance that 16S rRNA genes are conservative markers for bacterial diversity (Sloan et al. 2007) inject further uncertainty into estimates of microbial diversity and community structure.

As a consequence, determining the result of heavy metal contamination on the structure of a microbial soil population is not straightforward. Using molecular approaches to identify the impact of heavy metal addition on microbial diversity and function may be limited in that they lack the sensitivity to identify small populations whose activity may be of high functional significance. Recent estimations of the impact of metal pollution on the bacterial populations of soil suggests that they reduce diversity by 99.9% (Gans et al. 2005) affecting rarer taxa most especially. Functional studies that seek to identify how heavy metals impact on community function also have limitations. For example, measures of changes in total biomass, respiratory rates, enzyme activities and substrate utilisation profiles can demonstrate significant responses to heavy metal addition (Kelly et al. 1999; Lazzaro et al. 2006a), and can act as robust biological indicators of the ecological health of the soil (Dawson et al, 2007). However, they do not necessarily correspond to changes in specific microbial processes that are important in maintaining soil fertility. Despite these methodological problems this remains an area of significant research activity, as improved tools become available the impact of heavy metals on soils may be more clearly elucidated, however, the heterogeneity and complexity of soil and soil communities make this a formidable challenge.

The Impact of Environmental Stress on Microbial Communities in Soil

The soil is a complex and heterogeneous environment and as yet we have no clear idea as to its phylogenetic and functional diversity (Curtis et al 2006) . In order to understand the impact of anthropogenic pollution on soil health a greater understanding of the relationship between the soil community, the pollutant and the physical and chemical parameters of the soil are essential. The microbial community is a key driver in the cycling of nutrients and energy through the system and is therefore a vital component in determining the resilience of the soil. Soil resilience can be defined as the ability of the soil to recover after a disturbance, this term has been subdivided into resistance (to disturbance) and resilience (to disturbance) that in combination define soil ecosystem stability (Griffiths et al. 2001).

Recent studies on the resilience of soil microbial communities have been directed at testing how environmental stresses affect the functional stability of the ecosystem. Two theories have been postulated. Firstly, that systems under stress are more stable than those which are not, as they have acquired physiological adaptations, that enable them to maintain function the face of additional stresses. The second states that non-stressed systems have a greater stability as they have more resources at their disposal that can be deployed to maintain ecosystem function in the face of stress (Tobor-Kaplan et al. 2005). These hypotheses were tested by Griffiths et al. (2001) using grassland, agricultural and industrial soil and Kuan et al. (2006) with grassland soils subject to various disturbances (biocide, sewage sludge application). Both studies challenged the soils with heat and Cu stress and measured the effects on soil functional stability. In each study, less stressed soils had greater stability supporting the hypothesis that non stressed soils have more resistance to stresses that those previously exposed to environmental challenge.

The work described above illustrates a growing awareness amongst microbial ecologists that the study of complex microbial systems in soil may benefit from the use of applied classical ecological theory to enable them to frame testable hypotheses on the role of phylogenetic and functional diversity on soil ecosystem functioning (Prosser et al. 2007). There has been significant interest in applying ecological theory to help describe the relationship between ecosystem sustainability and biodiversity, a number of studies have highlighted a positive correlation between ecosystem functioning and biodiversity or the insurance hypothesis of biodiversity. This theory states that diverse systems are more stable (Wardle and Zackrisson, 2005). However, there are few studies that have addressed this relationship within soil microbial communities. Wertz et al. (2006) showed that the dilution of soil diversity by four orders of magnitude had no appreciable impact on carbon mineralisation, denitrification and nitrification. This data suggests that the huge biodiversity of soil makes the functioning of this ecosystem largely insensitive to the erosion of biodiversity, even for specific functions such as nitrification, so long as all functional groups are not eliminated and their abundance can be recovered. The corollary of this observation is that anthropogenic pollution that affects soil microbial diversity should not affect ecosystem functioning. However, studies in the impact of disturbance on soil ecosystems have reported a significant loss of diversity and effects on the functional response of the ecosystem (Lazzaro et al. 2006b; Griffiths et al 2001; Kuan et al. 2006) this suggests that the qualifications relating to the persistence of all functional groups and the recovery of abundance are frequently not met in stressed soil. Superficially these observations that reduced biodiversity negatively affect ecosystem functioning would appear to lend support to the insurance hypothesis of biodiversity. However, other factors could also be significant, such as the energy costs required to cope with stress, which may be accommodated more readily by unstressed communities were maintenance energy is lower (Tobor-Kaplan et al. 2005).

Heavy Metals in Soil

The broad responses to environmental stress observed in soil communities highlights some of the important issues that need to be addressed in resolving how and why soil microbial ecosystems respond in the ways we observe to disturbance. From here on we shall draw more specifically on the literature that deals with the particular impacts of heavy metals on the soil ecosystem. Almost ten years ago the impact of heavy metals on microorganisms and microbially driven processes in agricultural soils was extensively reviewed (Giller et al 1998). In this study, the authors identified that there was an enormous variability, up to two orders of magnitude, in literature on the concentrations of heavy metals that exerted toxicity on various soil microbial communities. Subsequent to the publication of that review, there has been the widespread application of molecular techniques to study the impact and effects of heavy metal contamination on soil communities. However, many of these studies have simply confirmed the earlier work identifying the variability of the response of soil communities to heavy metal application (Feris et al. 2004; Griffiths et al 2005). It is clear that to understand heavy metal pollution of soil, a multi-disciplinary approach is essential, embracing the interactions between the biological, chemical and physical parameters within the soil. Perhaps the most significant factors to understand are those that modify the toxicity of heavy metals by affecting their bioavailability and the differential sensitivity of the microbial community exposed to them.

The bioavailability of heavy metals in soils determines what impact they will have on soil communities and contributes to the high degree of variability between soils, of heavy metal concentrations that exert toxicity. Differences in bioavailability may arise for a number of reasons and may reflect a range of diverse and heterogeneous interactions between heavy metals the soil matrix and the microbial biomass. The bioavailability of heavy metals is dose dependent (Almås et al. 2004), however, it has been shown that the total concentration of a heavy metal in a soil is not a reliable indicator of its bioavailability (Prokop et al. 2003). The toxicity of a metal is dependent on how it speciates in the soil (Knight and McGrath, 1995) with the free metal form being the most biologically active. Determining what fraction of the heavy metal in a soil is bioavailable has been measured using a range of techniques, in the case of Cd, it has correlated well with the fraction that could be extracted using chelating agents such as DPTA (Prokop et al. 2003). However, another study suggested that soil pore water concentration may give the best prediction of metal toxicity to soil communities (Almås et al. 2004).

The composition of the soil matrix can also influence the toxicity of heavy metals. For example, in soils with clay contents less than 50%, heavy metals exhibited higher inhibiton of microbial activity than those with higher proportions of clay (Lazzaro et al. 2006b). This may reflect the observation that clay minerals such as sodium and calcium bentonite and zeolite have been shown immobilise and decrease the water extractable concentration of heavy metals including Cd, Zn, Cu and Ni (Usman et al. 2005). Similarly, the organic carbon content of the soil can also exert an significant effect on the availability of heavy metals, Cd has been shown to adsorb strongly to organic carbon (Prokop et al 2003), therefore, one would expect the proportion of these two components of the soil to exert significant effects on the bioavailability of heavy metals in a soil. Recent work has demonstrated the significance of the heterogeneity of the soil environment across small (a few centimetre) distances. Most studies have monitored bioavailability using homogenised bulk soil, however, it has been shown that the spatial heterogeneity of soil matrices can result in a 30–fold differences in metal concentrations at sites less than 1 cm apart (Becker et al. 2006). Given that soil microbial communities demonstrate high phylogenetic and functional heterogeneity (Fierer et al. 2005; Rondon et al. 2000), large differences across relatively short distances may be explained at least in part by these steep gradients in heavy metal concentrations. The bioavailability of heavy metals also changes with time reflecting changes in the soil environment often driven by microbial activity, the metabolism of organic substrates by the microbial flora of the soil has been shown to increase the concentration of dissolved organic carbon content in soil and reduce the pH which enhanced both metal mobility and bioavailability. Concurrently, the adsorption to biomass and the sedimentation with biologically generated phosphate or other anions will immobilise heavy metals (Wu et al 2006).

The differential response to heavy metals by microbial communities, in the presence of equivalent bioavailable concentrations reflect the different sensitivities of these communities to heavy metal exposure (Giller et al. 1998). In many studies the presence of high concentrations of heavy metals has been associated with a decrease in the biomass (Wang et al. 2007; Liao et al, 2007; Vaasquez-Murrieta et al. 2006; Li et al. 2006; Frey et al. 2006), indeed some data suggests that heavy metals even in relatively low concentrations can depress the microbial biomass even during long term exposure and this is accompanied by changes in the structure of the microbial community (Witter et al 2000). Decreases in biomass may relate to the reduction in fungal, particularly mycorhizzal biomass as plant roots are frequently absent in highly contaminated soils, particularly forest soil are undisturbed (Frostegård et al. 1993). However, such effects are not always observed and in a number of field based studies no significant effects of heavy metals on biomass have been noted (Griffiths et al. 2005; Ranjard et al. 2006).

In addition, heavy metals have had measurable impacts on a range of other parameters used to measure microbial activity, such as the production of carbon dioxide, which correlates with the basal respiration rate of the soil microbes, and enzymatic activities that are involved in key metabolic cycles or substrate utilisation profiles. The effect of high concentrations of heavy metals on basal respiration rates and enzymic activities have also produced conflicting data in the literature. Enzyme activities have been shown to be significantly decreased (Wang et al, 2007; Kandeler et al. 2000), whereas, other studies report no significant changes in soil contaminated with Zn and Cu compared to controls (Niklínska et al. 2006).

A number of studies have, in addition, to identifying changes in biomass and functional activities of the soil community (Abaye et al. 2005) also reported that these responses were frequently accompanied by changes in the microbial community structure (Giller et al. 1998). It is not clear whether these changes are related to changes in the functional activity of the community or whether they represent adaptation of the community to the presence of heavy metals (Griffiths et al. 2001). The overarching issue is whether we can apply a more profound understanding of the impact of heavy metals to the microbial communities in soil to more effectively manage the toxicity and environmental degradation that they cause.

Most work has focussed on how the diversity of microbial community structure changes in response to heavy metals. Ecological theory tells us that environmental stress reduces biodiversity and intuitively, one would expect that the presence of heavy metals in soils will have a negative impact on the diversity of the microbial community such responses have been observed (Li et al. 2006; Joynt et al. 2006) but not in all cases (Feris et al 2003). Moreover, it is logical to assume that the presence of a toxic stress such as heavy metals will also lead over time to the adaptation of the community to tolerate that stress, some studies have seen such an increase in heavy metal resistance among microbial populations (Schmidt et al. 2005; Frey et al. 2006), however, others have reported no such development of tolerant communities (Feris et al. 2003). It has been pointed out that the duration of the exposure that is being studied will affect the response on the microbial community, laboratory studies tend to be short term, whereas field studies tend to have significantly longer durations (Bååth et al. 2005). Short term addition of heavy metals soils have been reported to have quite different consequences to long term exposure. Renella et al. (2002) studied the impact of the addition of individual and combination additions of Cu, Zn and Cd salts to arable and grassland soils. They found that reductions in microbial biomass and respiration rate were variable small (5-12 %) compared to the effects of chronic exposure which gave reductions in these parameters of up to 50 % (Brookes and McGrath 1984). In summary, the community structure of heavy metal contaminated soils, particularly as a result of long term exposure, appears to change significantly, although this does not necessarily mean the diversity will decrease (Lorenz et al 2006) or that the community will be more tolerant of heavy metals (Feris et al. 2003; Giller et al 1998; Lorenz et al 2006). Often these changes have been related to specific taxa, such as Actinobacteria and fungi becoming more prevalent (Kelly et al 2003; Hery et al. 2003), however, the variability and conflicting data in the literature suggest that such changes are affected by a range of inter-related variables including the nature of the soil matrix, its chemistry and the composition of the soil flora. Therefore, each soil should be considered on a case by case basis.

The Response of Actinomycetes to Heavy Metals in Soil

Actinobacteria are a broad taxonomic group within the domain *Eubacteria.* This division has been divided into six orders and fifteen suborders (Stackebrandt et al. 1997) that embraces a large number of genera. The group describes a diverse phylogenetic and functional assemblage of bacteria that share a high proportion of guanine and cytosine nucleotides in their DNA, greater than 50%, and sequence similarities of their 16S ribosomal ribonucleic acid (rRNA). These bacteria are also described in the literature as high G+C Gram positives and Actinomycetes, names that have no standing in the nomenclature of bacterial classification (Eusby 2007) but remain widely used. They are of significant biotechnological interest as some taxa within the group, notably the within the genus *Streptomyces*, produce secondary metabolites such as the antibiotics erythromycin and tetracycline that are of great economic and pharmaceutical significance (Paradkar et al. 2003). In addition, they are of interest from the evolutionary perspective and recently, this group of bacteria have been proposed to belong to a novel paraphyletic division of the Eubacterial domain called the Posibacteria from which the Eukaryota and Archaebacteria emerged (Cavalier-Smith, 2002).

Within the Actinobacteria there is huge phlyogenetic and functional diversity, and individual taxa will perform a variety of, probably quite diverse, functional roles within soil. However, their ubiquity in soil, their metabolic diversity and their significance in ecology, biotechnology, and human and veterinary medicine make them a useful candidate taxa to study to determine how bacterial populations responds to the perturbation of the soil environment by the presence of heavy metals.

Actinobacteria Resistant to Multiple Heavy Metals

Bacteria are thought to tolerate metals by means of several systems. These include precipitation at the cell surface, biosorption to extracellular polymeric substances (EPS), including polysaccharides, and proteins, or to the cell surface (Valls *et al.*, 2000; Kefala *et al.*, 1999). There are also a number of active heavy metal efflux systems which are widely distributed amongst bacteria, such as the mercury efflux (Mer) proteins and P-type ATPases that transport heavy metals across the cellular membrane (Neis, 2003). Actinobacteria have been shown to employ both avoidance of heavy metals by biosorption (Rho and Kim, 2002) and resistance through their active efflux (Schmidt et al. 2005). For example, *Frankia* strains were shown to exhibit resistance to lead and copper through binding to an extracellular matrix composed of polysaccharides and proteins that prevented metal ions from entering the cell (Richards *et al.*, 2002). In contrast, Ravel *et al.,* (1998) isolated *Streptomyces* spp. from polluted sediment samples that demonstrated high levels of mercury resistance. This was mediated through the expression of a plasmid borne mercury efflux (Mer) system. Similarly, evidence for high affinity nickel transporters has been demonstrated in this genus (Amoroso et al. 2000)

Actinobacteria in Heavy Metal Contaminated Soil

The significance of Actinobacteria within the microbial communities of soils contaminated with heavy metals remains unclear, Gremion et al (2003) studying the bacterial community of a soil contaminated with Cd, Cu and Zn showed that Actinobacteria dominated clone libraries derived from both the bulk soil and the rhizosphere of *Thlaspi caerulescens* a heavy metal hyperaccumulator. Moreover, 70% of the Actinobacterial clones belonged to the genus *Rubrobacter*, a widespread taxa in soils, that has as yet few cultured representatives (Holmes et al. 2000). Similarly, a study on an agricultural soil amended with sewage sludge containing elevated levels of Zn produced a clone library in which the most common OTUs were high G+C Gram positive clones belonging to the *Rubrobacter* genus (Moffett et al. 2003). High abundance of Actinobacteria have also been found in soils that contain naturally high concentrations of heavy metals such as Ni, Cr and Co, culture based studies identified 80% of the bacteria recovered as belonging to this group (Amir and Pineau 1998) and actinobacterial sequences in clone libraries from nickel mine spoils comprised 96 % of the community (Hery et al, 2005). Studies using PFLA techniques also showed that although total biomass decreased on these soils compared to non-serpentine soils the proportion of actinobacterial marker fatty acids to be significantly increased (DeGrood et al. 2005).

However, the abundance of Actinobacteria is probably determined by a range of factors in the soil. For example, studies on the bacterial communities of rhizosphere soil from grain legumes indicated Gram positive organisms made up almost 45% of the community derived from the analyses of a clone library. Furthermore, Actinobacteria contributed between 70-95% of the clones from this fraction. The authors speculated that the predominance of Actinobacteria could be attributed to the timing of sampling which occurred after fruiting of the legumes. At this point the root exudation would diminish resulting in organisms that adopted a K-strategy predominating (Sharma et al. 2005). They have also been observed to increase as a result of remediation of contaminated soils, such changes have been linked to to increases in the total N, pH and porosity of the soil as it is re-vegetated (Zhang et al. 2006). Actinobacteria have also been shown to be dominant within particular microenvironments within the soil. A study of the bacterial communities found inside and outside of soil micro-aggregates in disturbed and undisturbed soils showed them to be distinctly different, while *Proteobacteria* clones were prevalent in libraries derived from whole microaggregates. The community inside micro-aggregates, in undisturbed soil, had a high abundance of Actinobacteria that consisted entirely of taxa from the order *Actinobacteridae*. Whereas, in micro-aggregates from disturbed soil, Actinobacteria were dominant but 53% of the clones were attributed to the order *Rubrobacteridae* (Mummey and Stahl, 2004). This work demonstrated that there is a stratification of bacterial taxa in relation to the communities inside and outside soil micro-aggregates, furthermore, the inner micro-aggregate community was also significantly different between different soil types, indicating the actinobacterial community structure is affected by soil disturbance even in these sequestered environments.

In contrast to the assertion by Gremion et al (2003) a number of studies on the impact of heavy metal contamination on bacterial community structure have reported that there was a significant decline in the contribution of Actinobacteria to the bacterial community. In a forest soil contaminated with cadmium, copper zinc and lead there was a significant decrease in the Actinobacterial marker using a PFLA protocol compared to control soils. The microbial community also showed more general changes in response to heavy metal amendment, one that was more tolerant of heavy metals but with a compromised functional activity (Frey et al. 2006). Soils subjected to amendment with zinc in a microcosm study, and those from a soil in the vicinity of a Zn smelter also showed significant changes in the microbial community that included a decrease in the Actinobacteria (Kelly et al. 1999; Kelly et al. 2003). However, in the former case there was no significant change in the culturable bacterial population or the microbial biomass of the soil compared to the control from the vicinity

In this study we have examined the bacterial communities present in two soils subjected to long-term heavy metal contamination, principally lead and zinc, as a result of historical lead mining activity. The first site was a metallophytic grassland, a habitat that contains a high diversity of flora capable of tolerating heavy metals in the soil (Balabane et al. 1999). The site had been previously surveyed for vegetation coverage (Leonard, 2001). Here we characterised the extractable lead and zinc from the soil in eight distinct areas defined by the vegetation. Heavy metal concentrations differed significantly across these areas and we characterised the bacterial soil community in each to understand how the community structure differed under the different metal concentrations. The second site was a spoil heap from an abandoned lead mine. Here we were interested in seeing how a bacterial community that had developed under high concentrations of heavy metals responded to additional metal stress. In particular whether such communities exhibited higher or lower stability to additional environmental disturbance.

Table 1. The differentiation of the Garrigill field site into eight distinct areas based on the most abundant plants, lichens and Bryophytes in each after Leonard (2001)

Area	Vegetation
1	*Festuca ovina* (abundant); *Viola lutea**
2	*Gunneracea*; Scrub (*Crataegus monogyna)*
3	*Festuca ovina* (abundant), *Plantago lanceolata, Hyloconium splendens*
4	*Festuca ovina* (sparse); *Plantago lanceolata, Hyloconium splendens*
5	*Festuca ovina* (sparse); *Crepis aurea*
6	*Minuartia verna*, Thalspi caerulescens*, Peltigera leucophlebia**
7	*Armeria maritime*, Minuartia verna*, Thalspi caerulescens**
8	Mixed woodland (Pine; Beech); *Ulex europaeus*

* Species diagnostic of a metallophytic grassland community.

MATERIALS AND METHODS

Field Sites and Sampling Strategy

Two sites were used in this study both were metallophytic grassland sites (Balabane et al, 1999) in the Pennine uplands of northern England. The Garrigill site, located at grid reference 714 435 (OS Landranger 87) has thin, poorly sorted soil, with sparse but species rich vegetation, comprising of grasses, predominantly sheep fescue (*Festuca ovina*) and species typical of metallophytic grasslands (Table 1). The site is roughly triangular with distinct boundaries of the South Tyne to the south, a dry stone wall to the north and a tributary to the river defining the eastern boundary (Figure 1). The site has been extensively grazed by rabbits. Heavy metals on the site has arrived through alluvial deposition via the South Tyne river as a consequence of intensive lead and zinc mining in the nineteenth and early twentieth centuries in the North Pennines (Macklin *et al* 1997). The site represents a river shingle in the floodplain of the South Tyne that has, due to decreased flooding since the turn of the nineteenth century, become vegetated. The community has developed over the last 150 years and has areas of grassland, scrub and some mixed deciduous and coniferous woodland. The vegetation of the site was surveyed in an earlier study (Leonard, 2001) that identified eight areas (Figure 1).

The site was divided into 8 areas on the basis of the vegetation cover (Leonard, 2001). To sample each area a transect 3 m wide was plotted across the site that covered all areas except area 4. A 1 m^2 quadrat divided into 10 squares was then placed along the transect within each area. Each of the 8 areas was then randomly sampled from 3 squares within the quadrat by collecting soil cores from the top 10 cm of topsoil in May 2005. Area 4 was sampled identically.

The second site at Bardon Mill, grid reference 779 644 (OS Landranger 87) was derived from an abandoned mine spoil heap. The soil had very sparse vegetation, including grasses, however, most of the site had no vegetation cover. Mining occurred from the mid sixteenth century to the early twentieth century, therefore the site is no more than a few hundred years old. Bardon Mill soils were collected from non-vegetated areas by collecting top soil from the top 10 cm of soil. The area was much more limited and homogeneous than the Garrigill site. Therefore, triplicate samples of topsoil were excavated from the non-vegetated area of the site.

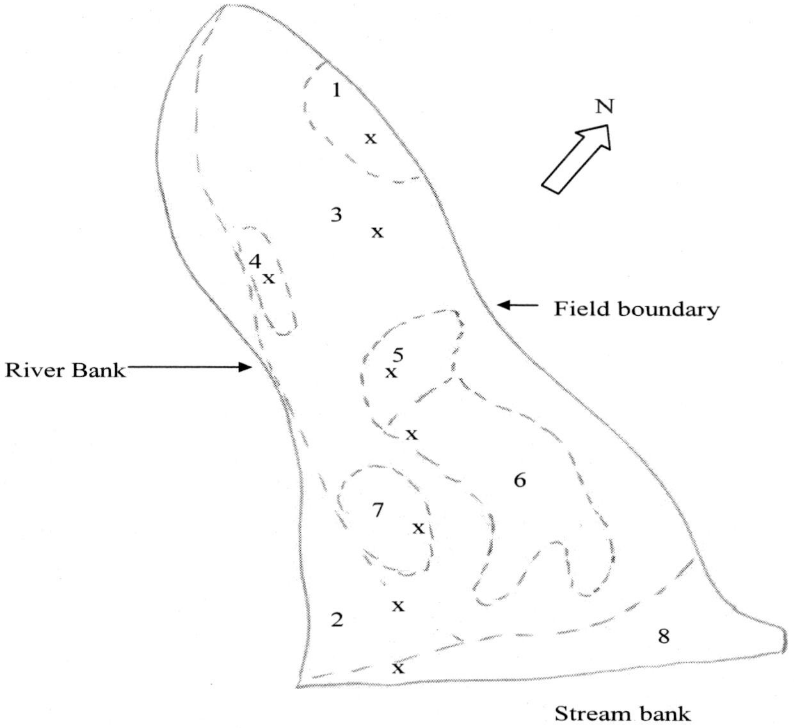

Figure 1. A plan of the Garrigill site showing the eight areas identified by vegetation survey (Leonard, 2001). The position of the 1 m^2 quadrats used for sampling are marked by 'x' in each area. All except that for area 4 were taken along a transect 3 m wide.

Immediately after collection soil was transported to the laboratory and stored at 4° C until required. Prior to being used in microcosms, soil was sieved (2mm) and air dried overnight. When required, soils were sterilised by autoclaving the soil (121° C, 15 mins) on three consecutive days with incubations at 37° C in between and a final incubation at 100° C for 30 min to dry (Baxter et al. 2006).

Analyses of Soil Samples

Total metal content was determined with duplicate soil samples (0.5 g). Samples were air dried and digested in 10 ml of boiling *aqua regia* (HNO_3: HCl, 1:3, v/v).Once the sample solutions were cool, they were filtered through Whatmans No. 1 filter papers to remov any insoluble sediment or organic material. The solution was then made up to 100 ml with distilled water. Sample were tested for total metal content by Flame Atomic Absorption Spectoscopy (Flame Atomic Absorption Spectroscope) (Perkin Elmer / AAnalyst 100), with "Calib" software. Samples were calibrated against Pb and Zn solutions.

Carbon and Nitrogen analyses were commercially done with duplicate samples of soil (1g) using a Carlo Erba 1108 Elemental Analyser controlled with CE Eager 200 software, run in accordance with the manufacturer's instructions and weighed using a Mettler MT 5 Microbalance by the Advanced Chemical and Material Analysis, University of Newcastle-upon-Tyne, Newcastle, UK.

The contamination index (CI) was calculated for Pb and Zn in the Garrigill site soils according to the method of Feris et al. (2003). The index measures the contamination in the soil relative to the most uncontaminated area in the site (area 8) according to the formula

$$CI = \sum\{\log [HM_n / HM_{n\ in\ control}]\} / \text{Number of metals in index}$$
$$n = Pb \text{ and } Zn$$ (Feris et al. 2004)

Soil Microcosms

For each treatment, namely, controls containing unmodified soil and test soil with additional heavy metal added, three microcosms, consisting of 500 g of soil in each replicate, were established in 600ml glass beakers. The soil was rewetted to a moisture content of 10% (w/w, dry weight basis). To increase the heavy metal stress of the soils to five times that of the background levels $PbNO_3$ and $ZnSO_4$ were added to the soil in aqueous solution. A carrier system was used to ensure consistent metal application. Appropriate concentrations of $PbNO_3$ and $ZnSO_4$ dissolved in dH_2O were applied to 1 g aliquots of sterile soil subsequently these were incubated overnight to allow evaporation of the water. The treated aliquot was cut into the rewetted soil by thorough mixing, giving soil microcosms with final concentrations of $PbNO_3$ and $ZnSO_4$.

Samples (1g) of soil were taken from each microcosm for DGGE analysis. Soil water content was measured and found to be constant during the experiment. Initial samples (day 0) were taken immediately after the microcosms were set up. Between samplings microcosms were incubated at 18° C in a water-saturated atmosphere (Rasmussen and Sorensen 2001).

DNA Extraction and PCR Amplification of 16S rRNA Genes from Soil Microcosms

DNA was extracted from 1g of soil with an UltraClean Soil DNA kit (MoBio) and stored at –20°C. PCR amplification of V3 region of the bacterial 16S rRNA gene was performed using Primer 1 (5'-CCTACGGGAGGCAGCAG-3') and primer 2 (5'-CGCCCGCCGCGC GCGGCGGGCGGGGCGGGGGCACGGGGGGCCTACGGGAGGCAGCAG-3') (Muyzer et al., 1993). Amplification of DNA was performed with 0.5 μM each primer, 2 x *Pfx* amplification buffer, 1 μl template DNA, 0.3mM each dNTP, 1mM $MgSO_4$, 1 x PCR enhancement solution, 500 μg/ml BSA and 1.25 U Platinum *Pfx* DNA polymerase (Invitrogen) in a 50 μl PCR reaction mixture. The thermocycling conditions used were as follows: 3 min. at 94°C (1 cycle), 94°C for 1 min. (10 cycles), 60°C for 1 min. (with reduction of 1°C every cycle), 68°C for 1 min. (10 cycles), then 94°C for 3 min., 50°C for 1 min. and 68°C for 1 min. (25 cycles) with a final extension step of 68°C for 10 min (Baxter &

Cummings 2006). All reactions were carried out using the Eppendorf Mastercycler Gradient PCR machine. PCR products were analysed by agarose (1.5% w/v) gel electrophoresis and ethidium bromide staining.

Reverse Transcription PCR (RT-PCR)

Bacterial RNA was extracted from soil using with an Ultraclean™ Microbial RNA Kit (MoBio) according to the manufacturer's instructions, and stored at -20° C.

RNA was amplified using the Omniscript RT kit (Qiagen). RT PCR was performed in a final volume of 20µl, containing 5µl RNA, 10U RNAseOUT (Invitrogen), 10µM Random Hexamers (Qiagen), 0.5µM each dNTP, 10U Omniscript reverse transcriptase, and 1x Buffer RT. The mixture was incubated at 37° C for 60min then stored at -20° C.

Selective amplification of Actinomycetes was performed using Primer F243 (5'-GGATGAGCCCGCGGCCTA-3') and primer R513 (5'-CGGCCGCGGCTGCTGGC ACGTA-3') (Heuer *et al.*, 1997). Amplification of soil Actinomycete DNA was performed as previously described by Heuer *et al.* 0.5µM each primer, 1µl of template DNA, 2 x *Pfx* amplification buffer, 1mM $MgSO_4$, 1 x PCR enhancement solution, 25µg BSA, and 1.25U Platinum *Pfx* DNA polymerase (Invitrogen). PCR conditions were 5 min at 94° C; 35 cycles 1 min at 94 ° C, 1 min at 63 ° C, and 2 min at 72° C; and a final extension for 10 min at 72°C.

PCR products were analysed by agarose (1.5% w/v) gel electrophoresis and ethidium bromide staining.

Denaturing Gradient Gel Electrophoresis

Denaturing Gradient Gel Electrophoresis (DGGE) analysis was performed using the D-Code DGGE system (Bio-Rad) as described by Baxter & Cummings (2006). For analysis of bacterial populations, PCR products were loaded on to polyacrylamide gels (12%) with a denaturant gradient ranging from 35-55% denaturant (with 100% denaturant corresponding to 7M urea plus 40% v/v formamide). Gels were run at 60° C for 10.5 hours at 80 V. Gels were stained in SYBR Green I (1:10000 dilution in 1x TAE buffer; Molecular Probes) for 30 min. and viewed with UV transillumination and photographed using the Gel Doc 2000 gel documentation system (Bio-Rad).

Statistical Analysis of DGGE Profiles

For each treatment, PCR products, derived from DNA extracted from 3 independently derived communities, were analysed by DGGE. The patterns obtained from these replicates were highly reproducible. To facilitate sample processing, PCR products were pooled for each treatment. The pooled PCR products were then subjected to triplicate DGGE analyses to confirm that the band patterns were reproducible with those derived from the independently derived community replicates (Sigler and Turco, 2002; Brodie *et al.* 2003).

DGGE profiles were analysed using Quantity One 4.1.1 software (Bio-Rad). The lane background was subtracted by using a rolling disc algorithm and the lanes were 'normalised' to ensure all lanes contained the same amount of total signal. Lane profiles were corrected for differences in migration rate by manually assigning R_f lines to marker lane bands (Griffiths et al. 2003). To facilitate the efficient handling of samples and data, each sample from each replicate were screened with PCR-DGGE to assess replicate variability in DGGE profiles. When the reproducibility of PCR profiles within a treatment had been ensured, all PCR products derived from a common treatment were then pooled for the final analysis in order to accommodate the electrophoresis of samples on a single gel.

Band patterns were analysed using three methods. Correspondence analysis (CA) was performed to visualise the relationships among band patterns using Statistica 6.0 software (StatSoft Inc.). Cluster analyses of band patterns were also performed with the unweighted pair group method using averages (UPGMA) based on the Morisita index of Similarity. Band pattern data was analysed using Raup and Crick's index of similarity The Raup and Crick index of similarity was used to test if similarities observed between soil cultures profile patterns were significantly higher or lower than would be observed by chance (Rowan et al. 2003). Raup and Crick and cluster analyses were calculated using the PAST statistical analysis software.

RESULTS

The effect of additional heavy metal stress on bacterial communities in soils from lead mining spoil.

To study how bacterial community structure responded to the input of additional heavy metals in contaminated soils, microcosm experiments were performed using top soil from the Bardon Mill site. The concentration of extractable Pb and Zn in this soil were determined to be ~3 mg g^{-1} and 1 mg g^{-1} respectively. In experimental microcosms these levels were supplemented with $PbNO_3$ and Zn $ZnSO_4$ to give an increase in the concentration of these two metals to 15mg g^{-1} $PbNO_3$ and 5 mg g^{-1} $ZnSO_4$. Subsequently, the impact of these additions were monitored over a 51 day period and compared with control microcosms that had received no additional heavy metal inputs, using DGGE analyses of the bacterial community.

Samples of soil (1 g) were collected immediately after setting up the microcosms (day 0) and then on days 2, 7, 10, 16, 23, 30, 37, 44 and 51. Total DNA was extracted from each and amplified with universal eubacterial (Eub)primers and primers to amplify actinobacteria 16S rRNA genes. Amplified products from each microcosm were run on DGGE gels (Figure 2). Visual comparison of the gels showed that there were qualitative and quantitative differences between the control microcosms and those that had been supplemented with additional lead and zinc. The samples of community DNA amplified with universal Eub primers showed that across the time course of the experiment (51 days) the diversity calculated using the Shannon Weiner index, within each treatment did not change significantly. However, when the control and heavy metal amended microcosms were compared there was a significant ($P > 0.05$) decrease in the diversity of the soil that had received additional heavy metal input compared to the control microcosm (Table 2).

In relation to the actinobacterial community structure, control microcosms showed a decrease in the diversity after the initial sample, but subsequently the H' values remained similar throughout the course of the experiment (Table 2). In the amended microcosm, there was no significant change over the course of the experiment and the mean diversities measured between control and amended soils was not significantly different (Table 2). In contrast to the reduced diversity in amended soils determined using V3 primers sets.

The DGGE band patterns generated at each time point were subjected to statistical analyses to identify any measurable changes in the community structure across the time course of the experiment and in response to heavy metal addition. Cluster analyses using the UPGMA algorithm of lane band patterns produced using the eubacterial primers (Figure 3 and Figure 4) were performed. The eubacterial community in non-supplemented soil gave two clusters one consisting of the earlier samples (days 0 to16), and the second comprised of the later samples. The similarity between the two clusters was about 80%. A similar partition of the profiles was also seen in the microcosm augmented with fivefold more Pb and Zn (Figure 3). Although these clustering patterns were observed in both microcosm treatments, the similarity between the clusters was less pronounced (73 %) in the microcosm soil supplemented with additional Pb and Zn. The use of the Raup and Crick (S_{RC}) index to quantify the relationships between the community profiles indicated that the eubacterial communities changed over time, irrespective of the presence of additional heavy metal and became significantly similar to each other (Table 3). However, additional heavy metal appeared to inhibit this drift toward significant similarity, so for example, in microcosms without added Pb and Zn, the profiles from days 23 to day 51 showed a similarity to each other suggesting greater than could be explained by chance matching of bands. This is indicative of deterministic selection was acting on the community (Table 3a). In contrast, this trend towards similar population profiles was much less pronounced, occurring from day 37 onwards, in microcosms amended with fivefold more Pb and Zn added.

The actinobacterial profiles gave less clear time related clusters. In non supplemented microcosm soil two distinct clusters were seen, however, one consisted of the profiles derived from days 23 and 30, the other cluster contained the remaining profiles that appeared interdigitated, with the exception of the day 0 profile, the similarity between the two was 73%, in addition. The day 0 profile formed an outlier in this second cluster, being 79% similar to the other profiles within it (Figure 4a). In supplemented soils the first two samples points (days 0 and 2) gave a distinct cluster, the remaining profiles formed a second but the similarity between the two was 91.6 %. In contrast to the eubacterial profiles, it appeared that adding additional heavy metal to the soil increased the similarity between profiles. The Raup and Crick analyses showed actinobacterial communities under both treatments showed a drift towards similar community profiles, such that the comparison of profiles from day 7 onwards under both treatments shows high, frequently significant, similarity between the community profiles (Tables 3c,d).

Table 2. A comparison of the Shannon-Weiner diversity indices (H') of the Eubacterial (Eub) and Actinobacterial (Act) communities in microcosms derived from the Bardon Mill field site

		H'		
	Eub	Eub x5	Act	Act (x5)
day 0	3.81	2.68	3.28	2.92
day 2	3.71	2.57	3.01	3.01

Table 2. Continued.

	H'			
	Eub	Eub x5	Act	Act (x5)
day 7	3.88	2.81	2.97	2.99
day 10	3.83	3.03	3.01	2.96
day 16	3.88	2.75	2.91	3.03
day 23	3.90	2.89	2.70	2.97
day 30	3.84	2.96	2.86	3.04
day 37	3.74	2.96	2.78	3.09
day 44	3.79	2.90	2.85	3.00
day 51	3.78	3.00	2.94	2.95
Mean	**3.82 ± 0.06**	**2.86 ± 0.15**	**2.93 ± 0.16**	**3.00 ± 0.05**

Microcosm soil was either left untreated or amended with a fivefold increase in Pb and Zn (x5) concentrations.

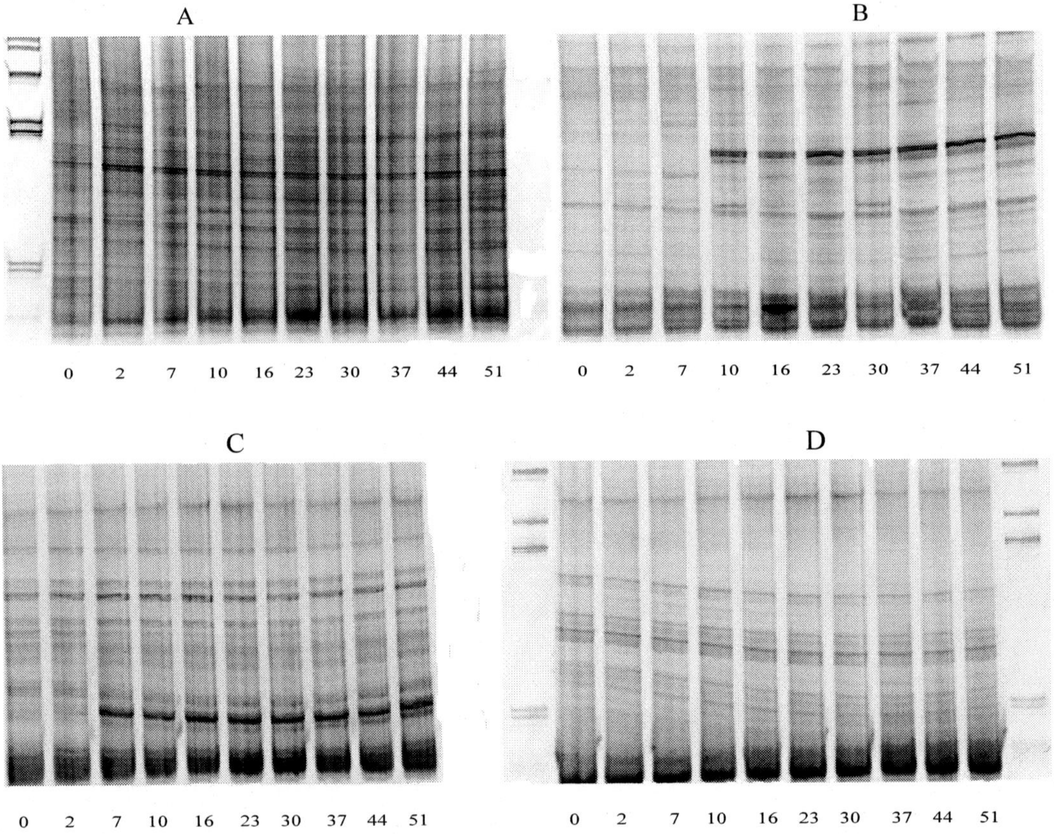

Figure 2. 16S rDNA DGGE analysis of bacterial communities in soil microcosms derived from the Bardon Mill field site. A. Eubacterial profile from non-amended soil. B. Eubacterial profile from soil spiked with a fivefold increase in Pb and Zn. C. Actinobacterial profile from non-amended soil. D. Actinobacterial profile from soil spiked with a fivefold increase in Pb and Zn. Each gel was produced from pooled DNA derived from triplicate microcosms after the reproducibility of the banding patterns was confirmed between replicates, and replicates and pooled DNA. Lane 1: day 0 profile; Lane 2: day 2; Lane 3: day 7; Lane 4: day 10; Lane 5: day 16; Lane 6: day 23; Lane 7: day 30; Lane 8: day 37; Lane 9: day 44; Lane 10: day 51.

Table 3. A comparison of similarity values (S_{RC}) for the community profiles from Bardon Mill soil microcosms over a 51 day incubation

	day 0	day 2	day 7	day 10	day 16	day 23	day 30	day 37	day 44	day 51
A										
day 0	1	**1**	**0.98**	0.74	0.56	0.84	0.77	0.94	0.94	0.43
day 2		1	1	**0.99**	**0.95**	**0.96**	0.89	**0.96**	0.81	0.79
day 7			1	0.79	0.92	0.97	0.92	0.53	0.56	0.33
day 10				1	**0.97**	0.67	0.49	0.88	0.74	0.34
day 16					1	**0.98**	0.93	0.90	0.57	0.76
day 23						1	**1**	**1**	**0.99**	**0.99**
day 30							1	1	1	1
day 37								1	1	1
day 44									1	1
day 51										1
B										
day 0	1	**0.97**	0.76	0.34	0.84	0.89	0.92	0.90	0.84	0.84
day 2		1	**0.99**	0.32	**0.99**	**0.98**	0.63	**0.98**	**0.99**	0.8
day 7			1	0.50	**0.98**	0.82	0.80	0.80	0.92	0.34
day 10				1	0.82	0.28	0.28	**0.002**	0.24	0.33
day 16					1	0.98	0.86	0.64	0.84	0.14
day 23						1	0.99	0.99	0.88	0.51
day 30							1	0.93	0.89	0.54
day 37								1	**0.99**	**0.97**
day 44									1	**0.99**
day 51										1
C										
day 0	1	0.59	**0.06**	0.27	**0.05**	0.28	0.16	0.34	**0.02**	0.22
day 2		1	0.79	0.63	0.46	0.89	0.87	0.94	0.65	0.95
day 7			1	**0.99**	0.8	0.77	0.80	**0.97**	0.99	0.96
day 10				1	**0.98**	0.88	0.88	**0.99**	0.99	0.99
day 16					1	**0.99**	0.9	**1**	1	0.97
day 23						1	**1**	**1**	0.98	0.99
day 30							1	**0.99**	0.95	0.98
day 37								**1**	1	1
day 44									1	0.99
day 51										1
D										
day 0	1	0.92	0.69	0.88	0.56	0.61	0.50	0.28	0.40	0.60
day 2		1	**0.96**	**0.99**	0.66	**0.97**	0.64	0.45	0.83	0.93
say 7			1	**0.99**	**0.99**	**0.99**	0.91	0.70	**0.96**	0.99
day 10				1	**0.98**	**0.99**	**0.99**	0.91	**0.95**	**0.99**

Table 3. Continued.

	day 0	day 2	day 7	day 10	day 16	day 23	day 30	day 37	day 44	day 51
day 16					1	**0.97**	0.89	0.84	0.88	**0.97**
day 23						1	**0.99**	**0.98**	**1**	**1**
day 30							1	1	**0.97**	**0.99**
day 37								1	**0.95**	0.91
day 44									1	**1**
day 51										1

A. Eubacterial communities B. Eubacterial communities amended with a fivefold increase in Pb and Zn concentrations. C. Actinobacterial communities D. Actinobacterial communities mended with a fivefold increase in Pb and Zn concentrations. 0.95> S_{RC}> 0.05 similarity no greater than expected by chance (null hypothesis upheld), S_{RC}< 0.05 significant dissimilarity, S_{RC} > 0.95 significant similarity (null hypothesis fails).

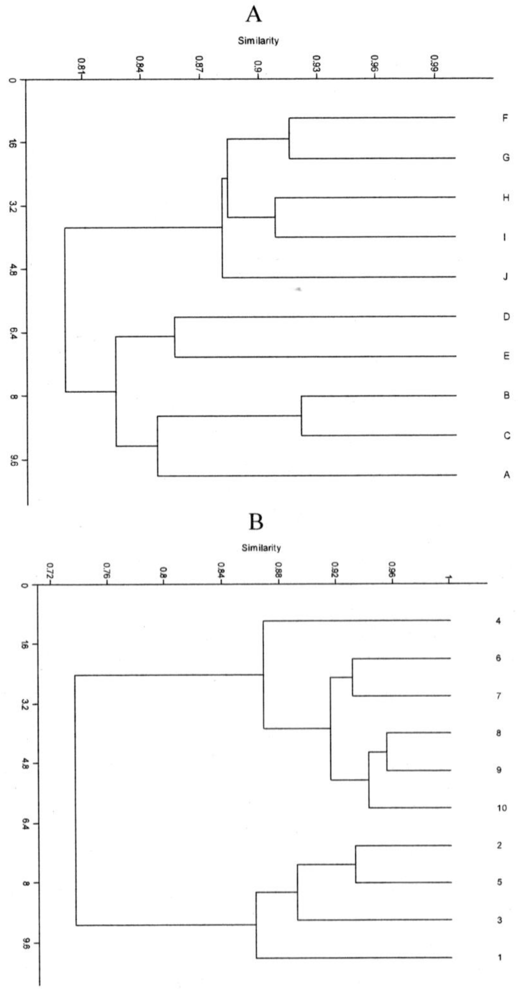

Figure 3. Cluster analyses using UPGMA algorithm of DGGE profiles from microcosm soils. A. Eubacterial profiles from non-amended soil B. Eubacterial profiles from soil spiked with a fivefold increase in Pb and Zn. Numbers refer to the day that the soil was sampled.

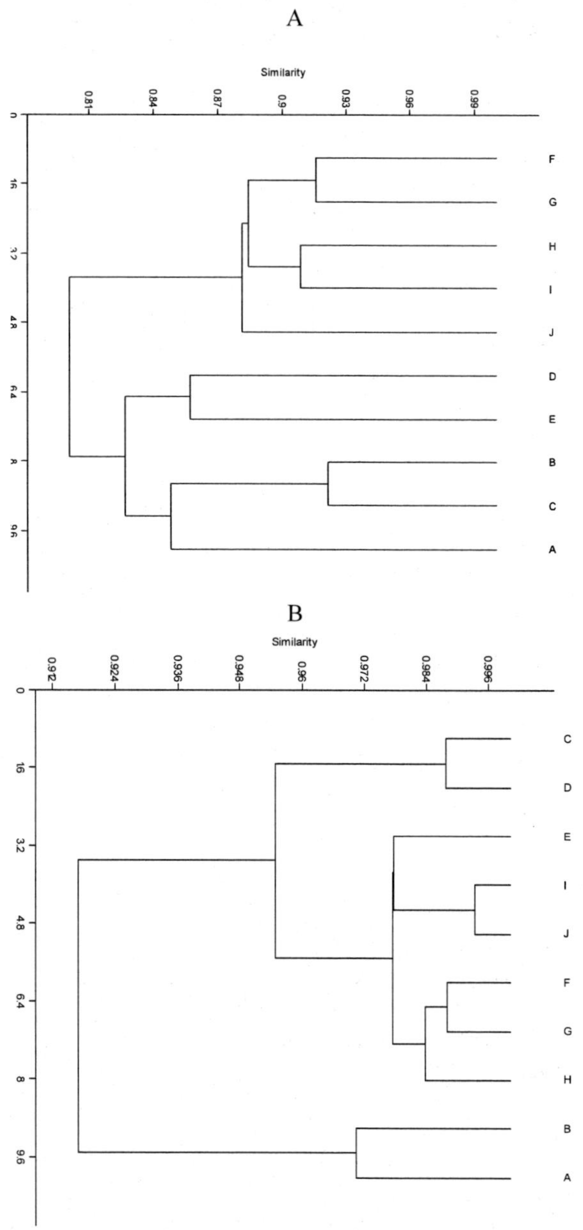

Figure 4. Cluster analyses using UPGMA algorithm of DGGE profiles from microcosm soils. A. Actinobacterial profiles from non-amended soil. B. Actinobacterial profiles from soil spiked with a fivefold increase in Pb and Zn. Numbers refer to the day that the soil was sampled.

Bacterial Community Analyses from Metallophytic Grasslands of the South Tyne Valley

The Garrigill field site was located on the bank of the upland river the South Tyne in the Pennine uplands of northern England. The site had been subjected to a vegetation survey (Leonard, 2001) that had divided the site into eight areas on the basis of predominant vegetation and its coverage (Figure 1). The majority of the site was considered to be a metallophytic grassland that included heavy metal tolerant indicator species, however, areas 2 and 8 on the SE border of the site were predominantly woodland and scrub (Table 1). The analyses of the bacterial community was undertaken by sampling top soil from each of the 8 areas, soil was characterised in terms of its composition (sand 84.2%, silt 10.5 % and clay 5.3 %) total extractable lead and zinc content and total organic carbon and nitrogen (Table 4).

The amounts of Pb and Zn in the soil was found to vary significantly across the site. Area 1 had the highest concentration of heavy metals and the wooded area 8 the lowest (Table 4). The contamination index (CI) was determined with respect to Pb and Zn for each area (Feris et al., 2003).

We extracted RNA from soil samples to determine the diversity of the metabolically active eubacterial and actinobacterial communities. The metabolically active eubacterial community amplified (Figure 5a) produced a limited number of bands. The diversity calculated with the Shannon Weiner (H') index was correspondingly low (Table 4). In the most contaminated area 1 only a single band was visible on the gel and the H' value was significantly ($P>0.05$) lower than that of the other areas. The limited diversity was not consistently linked to the contamination index, the most diversity was found in area 5 (CI = 0.53) and this was significantly higher than that of the other areas in the field site. The soil in this area had the highest organic C and N content (8 %) and these elements were also shown to be relatively high in areas 6 and 7 and to a lesser extent area 2 which exhibited the next highest eubacterial diversities (Table 4). Conversely, diversity was lowest in 1, 4 and 8 which had the lowest amounts of measured organic C and N. While area 8 had the lowest CI, that area of 4 was amongst the highest.

Table 4. Characterisation of the soil form the 8 areas of the Garrigill field site in terms of heavy metal, organic C, total N content, and the diversities of the metabolically active eubacterial and actinobacterial communities derived from the DGGE profiles (Fig. 5)

	Extractable Metals (ppm)			Elemental analyses		Shannon Weiner index (H')	
	Pb	Zn	CI	N (%)	C (%)	Actinobacteria	Eubacteria
Area 1	5285	1617	0.59	0.23	1.76	3.82	0.01
Area 2	1075	621	0.04	0.26	2.23	3.82	1.16
Area 3	1299	638	0.09	0.26	2.19	3.4	0.53
Area 4	3685	1087	0.43	0.18	1.12	3.46	0.34
Area 5	1356	979	0.19	0.73	8.00	3.55	2.76
Area 6	2918	1746	0.48	0.52	6.12	3.55	0.96
Area 7	3924	1984	0.57	0.42	6.19	3.03	0.85
Area 8	909	613	0	0.12	1.05	3.08	0.5

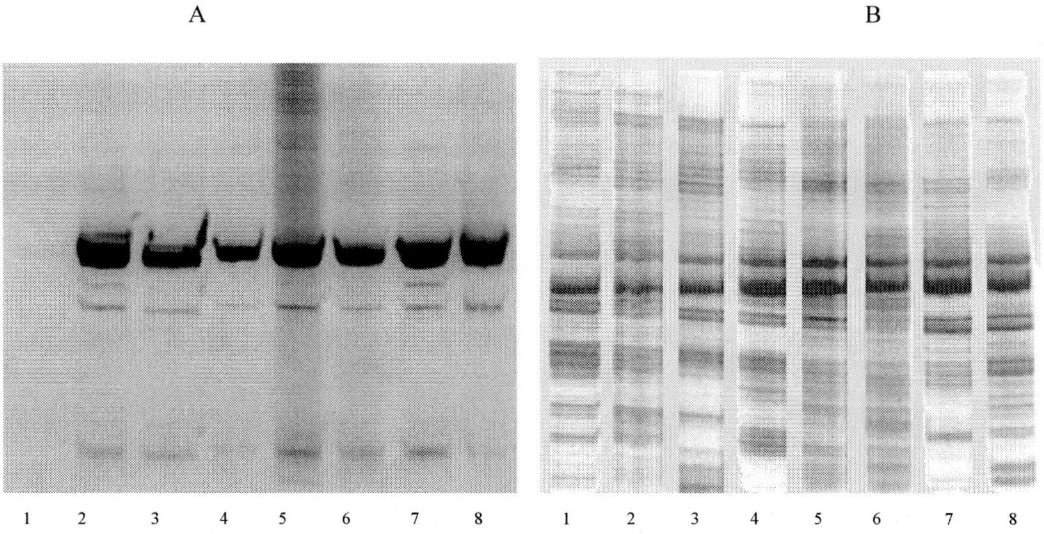

Figure 5. 16S rRNA DGGE analysis of bacterial communities in soil from the Garrigill site A. Eubacterial profiles. B. Actinobacterial profiles. Each gel was produced from pooled DNA derived from triplicate 1 g soil samples after the reproducibility of the banding patterns was confirmed between replicates, and replicates and pooled DNA. Lane 1: area 1 profile; Lane 2: area 2; Lane 3: area3; Lane 4: area 4; Lane 5: area 5; Lane 6: area 6; Lane 7: area 7; Lane 8: area 8. Areas correspond to those described in Figure 1.

Using more specific primers to amplify the metabolically active Actinobacteria community produced a large number of bands and a high diversity index in each area (Figure 5b). In contrast to the community identified using the eubacterial primers area 1, with the highest CI, also had the highest diversity of any area on the site. However, as with the eubacteria there was no clear cut relationship between the CI and the bacterial diversity. Similarly, the amount of increased organic C and N in the soil did not give consistent responses in actinobacterial diversity (Table 4).

Correspondence analyses of the DGGE band patterns of eubacterial and actinobacterial communities for each area enabled visualization of the relationship between the bacterial communities from the different areas (Figure 6). Correspondence analyses is a multivariate technique that enables the transformation of a complex data set by constructing a new set of variables that are linear combinations of the original variables, i.e. the 16S rDNA gene bands and the soil from the 8 areas of the site. For the total bacterial analyses using the V3 primers the samples with the lowest CI (areas 2, 3 and 8) clustered very closely on the first axis that explained 48.5 % of the inertia but were separated across the second axis that explained almost 23% of the inertia Figure 6a). Areas 3 and 8 also formed a tight cluster with areas 4, 7 and 6 across both axes, areas 6 and 7 had amongst the highest organic C and total N content, however, area 4 in contrast had low values but all sites shared a similar CI (Table 4). Area 5 that had the significantly highest diversity was separated from all other profiles across the first axis, and area 1 that had the highest CI lay between area 5 and the remaining profiles along the first axis.

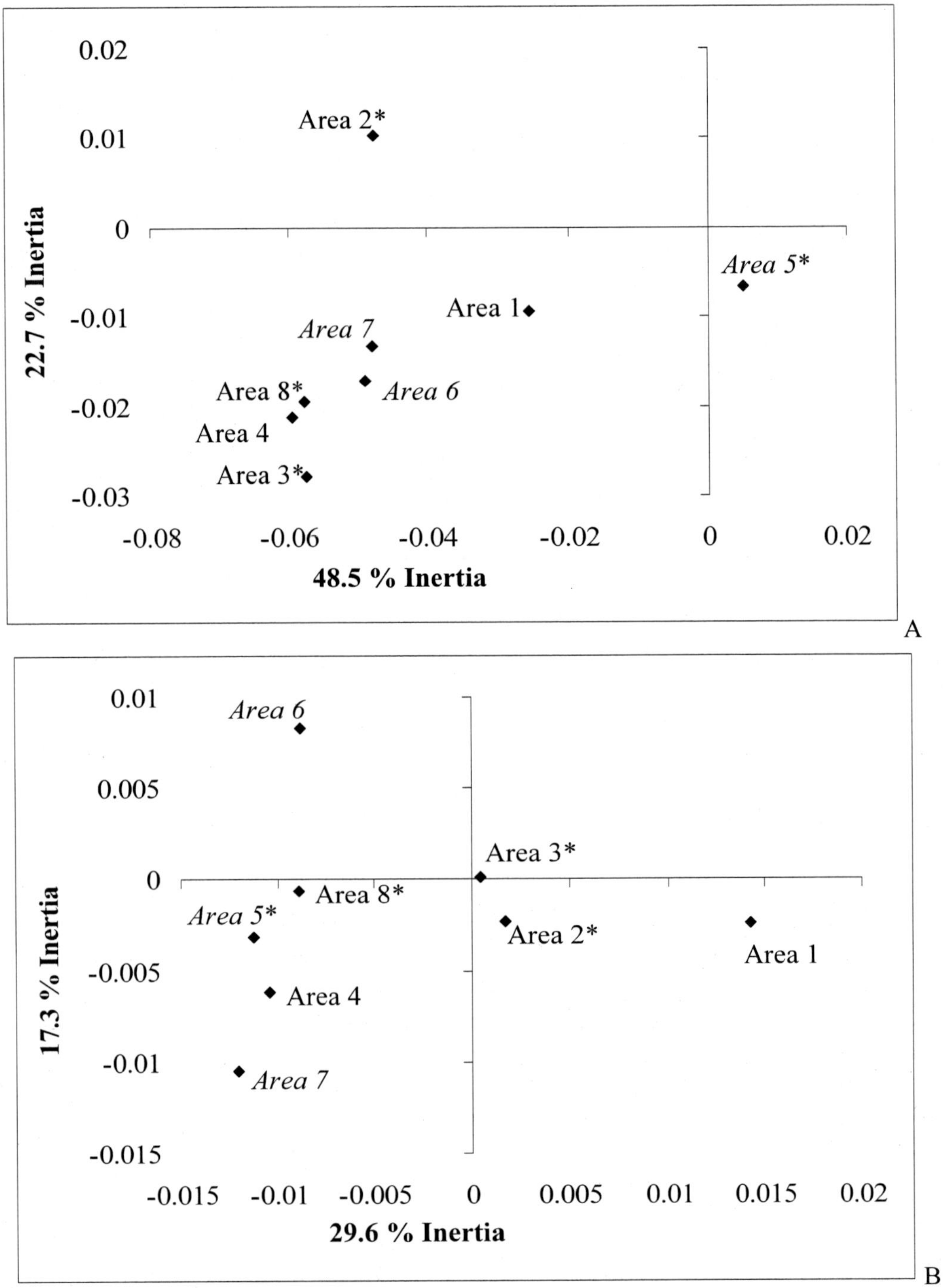

Figure 6. Correspondence analysis of the 16S rRNA DGGE community profiles from the 8 areas within the Garrigill field site. A. profiles of Eubacterial communities. B. Profiles of Actinobacterial communities. Those areas italicised have soil with high organic C and total N content. * Indicates a low contamination index (see Table 4).

The profiles of the metabolically active Actinobacteria indicated that the profile from area 1 with the highest CI was distinct from the other profiles across the first axis that accounted for 29.6% of the inertia (Figure 6b). In addition areas 2 and 3 formed a tight group, on both axes, these soil samples exhibited, along with the woodland area 8, the lowest CI and moderate to low levels of organic C and total N. The latter area was separated form these profiles along the first axis but clustered closely with them along the second. Indeed all the low CI soil profiles (2, 3, 5 and 8) formed a discrete group along axis two bit were separated along axis 1 (Figure 6b). The profiles 4, 5, 6, 7 and 8 were very tightly clustered across the first dimension but were separated across the second axis that described over 17% of the inertia in the sample (Figure 6b).

Raup and Crick analyses was undertaken to determine whether any of the profiles across the site shared a statistically meaningful relationship (Table 5). the analyses of the profiles determined from the metabolically active eubacteria indicate that two sites areas 3 and 5 have bacterial community profiles that are significantly different to each other (Table 5a), the CI for these two areas are both on the lower side of the range for the site but have a significant difference in their organic C and total N content (Table 4). The vegetation at both sites was predominantly grass (*F. ovina*), but in area 5, the coverage of grass was sparse and typical heavy metal tolerant plant species of a metallophytic grassland were evident (Table 1). The community profile of area 4 was significantly similar to that of areas 6, 7 and 8 suggesting a deterministic selection for these populations (Table 5a), vegetation between areas 4, 6 and 7 was similar, with comparable CI values for each soil, but area 4 had significantly lower amounts of organic C and N present in the soil. In contrast, area 8 was woodland and had the lowest CI index on the site (Table 4). Areas 2 and 7 showed significant similarity in population profiles (Table 5). However, the organic C and N content of the soil and the vegetation was distinct between these two sites, area 2 being largely scrub, whereas area 7 had a large coverage within the quadrat of typical metallophytic flora, however the two sites were physically linked (Figure 1).

Table 5. A comparison of similarity values (S_{RC}) for the metabolically active community profiles from soil derived from the 8 areas (Figure 1) of the Garrigill field site

A

	Area 1	Area 2	Area 3	Area 4	Area 5	Area 6	Area 7	Area 8
Area 1	1	0.7725	0.88	0.865	0.5875	0.755	0.855	0.8725
Area 2		1	0.2675	0.92	0.1425	0.375	**0.995**	0.865
Area 3			1	0.605	**0.005**	0.2525	0.46	0.55
Area 4				1	0.8225	**0.985**	**0.99**	**0.9875**
Area 5					1	0.2025	0.6125	0.4725
Area 6						1	0.925	0.8
Area 7							1	0.92
Area 8								1

Table 5. Continued.

B

Area 1	1	**0.9775**	**1**	0.3825	0.1325	0.41	0.565	0.81
Area 2		1	**0.975**	0.305	0.405	0.375	0.345	0.8925
Area 3			1	0.6025	0.315	0.7975	0.64	0.6275
Area 4				1	0.6175	0.7475	**0.985**	0.3575
Area 5					1	0.705	**0.995**	0.4375
Area 6						1	0.6425	0.8675
Area 7							1	0.835
Area 8								1

A. Eubacterial communities;

B. Actinobacterial communities. 0.95> S_{RC}> 0.05 similarity no greater than expected by chance (null hypothesis upheld), S_{RC}< 0.05 significant dissimilarity, S_{RC} > 0.95 significant similarity (null hypothesis fails).

The profiles derived for the metabolically active Actinobacteria showed that areas 1, 2 and 3 had statistically similar community profiles (Table 5b), while areas 1 and 3 shared common features in terms of vegetation coverage of predominantly grass and heavy metal tolerant flora, area 2 was in contrast scrub (Figure 1). Organic carbon and N content of the soils were similar, and while areas 2 and 3 shared similar CI values, amongst the lowest on the site, area 1 was the most contaminated soil. The profile from area 7 was significantly similar to that of areas 4 and 5 (Table 5b). Each of these areas shared a moderate to high CI value, however the vegetation cover and organic carbon content of the soil was different in each area.

CONCLUSION

There have been many studies investigating the impact of long term heavy metal addition on the community structure on soil bacterial populations. However, this is the first report to our knowledge of the bacterial community of the metallophytic grassland of the Pennines. These habitats were formed as a result of a short period mining activity between the mid-eighteenth and mid twentieth centuries. They comprise of spoil heaps from abandoned mine workings and also alluvial deposits, on the flood plains of the local watercourses that have been contaminated by water from abandoned mine workings. They represent a unique habitat, containing a highly diverse flora, many nationally scarce within the UK, that are described as metallophytes (high metal tolerance). These rare habitats are increasingly threatened by natural events, such as succession by birch and gorse, and a range of anthropogenic activities, including, farming, spoil reclamation, gravel extraction and leisure uses. Our aim was to investigate the stability of the bacterial soil communities associated with these sites and to characterise the bacterial community structure under such a metallophytic grassland subjected to different heavy metal concentrations and vegetation cover.

The first soil studied was obtained from a historical lead mine spoil heap at the Bardon Mill site, The hypothesis we wished to test was whether a bacterial community previously exposed to high heavy metal concentrations would tolerate additional heavy metal, due to adaptation and functional changes within the community predisposing it to cope with the additional stress (Odum, 1981). We predicted that additional heavy metal addition to the soil would not lead to a significant decrease in on the community diversity and structure, compared to the control microcosm if the hypothesis was correct.

Visual inspection of the DGGE gels derived from the microcosm soils suggested that using universal eubacterial (Eub) primers the impact of a fivefold increase in Pb and Zn was to reduce the diversity of the bacterial soil community. This was quantified by determining the Shannon–Weiner index which demonstrated a significant decrease in the diversity of the soil when Pb and Zn concentrations were increased fivefold (Table 2). In contrast, the actinobacterial community in microcosms supplemented with additional Pb and Zn , showed no significant change in diversity. These observations are of interest, because the decrease in diversity determined with the Eub primer set suggests that the community is not able to tolerate additional heavy metal stress. Caution must be exercised in the interpretation of DGGE data as the primers used may have a significant effect on the profiles obtained. Often PCR primers can show a bias towards particular sequences resulting in differential amplification (Joynt et al 2006). This can lead to underestimations of diversity because the more abundant sequences are preferentially amplified masking sequences of lower copy number. Nevertheless, DGGE with Eub primers can demonstrate whether environmental disturbance does cause gross changes in community structure.

The results of our experiment examining the eubacterial populations support the observation by Tobor-Kaplan et al. (2005) that ecosystem stability to heavy metals is not enhanced by long term adaptation to this stress, because diversity significantly declined in the soils challenged with additional Pb and Zn (Table 2). Several other studies have found similar reductions in bacterial diversity as a result of heavy metal addition (Sandaa et al. 1999, Joynt et al. 2006; Wu et al. 2006). However, this response is not universal as for example, Lazzaro et al. (2006b) found that diversity was not significantly reduced in forest soils and perhaps more significantly, in a study of a hyporheic (sediment) communities from heavy metal contaminated watercourses no correlation between microbial diversity and metal concentration was apparent (Feris et al. 2003).

In contrast to the eubacterial populations, our Actinobacteria community data was more in line with these latter studies. The Shannon-Weiner diversity indices did not change significantly between the two soil treatments. This suggests that the Actinobacteria taxa are not negatively affected by the additional heavy metal stress. There are conflicting reports of the response of Actinobacteria to heavy metal input into soil. A number of studies have reported that its abundance and diversity decreases in the presence of heavy metal contamination in soils (Frey et al. 2006; Lorenz et al. 2006; Kelly et al. 1999). Whereas, our data supports the observations that they are a significant and diverse element of the bacterial community in soils that have high natural concentrations of heavy metals or has been subjected to long term contamination (DeGrood et al. 2005; Gremion et al. 2003).

The impact of increasing the heavy metal concentration on the bacterial community structure was explored using cluster analyses of the DGGE profiles derived from the microcosm experiments (Li et al. 2006; Kandeler et al. 2000). This analyses indicated that the similarity of the eubacterial profiles showed a time dependency with two clusters, one consisting of profiles from early sample times, and the second comprised of the later time points of the experiment. A similar pattern was observed in microcosms irrespective of the addition of extra Pb and Zn but the difference between early and later clusters was higher than in controls (Figure 3). The S_{RC} values showed these changes were significant and suggested a deterministic selection (Rowan et al 2003) was acting on the soil communities in the microcosms (Table 3a and 3b). Further heavy metal addition seemed to reduce this deterministic selection. In comparison, the actinobacterial community profiles in the control microcosms showed less time related changes in community profiles. The S_{RC} profiles indicated the addition of heavy metals increased the frequency of significant similarities occurring between profiles. The implication of these data are that the Actinobacteria are subject to deterministic selection in the microcosm and that adding additional heavy metals increases this selection pressure (Figure 4a).

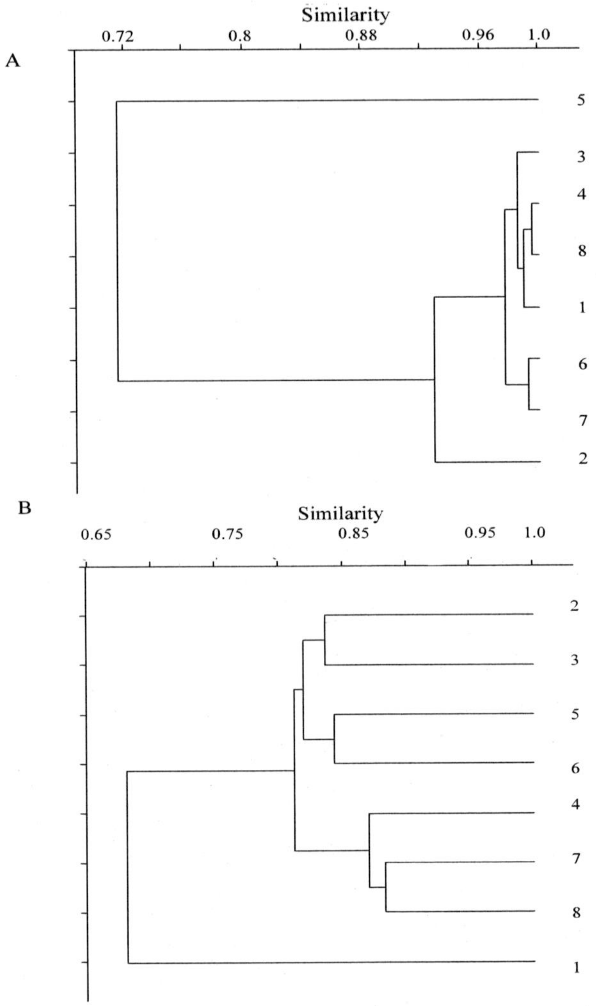

Figure 7. Cluster analyses using UPGMA algorithm of DGGE profiles from the 8 areas sampled on the Garrigill field site. A. Eubacterial community profiles. B. Actinobacterial profiles. The numbers on each branch correspond to the areas of the site from which the profile was derived (Figure 1).

Our data suggests that the short term response to additional heavy metal exposure did affect the eubacterial and actinobacterial community structure in the microcosms, but not dramatically. Renella et al. (2002) showed related modest changes to community parameters in short term exposure experiments. It has been widely reported that heavy metals can change the community structure of bacterial populations in soil. Even prior to the widespread application of molecular techniques changes in the genetic structure of communities exposed to long term heavy metal exposure had been noted (Giller et al. 1998). Frey et al. (2006) analysed the impact on community structure of heavy metal contamination and noted a shift in the community structure measured using both PLFA and T-RFLP and that the resulting community had a reduced functional diversity compared to control soils. They speculated that the changes caused by the heavy metal application were to shift to more metal tolerant species. However, the work of Feris et al. (2003) does not support this they found no correlation between heavy metal concentration and reduced diversity or biomass, nor any indication of a more metal tolerant community being selected for. There is data on the specific changes of soil microbial communities subjected to heavy metal stress, and once more these studies present conflicting data. For example, PFLA and DGGE data has indicated that fungi, Proteobacteria and Gram negative increase under such conditions, whereas, Actinobacteria and Firmicutes decrease (Lorenz et al. 2006; Kelly et al. 1999; Hinojosa et al, 2005). However, Bååth et al (2005) showed in historically contaminated soil that using PFLA that Gram positive bacteria where more prevalent and that fungi decreased in Pb contaminated soils. It is of interest that many reports identify a decrease in Actinobacteria on the addition of heavy metals to soil, we did not measure the change in actinobacterial biomass, however, Lorenz et al. (2006) identified a decrease in the respiratory quinones associated with this group in soil exposed to arsenic and cadmium. However, their DGGE analyses, while also showing a reduction in the number of bands present (and by inference the diversity of the community), indicated that bands were lost predominantly from the upper part of the gel, these would be low G+C taxa. Therefore, it may be worth undertaking more focussed studies to identify whether when Actinobacteria are seen to decline whether this can be attributed to a decline in abundance but not in diversity. As in our study no significant decrease in actinobacterial diversity in contaminated soils was observed.

The second field site at Garrigill, was studied to characterise the microbial community across a metallophytic grassland site with a soil subject to differences in both metal concentration and vegetation cover. The soil contained quantities of several heavy metals, the most abundant being Pb and Zn, as the effects of heavy metals on biological communities co-vary and demonstrate a high co-linearity which would as a result confound any attempts to analyse the effects of individual metals (Feris et al. 2003). As a result the contamination index (CI) has been employed to overcome this and to express a relevant measure of metal contamination. The characterisation of the site into 8 areas had been done on the basis of the predominant vegetation cover, it is known that the microbial community in soil, particularly in the rhizosphere, can alter as a result of changes in the quantity and composition of root exudates that plants produce (Baudoin et al 2003). Therefore, it might be expected that the microbial communities will demonstrate significant differences as a result of this rhizodeposition of organic carbon across the site. A crude measure of organic carbon in the soil showed that, area 5 had the most organic C and total N in the soil. This area also had the highest diversity in the metabolically active eubacterial community (Table 4). The presence of high concentrations of heavy metals often inhibits plant growth and as a result will decrease soil organic C, the most contaminated area on the site had the lowest level of organic C in its soil and relatively sparse vegetation (Table 1; Table 4). However, other areas with high CI also have high levels of organic C present in their soils. In particular, areas 6 and 7 show high organic C and total N levels, the biodiversity among the eubacterial community are also higher than the average across the whole site in these areas. Both have significant coverage by metallophytic flora (Table 1) unlike area 1. metal tolerant plants are known to harbour heavy metal tolerant bacteria in their rhizospheres (Abou-Shanab et al. 2006), thererefore, heavy metal tolerant flora may enhance both soil C and biodiversity quite significantly. Analyses of the community profiles from the different areas showed a very complex relationship between the eubacterial and actinobacterial community profiles and the other parameters (CI, organic C and total N, vegetation cover) (Figure 6). A cluster analyses of the similarity of the profiles (Figure 7) demonstrates that the metabolically active eubacterial community could be divided into one cluster containing the single area 5 that was 72% similar to the other cluster which comprised of all the other area profiles whose similarity to each other was between 92-98 % (Figure 7). Area 5 had both the highest organic C and total N content in its soil and the highest diversity. In contrast, with the analyses of the actinobacterial communities diversities across the site were less variable, however, again 2 clusters were produced Area 1 with the highest diversity formed a cluster that was 66% similar to the second which contained all the other area profiles. The highest diversity of metabolically active Actinobacteria was therefore in soil with the highest heavy metal concentration and the lowest organic C and total N.

The results of the study on the metabolically active Eubacteria and Actinobacteria at the Garrigill site highlights the difficulties in studying bacterial soil populations *in situ.* A number of factors can influence the development of the metabolically active community, including environmental disturbance and resource availability. Our data indicate that the diversity of the active eubacterial population is strongly linked to the availability of organic C in the soil the highest organic C content equating to the highest diversity in area 5. The heavy metal concentration did appears to exert some effect, in that areas 1, and 4 with high CI values and had the lowest diversity, and area 6, and 7 also with high CI had much reduced community diversity compared to area 5 despite all three areas having soil with high organic C contents. However, Actinobacteria had the most diversity in area 1 with the lowest organic C content. The role of soil organic C in shaping the microbial community has received some attention, in metallophytic grasslands organic C levels are equivalent to those of other temperate grasslands but there is some evidence that plant material, particularly, from the metallophytic flora is only poorly degraded (Balabane et al. 1999). As a result the quality of the organic carbon the soil may be reduced and this favours those microorganisms that can access this refractile carbon, typically these are K-strategists such as Actinobacteria (Fierer et al. 2003). Across the Garrigill site it is possible that the quality of the organic C entering the soil is variable, so that in area 5 with low levels of heavy metal and few metallophytic hyperaccumulators the plant derived organic C is readily degradable by r-strategists adapted to acquiring labile carbon rapidly (Hery et al 2005). This would account for the high eubacterial diversity in this area. In contrast where organic C is low and heavy metal high as in area 1 (and to a lesser extent in areas 4, 6 and 7) Actinobacteria will be favoured over r-strategists as they are capable of accessing organic material returned to the soil by turnover of the metallophytic flora that may be sequestered with high levels of heavy metals (Balabane et al. 1999). That being said there are a number of studies that demonstrate Actinobacteria can be competitive in soils containing high levels of heavy metals even when labile organic carbon is present (Hery et al 2005), the high background levels of Pb and Zn in the Garrigill soils may enable the Actinobacteria to maintain a relatively high diversity across the site even in those areas where organic C appears plentiful.

In conclusion, this study has explored the stability and diversity of bacterial communities with an emphasis of the Actinobacteria a significant taxa within the soil flora, in historically contaminated soil from metallophytic grassland sites. We have demonstrated that the eubacterial communities when exposed to additional stress do not exhibit tolerance in terms of maintenance of the diversity of the community. The actinobacterial community, however, appears much more robust in terms of the community structure and diversity in the face of such additional stresses. Subsequently we explored the character of the metabolically active bacterial communities across a field site. Analyses of community structure showed that eubacterial diversity was low and strongly affected by available organic carbon and heavy metal content in the soil. In contrast, the actinobacterial community was much higher, and was lwess susceptible to differences in organic carbon across the site, but did appear to respond to the heavy metal concentration. Such data is important in underpinning our knowledge of scarce metallophytic grassland communities which continue to be threatened by leisure and agricultural activities. more widely, such studies provide important information on how microbial communities respond to heavy metals and can be used to help remediate and restore many of the contaminated sites around the world that have resulted from historical and current industrial activity.

REFERENCES

Abaye, D. A., Lawlor, K., Hirsch, P. R. & Brookes, P. C. (2005). Changes in the microbial community of an arable soil caused by long-term metal contamination. *European Journal of Soil Science*, *56*, 93-102.

Abou-Shanab, R. A. I., Angle, J. S. & Chaney, R. L. (2006). Bacterial inoculants affecting nickel uptake by Alyssum murale from low, moderate and high Ni soils. *Soil Biology & Biochemistry*, *38*, 2882-2889.

Almås, A. R., Bakken, L. R. & Mulder, J. (2004). Changes in tolerance of soil microbial communities in Zn and Cd contaminated soils. *Soil Biology & Biochemistry*, *36*, 805-813.

Amir, H. & Pineau, R. (1998). Effects of metals on the germination and growth of fungal isolates from New Caledonian ultramafic soils. *Soil Biology & Biochemistry*, *30*, 2043-2054.

Amoroso, M. J., Schubert, D., Mitscherlich, P., Schumann, P. & Kothe, E. (2000). Evidence for high affinity nickel transporter genes in heavy metal resistant *Streptomyces* spec. *Journal of Basic Microbiology*, *40*, 295-301.

Bååth, E. (1989). Effects of heavy metals in soil on microbial processes and populations (a review). *Water Air and Soil Pollution*, *47*, 335-379.

Bååth, E., Diaz-Ravina, M. & Bakken, L. R. (2005). Microbial biomass, community structure and metal tolerance of a naturally Pb-enriched forest soil. *Microbial Ecology*, *50*, 496-505.

Balabane, M., Faivre, D., van Oort, F. & Dahmani-Muller, H. (1999). Mutual effects of soil organic matter dynamics and heavy metals fate in a metallophyte grassland. *Environmental Pollution*, *105*, 45-54.

Bamborough, L. & Cummings, S. P. (2006). The impact of historical heavy metal contamination on the microbial community of upland pastures in the North Pennines. *Aspects of Applied Biology*, *80*, 95-100.

Baudoin E., Benizri E. & Guckert A. (2003) Impact of artificial root exudates on the bacterial community structure in bulk soil and maize rhizosphere. *Soil Biology & Biochemistry*, *35*, 1183-1192.

Baxter, J., Garton, N. J. & Cummings, S. P. (2006) The impact of acrylonitrile and bioaugmentation on the biodegradation activity and bacterial community structure of a topsoil. *Folia Microbiologica*, *51*, 591-598.

Baxter, J. & Cummings, S. P. (2006). The impact of bioaugmentation on metal cyanide degradation and soil bacteria community structure. *Biodegradation*, *17*, 207-217.

Becker, J. M., Parkin, T., Nakatsu, C. H., Wilbur, J. D. & A. Konopka, A. (2006). Bacterial activity, community structure, and centimeter-scale spatial heterogeneity in contaminated soil. *Microbial Ecology*, *51*, 220-231.

Brodie, E., Edwards, S. & Clipson, N. (2003), Soil fungal community structure in a temperate upland grassland soil. *FEMS Microbioogyl Ecology*, 45, 105-114.

Brookes, P. C. & McGrath, S. P. (1984). Effects of metal toxicity on the size of the soil microbial biomass. *Journal of Soil Science*, *35*, 341-346.

Bűrgmann H., Widmer F., Sigler W. V. & Zeyer J. (2003). mRNA extraction and reverse transcription-PCR protocol for detection of *nifH* gene expressed by *Azotobacter vinelandii* in soil. *Applied and Environmental Microbiology*, *69*, 1928-1935.

Cavalier-Smith, T. (2002). The neomuran origin of archaebacteria, the negibacterial root of the universal tree and bacterial megaclassification. *International Journal of Systematic and Evolutionary Microbiology*, *52*, 7-76.

Curtis, T. P., Head, I. M., Lunn, M., Woodcock, S., Schloss, P. D. & Sloan, W. T. (2006). What is the extent of prokaryotic diversity? *Philosophical Transactions of the Royal Society B-Biological Sciences*, *361*, 2023-2037.

Dawson, J. J. C., Godsiffe, E. J., Thompson, I. P., Ralebitso-Senior, T. K., Killham, K. S. & Paton, G. I. (2007). Application of biological indicators to assess recovery of hydrocarbon impacted soils. *Soil Biology & Biochemistry*, *39*,164-177.

DeGrood, S. H., Claassen, V. P. & Scow, K. M. (2005). Microbial community composition on native and drastically disturbed serpentine soils. *Soil Biology & Biochemistry*, *37*,1427-1435.

Duffus, J. H. (2002). "Heavy Metals" – A meaningless term? *Pure and Applied Chemistry*, *74*, 793-807.

Euzéby J. P. (2007) List of Prokaryotic names with standing in the nomenclature. http://www.bacterio.cict.fr. Accessed 18th August 2007.

Fierer, N., Schimel, J. P. & Holden P. A. (2003). Variations in microbial community composition through two soil depth profiles. *Soil Biology & Biochemistry*, *35*, 167-176.

Fierer, N., Jackson, J. A., Vilgalys, R. & Jackson, R. B. (2005). Assessment of soil microbial community structure by use of taxon-specific quantitative PCR assays. *Applied and Environmental Microbiology*, *71*, 4117-4120.

Feris, K. P., Ramsey, P. W., Rillig, M., Moore, J. N., Gannon, J. E. & Holbert, W. E. (2004). Determining rates of change and evaluating group-level resiliency differences in hyporheic microbial communities in response to fluvial heavy-metal deposition. *Applied and Environmental Microbiology*, *70*, 4756-4765.

Feris, K., Ramsey, P., Frazar, C., Moore, J. N., Gannon, J. E. & Holbert, W. E. (2003). Differences in hyporheic-zone microbial community structure along a heavy-metal contamination gradient. *Applied and Environmental Microbiology*, *69*, 5563-5573.

Frey, B., Stemmer, M., Widmer, F., Luster, J. & C. Sperisen, C. (2006). Microbial activity and community structure of a soil after heavy metal contamination in a model forest ecosystem. *Soil Biology & Biochemistry*, *38*, 1745-1756.

Frostegard, A., Tunlid, A. & Bååth, E. (1993). Phospholipid fatty-acid composition, biomass and activity of microbial communities from two soil types experimentally exposed to different heavy metals. *Applied and Environmental Microbiology*, *59*, 3605-3617.

Gans, J., Wolinsky, M. & Dunbar, J. (2005). Computational improvements reveal great bacterial diversity and high metal toxicity in soil. *Science*, *309*,1387-1390.

Giller, K. E., Witter, E. & McGrath, S. P. (1998). Toxicity of heavy metals to microorganisms and microbial processes in agricultural soils: A review. *Soil Biology & Biochemistry*, *30*, 1389-1414.

Gremion F., Chatzinotas A. & Harms H. (2003). Comparative 16S rDNA and 16S rRNA sequence analysis indicates that *Actinobacteria* might be a dominant part of the metabolically active bacteria in heavy metal-contaminated bulk and rhizosphere soil. *Environmental Microbiology*, *5*, 896-907.

Griffiths, B. S., Bonkowski, M., Roy, J. & Ritz, K. (2001). Functional stability, substrate utilisation and biological indicators of soils following environmental impacts. *Applied Soil Ecology*, *16*, 49-61.

Griffiths, B. S., Hallett, P. D., Kuan, H. L., Pitkin, Y. & Aitken, M. N. (2005). Biological and physical resilience of soil amended with heavy metal-contaminated sewage sludge. *European Journal of Soil Science*, *56*, 197-205.

Griffiths, R. I., Whitely, A. S., O'Donnell, A. G. & Bailey, M. J. (2003). Influence of depth and sampling time on bacterial community structure in an upland grassland soil. *FEMS Microbiology Ecology*, *43*, 35-43.

Hery, M., Nazaret, S., Jaffre, T., Normand, P. & Navarro, E. (2003). Adaptation to nickel spiking of bacterial communities in neocaledonian soils. *Environmental Microbiology*, *5*, 3-12.

Hery, M., Herrera, A., Vogel, T. M., Normand, P. & Navarro, E. (2005). Effect of carbon and nitrogen input on the bacterial community structure of Neocaledonian nickel mine spoils. *FEMS Microbiology Ecology*, *51*, 333-340.

Heuer, H., Krsek, M., Baker, P., Smalla, K. & Wellington, E. M. H. (1997). Analysis of actinomycete communities by specific amplification of genes encoding 16S rRNA and gel-electrophoretic separation in denaturing gradients. *Applied and Environmental Microbiology*, *63*, 3233-3241.

Hinojosa, M. B., Carreira, J. A., Garcia-Ruiz, R. & Dick, R. P. (2005). Microbial response to heavy metal-polluted soils: Community analysis from phospholipid-linked fatty acids and ester-linked fatty acids extracts. *Journal of Environmental Quality*, *34*, 1789-1800.

Holmes, A. J., Bowyer, J., Holley, M. P., O'Donoghue, M., Montgomery, M. & Gillings, M. R. (2000). Diverse, yet-to-be-cultured members of the Rubrobacter subdivision of the Actinobacteria are widespread in Australian arid soils. Fems *Microbiology Ecology*, *33*,111-120.

Hu, Q., Qi, H. Y., Zeng, J. H. & Zhang, H. X. (2007). Bacterial diversity in soils around a lead and zinc mine. *Journal of Environmental Sciences-China*, *19*, 74-79.

Joynt, J., Bischoff, M., Turco, R., Konopka, A. & Nakatsu, C. H. (2006). Microbial community analysis of soils contaminaed with lead, chromium and petroleum hydrocarbons. *Microbial Ecology*, *51*, 209-219.

Jackson, R. B., Fierer, N. & Schimel, J. P. (2007). New directions in microbial ecology. *Ecology*, *88*, 1343-1344.

Kandeler, E., Tscherko, D., Bruce, K. D., Stemmer, M.,Hobbs, P. J., Bardgett, R. D. & Amelung, W. (2000). Structure and function of the soil microbial community in microhabitats of a heavy metal polluted soil. *Biology and Fertility of Soils*, *32*, 390-400.

Kao, P. H., Huang, C. C. & Hseu, Z. Y. (2006). Response of microbial activities to heavy metals in a neutral loamy soil treated with biosolid. *Chemosphere*, *64*, 63-70.

Kefala, M. I., Zouboulis, A. I. & Matis, K. A. (1999). Biosorption of cadmium ions by *Actinomycetes* and separation by flotation. *Environmental Pollution*, *104*, 283-293.

Kelly, J. J., Häggblom, M. M., Tate, R. L. (1999). Changes in soil microbial communities over time resulting from one time application of ainc: a laboratory microcosm study. *Soil Biology and Biochemistry*, *31*, 1455-1465.

Kelly, J. J., Häggblom, M. M. & Tate, R.L. (2003). Effects of heavy metal contamination and remediation on soil microbial communities in the vicinity of a zinc smelter as indicated by analysis of microbial community phospholipid fatty acid profiles. *Biology and Fertility of Soils*, *38*, 65-71.

Knight, B., & McGrath S. P. (1995). A method to buffer the concentrations of free Zn and Cd ions using a cation-exchange resin in bacterial toxicity studies. *Environmental Toxicology and Chemistry*, *14*, 2033-2039.

Kozdroj, J., & van Elsas, J. D. (2001). Structural diversity of microbial communities in arable soils of a heavily industrialised area determined by PCR-DGGE fingerprinting and FAME profiling. *Applied Soil Ecology*, *17*, 31-42.

Kuan, H. L., Fenwick, C., Glover, L. A., Griffiths, B. S. & Ritz, K. (2006). Functional resilience of microbial communities from perturbed upland grassland soils to further persistent or transient stresses. *Soil Biology & Biochemistry*, *38*, 2300-2306.

Lazzaro, A., Hartmann, M., Blaser, P., Widmer, F., Schulin, R. & Frey, B. (2006). Bacterial community structure and activity in different Cd-treated forest soils. *FEMS Microbiology Ecology*, *58*, 278-292.

Lazzaro, A., Schulin, R., Widmer, F. & Frey, B. (2006). Changes in lead availability affect bacterial community structure but not basal respiration in a microcosm study with forest soils. *Science of the Total Environment*, 371, 110-124.

Leonard, J. (2001). A Study of metallophytic grassland in Weardale. MSc thesis. University of Northumbria.

Liao, M. & Xie, X. M. (2007). Effect of heavy metals on substrate utilization pattern, biomass, and activity of microbial communities in a reclaimed mining wasteland of red soil area. *Ecotoxicology and Environmental Safety*, *66*, 217-223.

Li, Z. J., Xu, J. M. Tang, C. X., Wu, J. J., Muhammad, A. & Wang, H. Z. (2006). Application of 16S rDNA-PCR amplification and DGGE fingerprinting for detection of shift in microbial community diversity in Cu-, Zn-, and Cd-contaminated paddy soils. *Chemosphere*, *62*, 1374-1380.

Lorenz, N., Hintemann, T., Kramarewa, T., Katayama, A., Yasuta, T., Marschner, P. & Kandeler, E. (2006). Response of microbial activity and microbial community composition in soils to long-term arsenic and cadmium exposure. *Soil Biology & Biochemistry*, *38*, 1430-1437.

Lunn, M., Sloan, W. T. & Curtis, T. P. (2004). Estimating bacterial diversity from clone libraries with flat rank abundance distributions. *Environmental Microbiology*, *6*, 1081-1085.

Moffett B. F., Nicholson F. A., Uwakwe N. C., Chambers B. J., Harris J. A. & Hill T. C. J. (2003). Zinc contamination decreases the bacterial diversity of agricultural soil *FEMS Microbiology Ecology*, *43*, 13-19.

Mummey, D. L., & Stahl, P. D. (2004). Analysis of soil whole- and inner-microaggregate bacterial communities. *Microbial Ecology*, *48*, 41-50.

Muyzer G., De Waal E. C. & Uitterlinden A. G. (1993). Profiling of complex microbial populations by denaturing gradient gel electrophoresis analysis of polymerase chain reaction-amplified genes coding for 16S rRNA. *Applied and Environmental Microbiology*, *59*, 695-700.

Mertens, J., Springael, D., De Troyer, I., Cheyns, K., Wattiau, P. & Smolders, E. (2006). Long-term exposure to elevated zinc concentrations induced structural changes and zinc tolerance of the nitrifying community in soil. *Environmental Microbiology*, *8*, 2170-2178.

Nakatsu, C. H., Carmosini, N., Baldwin, B., Beasley, F., Kourtev, P. & Konopka, A. (2005). Soil microbial community responses to additions of organic carbon substrates and heavy metals (Pb and Cr). *Applied and Environmental Microbiology*, *71*, 7679-7689.

Narang, R. & Dunbar, J. (2004). Modeling bacterial species abundance from small community surveys. *Microbial Ecology*, *47*, 396-406.

Nies D.H. (2003). Efflux-mediated heavy metal resistance in prokaryotes. *FEMS Microbiology Reviews*, *27*, 313-339.

Odum, E. P. The effects of stress on the trajectory of ecological succession. In Barrett, G.W. and Rosenburg R. (eds) Stress effects on natural ecosystems. London. J. Wiley & Sons; 1981; pp 43-47.

Paradkar, A., Trefzer, A., Chakraburtty, R. & D. Stassi, D. (2003). Streptomyces genetics: A genomic perspective. *Critical Reviews in Biotechnology*, *23*,1-27.

Prokop, Z., Cupr, P., Zlevorova-Zlamalikova, V., Komarek, J., Dusek, L. & Holoubek, I. (2003). Mobility, bioavailability, and toxic effects of cadmium in soil samples. *Environmental Research*, *91*,119-126.

Prosser, J. I., Bohannan, B. J. M. Curtis, T. P. Ellis, R. J. Firestone, M. K. Freckleton, R. P. Green, J. L. Green, L. E. Killham, K. Lennon, J. J. Osborn, A. M. Solan, M. van der Gast, C. J. & Young, J. P. W. (2007). Essay - The role of ecological theory in microbial ecology. *Nature Reviews Microbiology*, *5*, 384-392.

Ranjard, L., Echairi, A., Nowak, V., Lejon, D. P. H., Nouaim, R. & Chaussod, R. (2006). Field and microcosm experiments to evaluate the effects of agricultural Cu treatment on the density and genetic structure of microbial communities in two different soils. *FEMS Microbiology Ecology*, *58*, 303-315.

Rasmussen D. L. & Sørensen S. J. (2001). Effects of mercury contamination on the culturable heterotrophic, functional and genetic diversity of the bacterial community in soil. *FEMS Microbiology Ecology*, *36*,1-9.

Ravel J., Amoroso M. J., Colwell R. R. & Hill R. T. (1998). Mercury-resistant *Actinomycetes* from the Chesapeake Bay. *FEMS Microbiology letters*, *162*, 177-184.

Renella, G., Chaudri, A. M. & Brookes, P. C. (2002). Fresh additions of heavy metals do not model long-term effects on microbial biomass and activity. *Soil Biology & Biochemistry, 34*,121-124.

Richards, J. W., Krumholz, G. D., Chval, M. S. & Tisa L. S. (2002). Heavy metal resistance patterns of *Frankia* strains. *Applied and Environmental Microbiology, 68*, 923-927.

Rho, J. Y., & Kim, J. H. (2002). Heavy metal biosorption and its significance to metal tolerance of Streptomycetes. *Journal of Microbiology*, 40, 51-54.

Rowan, A.K., Snape, J.R., Fearnside, D., Barer, M.R., Curtis, T.P. & Head, I.M. (2003). Composition and diversity of ammonia-oxidising bacterial communities in wastewater treatment reactors of different design treating identical wastewater. *FEMS Microbiol Ecol, 43*, 195-206.

Sandaa, R. A., Torsvik, V. & Enger, O. (2001). Influence of long-term heavy-metal contamination on microbial communities in soil. *Soil Biology & Biochemistry, 33*, 287-295.

Sandaa, R. A., Torsvik, V., Enger, O., Daae, F. L., Castberg, T. & Hahn, D. (1999). Analysis of bacterial communities in heavy metal-contaminated soils at different levels of resolution. *FEMS Microbiology Ecology, 30*, 237-251.

Schmidt, A., Haferburg, G., Sineriz, M., Merten, D., Buchel, G. & Kothe, E. (2005). Heavy metal resistance mechanisms in actinobacteria for survival in AMD contaminated soils. *Chemie Der Erde-Geochemistry, 65*, 131-144.

Sigler, W.V. & Turco, R.F. (2002). The impact of chlorothalonil application on soil bacterial and fungal populations as assessed by denaturing gradient gel electrophoresis. *Applied Soil Ecology, 21*, 107-118.

Sharma, S., Aneja, M. K., Mayer, J., Munch, J. C. & Schloter, M. (2005). Characterization of bacterial community structure in rhizosphere soil of grain legumes. *Microbial Ecology, 49*, 407-415.

Sloan, W. T.,Woodcock, S., Lunn, M., I. M. Head, I. M. & Curtis, T. P. (2007). Modeling taxa-abundance distributions in microbial communities using environmental sequence data. *Microbial Ecology, 53*, 443-455.

Stacktbrandt E., Rainey F.A. & Ward-Rainey N.L. (1997). Proposal for a new hierarchic classification system, *Actinobacteria classis* nov. *International Journal of Systematic Bacteriology, 47*, 179-491.

Tobor-Kaplon, M., Bloem, A. J., Romkens, P. & de Ruiter, P. C. (2005). Functional stability of microbial communities in contaminated soils. *Oikos*, 111,119-129.

Usman, A., Kuzyakov, Y. & Stahr, K. (2005). Effect of clay minerals on immobilization of heavy metals and microbial activity in a sewage sludge-contaminated soil. *Journal of Soils and Sediments, 5*, 245-252.

Valls, M., Atrian, S., de Lorenzo, V. & Fernández, L.A. (2000). Engineering a mouse metallothionein on the cell surface of *Ralstonia eutropha* CH34 for immobilization of heavy metals in soil. *Nature Biotechnology, 18*, 661-665.

Vaasquez-Murrieta, M., Migueles-Gardunno, I., Franco-Hernandez, O., Govaerts, B. & Dendooven, L. (2006). C and N mineralization and microbial biomass in heavy-metal contaminated soil. *European Journal of Soil Biology, 42*, 89-98.

Wang, Y. P., Shi, J. Y., Wang, H., Lin, Q., Chen, X. C. & Chen, Y. X. (2007). The influence of soil heavy metals pollution on soil microbial biomass, enzyme activity, and community composition near a copper smelter. *Ecotoxicology and Environmental Safety, 67*, 75-81.

Wardle, D. A. & Zackrisson, O. (2005). Effects of species and functional group loss on island ecosystem properties. *Nature*, *435*, 806-810.

Wertz, S., Degrange, V. Prosser, J. I., Poly, F., Commeaux, C. Freitag, T., Guillaumaud, N. & Le Roux, X. (2006). Maintenance of soil functioning following erosion of microbial diversity. *Environmental Microbiology*, *8*, 2162-2169.

Witter, E., Gong, P., Baath, E. & Marstorp, H. (2000). A study of the structure and metal tolerance of the soil microbial community six years after cessation of sewage sludge applications. *Environmental Toxicology and Chemistry, 19*, 1983-1991.

Wu, S. C., Luo, Y. M., Cheung, K. C. & Wong, M. H. (2006). Influence of bacteria on Pb and Zn speciation, mobility and bioavailability in soil: A laboratory study. *Environmental Pollution, 144*, 765-773.

Zhang, C. B., Huang, L. N. Luan, T. G. Jin, J. & Lan, C. Y. (2006). Structure and function of microbial communities during the early stages of revegetation of barren soils in the vicinity of a Pb/Zn Smelter. *Geoderma, 136*, 555-565.

In: Soil Contamination Research Trends
Editor: Javier B. Dominguez, pp. 109-117
ISBN 978-1-60456-319-1

Chapter 4

CONTAMINATION OF VINEYARD SOILS WITH FUNGICIDE-DERIVED COPPER

Michael Komárek[a,*]**, *Vladislav Chrastný***[b,†]
[a]Department of Agrochemistry and Plant Nutrition, Czech University of Life Sciences Prague, Kamýcká 129, 165 21, Prague 6, Czech Republic
[b]Department of Applied Chemistry and Chemistry Teaching, University of South Bohemia, Studentská 13, 370 05, České Budějovice, Czech Republic

ABSTRACT

The repeated use of fungicides (e.g., Bordeaux mixture) resulted in the accumulation of Cu in vineyard soils. Due to the fact that Cu is strongly immobilized by soil organic matter and Fe, Mn, Al-(hydr)oxides, elevated Cu concentrations originating from fungicide applications were observed in superficial horizons of vineyard soils. The concentrations in these horizons often exceed 200 mg Cu kg^{-1} (in some cases > 1000 mg Cu kg^{-1}), which surpasses all regulatory limits set for agricultural soils. The highest Cu concentrations were found in the upper ~ 15 cm of the soil profile. Copper concentrations decrease with depth and distance from the treated plants. Abandoning vitriculture and changing to other agronomical practices (e.g., tillage, shift to other agricultural plants) can lead to the mobilization of Cu from such contaminated soils and thus present an important environmental concern. Furthermore, increased Cu concentrations (200–400 mg kg^{-1}) in vineyard soils proved to negatively influence growth of several plant species, especially in acidic soils with a low cation exchange capacity. This review presents an overview on the range and degree of Cu contamination of vineyard soils, together with the biogeochemistry of Cu in these soils and the sorption mechanisms involved.

* E-mail: komarek@af.czu.cz
† E-mail: vladislavchrastny@seznam.cz

INTRODUCTION

Unpolluted agricultural soils are an important precursor for quality crop and animal production. Contamination with metals reduces the quality of the soils and poses important environmental and toxicological threats. Metals in soils can be of different origin: while a relatively small portion originates from natural processes such as the weathering of bedrocks, a much larger content of metals originates from anthropogenic activities (e.g., industry, combustion of fossil fuels etc.) that significantly influence their biogeochemical cycles. Significant quantities of metals in agricultural soils originate from the application of sewage sludge, fertilizers and pesticides (Adriano, 2001). In the case of vineyard soils, increased Cu concentrations have been observed due to the intensive use of Cu-based fungicides (e.g., the Bordeaux mixture). These fungicides have been intensively used in Europe since the 18th century to control vine (*Vitis vinifera* L.) downy mildew caused by *Plasmopara viticola*. Increased Cu concentrations in soils can cause phytotoxicity and lead to groundwater pollution with this metal. This article presents an overview of literature published on Cu contamination of vineyard soils and its biogeochemistry.

BIOGEOCHEMISTRY OF COPPER IN SOILS

The average mean background Cu concentrations in unpolluted soils reach 30 mg kg^{-1}. The solubility of Cu, as of other metals, is greatly dependent on soil pH and will be most readily available at pH below 6 (Fig. 1; Adriano, 2001; Celardin et al., 2004; Boudesocque et al., 2007). Copper is mostly associated with soil organic matter (SOM) and Fe, Mn-(hydr)oxides, to a lesser extent with clay minerals (Parat et al., 2002; Bradl, 2004). This strong sorption and complexation makes it one of the least mobile trace metals in soils. In most cases, the sorption of Cu follows either the Langmuir or Freundlich isotherm (Bradl, 2004); with the Langmuir isotherm being a more suitable model (Yuan et al., 2007). The affinity of Cu to separate soil fractions decreases in the following order: Mn-(hydr)oxides > SOM > Fe-(hydr)oxides > clay minerals. However, in soils with a significantly higher content of Fe-(hydr)oxides, these can present the most important sink for Cu. The most abundant Fe oxides in soils include goethite (α-FeOOH), hematite (α-Fe_2O_3) and ferrihydrite ($5Fe_2O_3 \cdot 9H_2O$) (Ostergren et al., 1999). In highly organic soils, sorption on SOM by means of complexation especially with humic and fulvic acids present an important retention mechanism for Cu. However, SOM can influence the mobility of Cu by two different means: while particulate SOM will act as a sorbent for Cu, soluble SOM will actually complex Cu, increasing thus its solubility, especially at alkaline pH (Adriano, 2001). Soil carbonates proved to be another important factor controlling the mobility (and thus bioavailability) of Cu in soils. Ponizovsky et al. (2007) found that the activity of Cu in calcareous soils is to a great extent controlled by the surface precipitation of $CuCO_3$. This is especially important in alkaline soils containing high concentrations of carbonates. The retention of Cu in calcareous soils is associated with the release of Ca^{2+}, Mg^{2+}, Na^+ and H^+ into the soil solution at equimolar ratios (Ponizovsky et al., 2007). The solubility of Cu minerals in soils decreases in the following order $CuCO_3$ > $Cu_3(OH)_2(CO_3)_2$ (azurite) > $Cu(OH)_2$ > $Cu_2(OH)_2CO_3$ (malachite) > CuO (tenorite) > $CuFe_2O_4$ (cupric ferrite). Increasing the partial pressure of

CO_2 further decreases the solubility of the carbonates. Copper sulfates, such as $CuSO_4$ (chalcocyanite) and $CuSO_4{\cdot}5H_2O$ (chalcanthite), are highly soluble and require very high Cu concentrations to form in soils (Lindsay, 1979).

The presence of synthetic chelating agents significantly influences the behavior of Cu in the environment and its bioavailability (Nowack, 2002; Komárek et al., 2008). For example, high concentrations of EDTA (ethylenediaminetetraacetic acid) have been found in river waters which led to significant changes in metal speciation (Nowack, 2002). The adsorption of metals to the soil matrix is influenced, to a great extent, by the pH of the soil extract. Generally, metal adsorption increases with increasing pH values. The presence of chelating agents in soils results into the preferential formation of negatively charged complexes (e.g., $CuEDTA^{2-}$) and the adsorption behaviour of metals will be thus changed (Nowack, 2002). The complex CuEDTA ($\log K_{CuEDTA} = 20.5$) is most stable at pH 6–7, in acidic soils (pH < 5.7) Fe displaces Cu ($\log K_{Fe(III)EDTA} = 27.2$) and in alkaline soils (pH > 7.9) with high Ca concentrations the preferential formation of CaEDTA will occur ($\log K_{CaEDTA} = 12.4$) (Adriano, 2001; Martell et al., 2001; Nowack, 2002). The speciation of Cu in the soil solution in the presence of chelating agents plays a very important role in its bioavailability. According to some authors and to the Free Ion Activity Model (FIAM), complexes of metals with synthetic chelants are not bioavailable (Parker et al., 2001). The addition of chelants thus changes the mechanisms of heavy metal uptake. Without any chelant present, free Cu (as a micronutrient) will be mainly taken up according to the FIAM. With the addition of low chelant concentrations, free Cu will be chelated and, therefore, its uptake via active mechanisms (through a symplastic pathway) will be reduced. However, when high chelant concentrations are present, passive uptake of the complexes through the apoplasm via diffusion will occur (Nowack et al., 2006). This mechanism is used, for example, in chelant-enhanced phytoextraction of metals from contaminated soils (Komárek et al., 2008).

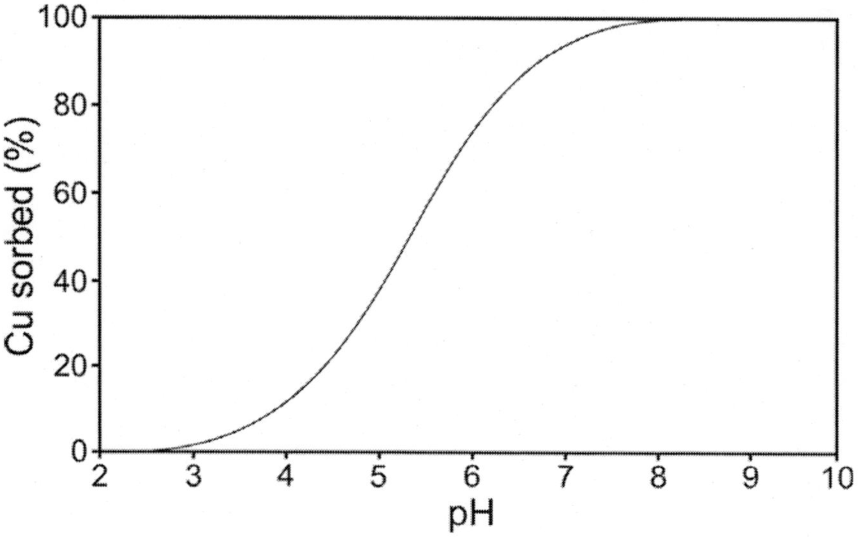

Figure 1. The influence of pH on Cu retention in a vineyard soil (redrawn after Boudesocque et al., 2007).

In order to determine the fractionation of metals (including Cu) in soils, several sequential extraction procedures have been proposed, most of them based on the classical sequential extraction by Tessier et al. (1979). These procedures allow to distinguish between separate operationally defined fractions of metals, such as exchangeable, acid extractable, reducible (Fe, Mn-(hydr)oxides), oxidizable (organic matter and sulphides), residual (silicate matrix). However, several critics of this approached have been published, mainly because of the redistribution and readsorption processes during the procedure and the possible absence of an extracted phase. All these drawbacks can lead to inaccurate results (Song et al., 1999). Nevertheless, sequential extraction procedures remain the most viable method for evaluation of chemical fractionation and mobility of metals in soils. Several one-step extraction procedures have been suggested for the determination of metal extractability and availability, among many H_2O, $CaCl_2$, $MgCl_2$, NH_4NO_3, EDTA, DTPA, HOAc, NH_4OAc etc. (Kennedy et al., 1997). The extracting solution should be chosen according to the specific soil characteristics and the results interpreted in accordance with the extractant used. Brun et al. (1998) recommend the 0.1 M $CaCl_2$ extraction as the best suitable for the determination of Cu bioavailability in neutral–acid soils due to a good correlation with Cu contents in wild growing plants. However, chemical-based methodologies do not give information about the actual binding form, but rather differentiate metals into separate operationally-defined fractions (e.g., exchangeable, reducible, oxidizable, residual etc.) (Gleyzes et al., 2002). Additional data about the binding forms and mineralogy can be obtained, for example, from X-ray diffraction (XRD), X-ray photoelectron spectroscopy (XPS), X-ray absorption near edge structure (XANES), extended X-ray absorption fine structure (EXAFS) and scanning microscope (SEM) analyses (Flores-Vélez et al., 1996; Boudesocque et al., 2007).

The phytotoxicity of Cu is the highest in acidic soils with a low cation exchange capacity. Chaignon et al. (2003) observed much higher Cu concentrations in tomato (*Lycopersicon esculentum* Mill.) roots grown on strongly acidic soils compared to calcareous or mildly acidic soils at the same Cu level. Root Cu concentrations are thus a good indicator of Cu bioavailability in soils. However, for a fast and simple prediction of Cu bioavailability, routine soil testing procedures, such as total Cu, $CaCl_2$- and EDTA-extractable Cu are sufficient (Brun et al., 1998; Chaignon et al., 2003). Increased metal concentrations in soils negatively influence the sustainability of agroecosystems. Earthworm populations are sensitive to tillage and Cu/Zn pesticides residues (Paoletti et al., 1998); therefore, they present useful reliable bioindicators of agroecosystem stability of vineyards. Significantly higher Cu concentrations have been observed in roots of several plants compared to shoots (Chaignon et al., 2003; Cattani et al., 2006). This can be explained by the fact that plants can, to some extent, actively control the uptake (lower pH of the rhizosphere compared to bulk soil) and translocation of Cu as a micronutrient. Furthermore, Cu is mainly fixed in the apoplasm due to its affinity to the negatively charged components of root cell walls and is not taken up into the root cells (Crawley, 1997). Copper uptake by plants is also dependent on phytosiderophores released by plant roots. Chaignon et al. (2002) observed that under Fe and Zn starvation, wheat plants (*Triticum aestivum* L.) acquired elevated concentrations of Cu due to the increased release of phytosiderophores.

COPPER CONTAMINATION OF VINEYARD SOILS

Vitriculture is a very important agricultural practice in many countries. In order to control vine diseases caused by fungi, different fungicides, especially Cu-based, such as the Bordeaux mixture ($CuSO_4$ + $Ca(OH)_2$), has been applied intensively to soils. Recently, other soluble Cu compounds (e.g., $CuSO_4 \cdot 3Cu(OH)_2$, Cu_2O, $Cu(OH)_2$, $CuCl_2 \cdot 3Cu(OH)_2$) have been used to replace the Bordeaux mixture. However, the application of Cu as a pesticide is not only limited to vineyards. Copper sulfate has been successfully applied as an algicide and molluscicide in water or even as a mordant or to fertilize Cu-deficient pastures (Merian, 1991). Nevertheless, the application of Cu-based fungicides increased significantly its concentrations especially in soils of vineyards (e.g., Flores-Vélez et al., 1996; Brun et al., 1998; Besnard et al., 2001), apple, citrus and coffee orchards (e.g., Adriano, 2001; Li et al., 2005; Loland and Singh, 2004), hop fields (e.g., Schramel et al., 2000) etc. For example, Arias et al. (2004) found that in Galicia, NW Spain, Cu concentrations exceeded 100 mg kg^{-1} in 60% of subsurface soil samples due to the application of Cu-based fungicides. Due to the fact that Cu is strongly immobilized by soil organic matter and Fe, Mn-(hydr)oxides, Cu concentrations in superficial horizons of vineyard soils often exceed 200 mg Cu kg^{-1} (Flores-Vélez et al., 1996; Brun et al., 1998; 2001; Pietrzak and McPhail, 2004). The highest Cu concentrations are therefore present in the upper 0–15 cm of the soil profile and the closest to the vine plants and Cu concentrations decrease with depth and distance from the treated plants (Magalhães et al., 1985; Pietrzak and McPhail, 2004; Wightwick et al., 2006; Rusjan et al., 2007). The low mobility of Cu in soils has been commonly accepted; however, Pietrzak and McPhail (2004) found > 50% of anthropogenic Cu (from pesticides) present in the so-called "active" fraction of the subsurface vineyard soil horizons (water extractable + exchangeable + adsorbed fraction). In France, the intensive use of the Bordeaux mixture led in some cases to an accumulation of Cu concentrations above 1000 mg kg^{-1} (Flores-Vélez et al., 1996). Extremely high concentrations (> 3000 mg Cu kg^{-1}) in Brazilian vineyard soils have been documented by Mirlean et al. (2007). These values are, so far, the highest published and indicate that the intensive use of pesticides in areas of wet subtropical climate results into significantly higher Cu concentrations in soils compared to drier areas. Table 1 summarizes data obtained from the literature on Cu concentrations in different vineyard soils. Statistical analyses (Pearson's correlation) showed that significant correlations were found only between exchangeable soil Cu (methods are summarized in Table1) and soil pH. Other soil parameters such as SOM correlate neither with total Cu nor exchangeable Cu (data from references in Table 1). This again highlights the importance of soil pH on Cu retention in soils. Warning and critical legislative limits valid in the EU set Cu concentrations in soils at 50 and 140 mg kg^{-1}, respectively (Council Directive 86/278/EEC, 1986). Moreover, the use of Cu-based pesticides can introduce other metals (Cd, Cr, Pb, Zn) into the soil (Mirlean et al., 2007) and the study of Robinson et al. (2006) revealed that vineyard posts treated with Cu-Cr-As can be another source of these risk elements in vineyard soils.

Table 1. Total and exchangeable Cu content in superficial horizons of vineyard soils. Total content is based on strong acid digestion (with HF) unless otherwise noted

Soil depth/cm	Total Cu content/ mg kg^{-1}	Exchangeable Cu content/ mg kg^{-1} (method used)	References
0–2	75–398	22-176 (Na_2-EDTA)	Chaignon et al., 2003
0–15	86–136	35-51 (Na_2-EDTA)	
2–15	73–288	22-111 (Na_2-EDTA)	
0–20	61–351	1.2-4.7 ($Mg(NO_3)_2$)	Parat et al., 2002
n.a.	n.a.	58-489 (2M HNO_3)	Celardin et al., 2004
0–2	323	n.a.	Flores-Vélez et al., 1996
2–5	209	n.a.	
5–28	127	n.a.	
0–1	9–249	n.a.	Pietrzak and McPhail, 2004
0–10	17–429	0.03-2.3 ($CaCl_2$)	Jacobson et al., 2005
0–20	12–30*	n.a.	Tack et al., 2006
0–10	125–603	49-207 (Na_2-EDTA)	Nóvoa-Muñoz et al., 2007
10–30	112–621	44-177 (Na_2-EDTA)	
0–15	14–251	2.3-95 (Na_2-EDTA) 0.1-9.2 ($CaCl_2$)	Brun et al., 1998; Brun et al., 2001
0–20	179–549	0.4-4.2 (H_2O)	Arias et al., 2005
0–5	37–3216	1.3-5.0 ($CaCl_2$) 7.3-607 (CH_3COONH_4) 19.5-1265 (Na_2-EDTA + CH_3COONH_4)	Mirlean et al., 2007
0–10	378	n.a.	Besnard et al., 2001

* Aqua regia extractable content.

It has been estimated that a single application of the Bordeaux mixture in the Champagne region, France, introduced 3–5 kg Cu ha^{-1} with 3–10 applications per year. However, only ~ 60% of Cu was retained in the soils, suggesting that losses through erosion must have occurred (Besnard et al., 2001). Because many vineyards are located on steep slopes, intensive erosion further influences Cu mobility in these soils and thus increases the risks associated with groundwater quality (Ribolzi et al., 2002). Among many, planting grass between vine rows has been suggested as an effective anti-erosion precaution. Organic soil amendments are often applied to increase soil fertility and decrease soil erosion by runoff. The most commonly used amendments are bark, vine shoots and compost. These amendments, however, influence Cu retention in soils through the introduction of particulate organic matter (POM) to soils. Copper may be adsorbed (immobilized) on the surfaces of POM or complexed (mobilized) with its reactive functional groups (Flores-Vélez et al., 1996; Besnard et al., 2001; Devez et al., 2005). As mentioned before, soil pH influences to a great extent the mobility of Cu in soils; therefore, in acidic vineyard soils (e.g., developed on granitic rocks) Cu can migrate downward more easily and present thus an important risk of ground water pollution. The main factors influencing the mobility of Cu in vineyard soils and the potential risk of groundwater pollution thus include: (i) tillage, (ii) soil pH and (iii) erosion (Besnard et al., 2001; Mirlean et al., 2007; Nóvoa-Muñoz et al., 2007).

Only in rare cases, vines suffered from high Cu concentrations in vineyard soils mainly due to the fact that vine plants are deeply rooting and the highest concentrations of Cu are present in the superficial soil horizons. However, Brun et al. (2003) found that increased Cu concentrations affected the phenology, growth and reproduction of some ruderal plant species, so it is possible to assume that plants with shallow roots are more susceptible to increased Cu concentrations. Therefore, changing the agricultural practices, such as replacing the vines with other shallow rooting crops or pasture plants, could lead to the contamination of the newly grown plants. Furthermore, increased Cu concentrations can, to a great extent, affect adversely the microbial activity in soils (Dumestre et al., 1999). Not only natural processes taking place in soils are negatively influenced, but the reduction of the activity of microorganisms and/or changes in microbial populations can lead to lower mineralization rates of organic xenobiotics, such as organic pesticides. Jacobson et al. (2005) found that increased Cu concentrations in vineyard soils influenced, to some extent, the persistence of the organic pesticide diuron in the soils and thus increased risks associated with groundwater pollution.

CONCLUSION

Polluted agricultural soils present a serious environmental concern. The contamination with organic and inorganic compounds reduces the quality of the soil and the sustainability of the agroecosystem. Copper originating from the intensive application of Cu-based fungicides belongs to the most important contaminants of vineyard soils. Copper concentrations in these soils often exceed regulatory limits and in some cases even phytotoxic levels. Despite the fact that Cu in soils is strongly immobilized by the soil sorption complex, there has been evidence that increased Cu can migrate through the soil profiles in vineyards and pose thus an important risk for groundwater quality. This is especially true for acidic soils, tilled soils and soils affected by intensive erosion. Therefore, stabilization strategies that lower the intensity of erosion (such as the application of organic amendments and vegetation cover) are needed for decreasing Cu runoff from vineyard soils via wind and water erosion. There have been only rare cases of Cu phytotoxicity symptoms observed on vine plants, mostly due to their deep rooting. However, precautions are needed when other crops with shallow roots are subsequently grown on such contaminated soils. Copper concentrations should be then lowered to acceptable levels through different remediation procedures.

REFERENCES

Adriano, D. C. *Trace Elements in the Terrestrial Environments: Biogeochemistry, Bioavailability, and Risks of Metals*, Springer-Verlag, New York-Berlin-Heidelberg-Tokyo, 2001; pp 533.

Arias, M., Lopez, E., Soto, B. *Agrochimica* 2005, *49,* 60-69.

Besnard, E., Chenu, C., Robert, M. *Environ. Pollut.* 2001, *112,* 329-337.

Boudesocque, S., Guillon, E., Aplincourt, M., Marceau, E., Stievano, L. *J. Colloid Interf. Sci.* 2007; *307,* 40-49.

Bradl, H. B. *J. Colloid Interf. Sci.* 2004, *277,* 1-18.

Brun, L. A., Maillet, J., Richarte, J., Herrmann, P., Remy, J. C. *Environ. Pollut.* 1998, *102,* 151-161.

Brun, L. A., Maillet, J., Hinsinger, P., Pépin, M. *Environ. Pollut.* 2001, *111,* 293-302.

Brun, L. A., Le Corff, J., Maillet, J. *Environ. Pollut.* 2003, *122,* 361-368.

Cattani, I., Fragoulis, G., Boccelli, R., Capri, E. *Chemosphere* 2006, *64,* 1972-1979.

Celardin, F., Trouillet, C., Tisiot, R. *Environ. Sci. Lett.* 2004, *1,* 225-227.

Chaignon, V., Sanchez-Neira, I., Herrmann, P., Jaillard, B., Hinsinger, P. *Environ. Pollut.* 2003, *123,* 229-238.

Crawley, M. J. *Plant Ecology* 2nd Ed., Blackwell Science, London, 1997; pp 736.

Devez, A., Gomez, E., Gilbin, R., Elbaz-Poulichet, F., Persin, F., Andrieux, P., Casellas, C. *Sci. Total Environ.* 2005, *348,* 82-92.

Dumestre, A., Sauvé, S., McBride, M., Baveye, P., Berthelin, J. *Arch. Environ. Con. Tox.* 1999, *36,* 124-131.

Flores-Vélez, L. M., Ducaroir, J., Jaunet, A. M., Robert, M. *Eur. J. Soil. Sci.* 1996, *47,* 523-532.

Gleyzes, C., Tiller, S., Astruc, M. *Trends Anal. Chem.* 2002, *21,* 451-467.

Jacobson, A. R., Dousset, S., Guichard, N., Baveye, P., Andreux, F. *Environ. Pollut.* 2005, *138,* 250-259.

Kennedy, V. H., Sanchez, A. L., Oughton, D. H., Rowland, A. P. *Analyst* 1997, *122,* 89R-100R.

Komárek, M., Tlustoš, P., Száková, J., Chrastný, V. *Environ. Pollut.* 2008, *151*, 27-38.

Li, W., Zhang, M., Shu, H. *Eviron. Sci. Pollut. Res.* 2005, *12,* 168-172.

Lindsay, W. L. *Chemical Equilibria in Soils*, Johh Wiley & Sons, New York-Chichester-Brisbane-Toronto, 1979; pp 222-237.

Loland, J. O., Singh, B. R. *Nutr. Cycl. Agroecosys.* 2004, *69,* 203-211.

Magalhães, M. J., Sequeiraa, E. M., Lucas, M. D. *Water Air Soil Poll.* 1985, *26,* 1-17.

Martell, A. E., Smith, R. M., Motekaitis, R. J. *NIST critically selected stability constants of metal complexes*, Version 6.0. NIST, Gaithersburg, MD, USA, 2001.

Merian, E. *Metals and Their Compounds in the Environment. Analysis and Biological Relevance*, VCH Weinheim, 1991; pp 1438.

Mirlean, N., Roisenberg, A., Chies, J. O. *Environ. Pollut.* 2007, 149, 10-17.

Nowack, B. *Environ. Sci. Technol.* 2002, *36,* 4009-4016.

Nowack, B., Schulin, R., Robinson, B. H. *Environ. Sci. Technol.* 2006, *40,* 5225-5232.

Ostergren, J. D., Brown, G. E., Parks, G. A., Tingle, T. N. *Environ. Sci. Technol.* 1999, *33,* 1627-1636.

Paoletti, M. G., Sommaggio, D., Favretto, M. R., Petruzzelli, G., Pezzarossa, B., Barbafieri, M. *Appl. Soil Ecol.* 1998, *10,* 137-150.

Parat, C., Chaussod, R., Lévêque, J., Dousset, S., Andreux, F. *Eur. J. Soil Sci.* 2002, *53,* 663-669.

Parker, D. R., Fedler, J. F., Ahnstrom, Z. A. S., Resketo, M. *Environ. Toxicol. Chem.* 2001, *20,* 899-906.

Pietrzak, U., McPhail, D. C. *Geoderma* 2004, *122,* 151-166.

Ponizovsky, A. A., Allen, H. E., Ackerman, A. J. *Environ. Pollut.* 2007, *145,* 1-6.

Ribolzi, O., Valles, V., Gomez, L., Voltz, M. *Environ. Pollut.* 2002, *117,* 261-271.

Robinson, B., Greven, M., Green, S., Sivakumaran, S., Davidson, P., Clothier, B. *Sci. Total Environ.* 2006, *364,* 113-123.
Rusjan, D., Strlič, M., Pucko, D., Korošec-Koruza, Z. *Geoderma* 2007, *141*, 111-118.
Schramel, O., Michalke, B., Kettrup, A. *Sci. Tot. Environ.* 2000, *263,* 11-22.
Song, Y., Wilson, M. J., Moon, H. S., Bacon, J. R., Bain, D. C. *Appl. Geochem.* 1999, *14,* 621-633.
Tessier, A., Campbell, P. G. C., Bisson, M. *Anal. Chem.* 1979, *51,* 844-850.
Wightwick, A., Mollah, M., Smith, J., McGregor, A. *Aust. J. Soil Res.* 2006, *44,* 711-717.
Yuan, S., Xi, Z., Jiang, Y., Wan, J., Wu, C., Zheng, Z., Lu, X. *Chemosphere* 2007, *68,* 1289-1297.

In: Soil Contamination Research Trends
Editor: Javier B. Dominguez, pp. 119-141

ISBN 978-1-60456-319-1

Chapter 5

SOIL POLLUTION AND RESPONSE OF MICROORGANISMS TO OXIDATIVE STRESS

Bystrik Polek
Institute of Molecular Biology, Slovak Academy of Sciences,
Bratislava, Slovak Republic

ABSTRACT

Many pollutants in the soils are subjected to oxidation processes, resulting in the formation of reactive intermediates, mainly radicals that react with oxygen producing the reactive oxygen species (ROS). In general, microorganisms are protected from ROS damage by several enzymes, including catalases. Model microorganisms, bacterial isolates, *Comamonas terrigena* N3H and *Comamonas terrigena* N1C selected from soils contaminated with crude oil, *Comamonas testosteroni* strain isolated from sludge of waste water and isolates of micromycetes *Aspergillus niger* from mine surroundings (As 1651 mg/kg, Sb 362 mg/kg respectively As 400 mg/kg) during the growth exhibited much higher level in total catalase activity than the same control species. Electrophoretic resolution of cell free extract of *C. terrigena* N3H exhibited two isozymes of catalase, catalase-1 and catalase-2 unlike to the control isolate, where only one band of heme catalase was detected. The isolate and the collection culture of *C. testosteroni*, grown in nutrient meat peptone medium, revealed only one band, the constitutively expressed heme catalase. *A. niger* isolates equally as the control culture, in the growth phase of developing hyphae and spore formation three catalase isozymes in the nitrate medium produced. The H_2O_2 – adapted variant of *C. terrigena* N3H, obtained by subculturings with gradually increasing concentration of oxidative stressor exhibited a much higher level of total catalase activity represented mainly by catalase-1, then the original isolate. Phenol peroxidation in *C. terrigena* N3H cells can be also largely attributed to the action of this isozyme, that is essential not only for the protection against oxidative stress, but its peroxidic activity can also play an important role in the degradation. The sequence analysis of *catalase-1* gene confirmed the presence of highly conserved catalase sequence motifs that in other strains (not exposed to pollution) could not be found. This fact can indicate a selective pressure of environment. Catalase expression in microbial isolates as a response to exogenously added stressors was dependent on individual strain, its growth phase and kind of the stressor. The isolates of *C. terrigena* exhibited much higher resistance to exogenous H_2O_2 (20, 40 mM) than collection cultures, mainly in the middle and late exponential growth phases. The results refer to complicated relations between biotic and abiotic effects that may directly affect the processes *in situ* environment.

INTRODUCTION

Soil Pollution and Oxidative Stress

Anthropogenic activity such as mining and smelting, combustion of fossil fuels and incineration of refuse is responsible for major releases of chemical compounds into the environment. Many of these pollutants are subjected to auto-oxidation or enzymatic oxidation, resulting in the formation of reactive intermediates, mainly radicals that react with oxygen producing the reactive oxygen species - ROS [1, 2]. Generally, the microorganisms are protected against the oxidative damage by means of several enzymes, including catalases, superoxide dismutases, gluthation-S-transferases and antioxidants occurring in the cells [3, 4]. Accurate regulation of enzymes activities, including a possible alteration in the expression of isozymes, may be important in surviving such stresses [5]. Catalases belong to enzyme group of oxidoreductases that have very high potential in protection of environment [6].

Catalases

On the base of the reaction abilities, catalases can be classified as: typical catalases, catalases-peroxidases and non heme catalases. All known catalases catalyze the decomposition of hydrogen peroxide into water and molecular oxygen (1). Besides the reduction, they are able also to peroxidate (2). Obviously, catalases are not only essential for the protection against oxidative stress, but could also play an important role in the degradation for instance of toxic phenolic compounds [7].

$$2H_2O_2 \longrightarrow 2\,H_2O + O_2 \qquad (1)$$

$$H_2O_2 + C_2H_5\,OH \longrightarrow CH_3CHO + 2\,H_2O \qquad (2)$$

For bifunctional catalases-peroxidases is characteristic lower catalase and higher peroxidase activity not only with 2-electron, for instance with aliphatic alcohols but especially with 1-electron substrates (ABTS-2, 2'-azino-bis (3-ethylbenzthiazoline- 6-sulphonic acid, guaiacol, L-DOPA). While catalases are tetrameric molecules with sole protoheme IX per one subunit, catalase-peroxidase is composed from two subunits. Only the N- end is catalytic active and combines catalytic properties both enzymes e.g. typical catalase and peroxidase [8]. These enzymes are involved into metabolic pathway of resistant polyaromatic compounds degradation, as well [9]. The strains selected from the extreme environments, among them belongs also the polluted soil, often own enzymes with the unusual structural forms and enzymatic properties [10, 11].

Typical catalases are found in both prokaryotes and eukaryotes. Catalase-peroxidases were detected mainly in prokaryotes cells, for instance *Escherichia coli* [12], *Legionella pneumophila* [13], *Mycobacterium* sp. PYR-1 [14], but also in eukaryotes, for instance in fungi *Septoria tritici* [15]. Nonheme catalases remove H_2O_2 that is product of pigmentation activity of tyrosinase [16].

Production of Catalases by Microbial Isolates from Polluted Environments

Soil bacteria of the family *Comamonadaceae* are known for their broad catabolic diversity in the degradation of various xenobiotics and are very often present in polluted environments. Isolate *Comamonas testosteroni* CPW301 was able to grow with phenol as the sole carbon source [17]. *Comamonas terrigena* N3H degraded dialkyl sulphosuccinates, a group of synthetic surfactants of commercial importance [18, 19]. This strain is able to cleave strongly resistant sulphonate bond that is included in the chemical structure of this surfactant.

The production of catalases was tested in different phases of growth cycle of bacterial strains: *C. terrigena* N3H and *C. terrigena* N1C, which were isolated from soils contaminated with crude oil [20]. *C. testosteroni* isolate from sludge of waste water treatment plant and collection cultures *C. terrigena* ATCC 8461 and *C. testosteroni* ATCC 11996. A gradual increase in total catalase activity in aging microbial cultures of all *Comamonas* spp. was observed (Figure 1). Growth of cells entering the stationary phase may cause an accumulation of endogenous oxidants, resulting in an increase in respiratory activity [5]. However, isolates, *C. terrigena* N3H, *C. terrigena* N1C and *C. testosteroni* selected from polluted environments, cultivated in reach MPB medium during the growth exhibited much higher level in total catalase activity than the same species from the culture collections.

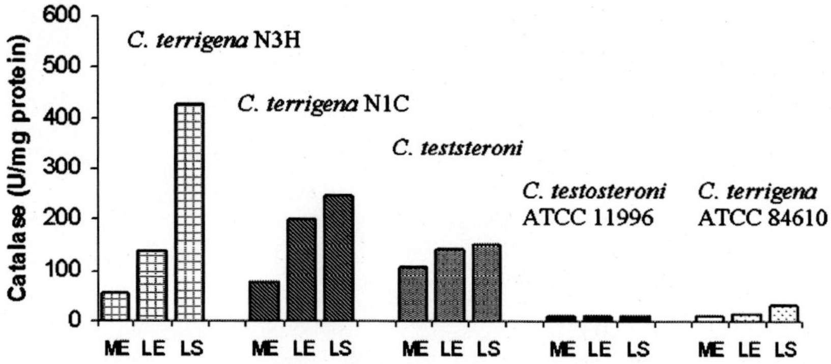

Figure 1. Changes in the level of specific catalase activity in crude extracts of *Comamonas* spp. during the growth. The bacterial cultures were grown in meat peptone broth medium and the catalase activities were determined in the middle exponential phase (ME), the late exponential phase (LE) and late stationary phase of growth (LS).

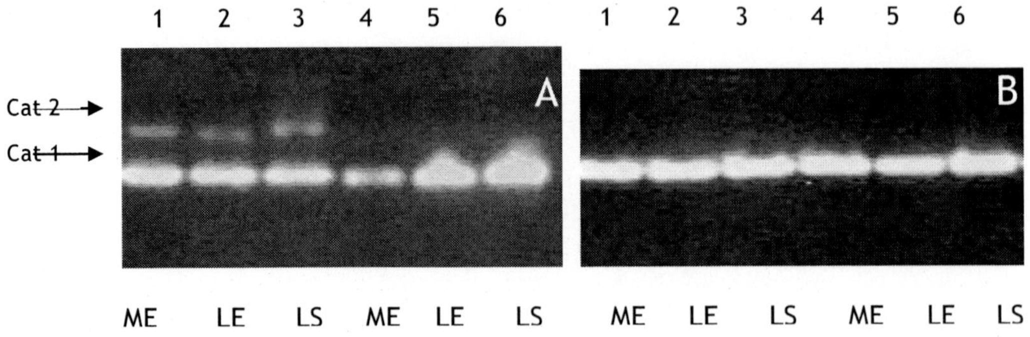

Figure 2. Native polyacrylamide gel electrophoresis (4-18 %) of catalases produced by *Comamonas* spp. (A) 1,2,3 - *C. terrigena* N3H; 4,5,6 - *C. terrigena* ATCC 8461. (B) 1,2,3 - *C. testosteroni*; 4,5,6 - *C. testosteroni* ATCC 11996. ME – the middle exponential phase, LE -the late exponential phase and LS – the late stationary phase of growth. The amount of cell free extract loaded on the gel was adjusted to reach the total catalase activity approximately 1U.

Electrophoretic resolution of cell free extract of *C. terrigena* N3H exhibited two isozymes of catalase, catalase-1 and catalase-2. The constitutively expressed catalase-1 is a dimeric protein with molecular weight of the monomer 55417 Da [21]. The induction of catalase-2 with a size equivalent to typical catalase was also detected. Unlike to the isolate, the electrophoretic pattern of the collection culture *C. terrigena* ATCC 8461 revealed only one band of heme catalase. *C. testosteroni* strains, the isolate and the collection culture, cultivated in meat peptone broth (MPB medium) revealed only one band, the constitutively expressed heme catalase (Figure 2).

In *E. coli* there exist two isozyme forms of catalase: hydroperoxidase I (HPI) and hydroperoxidase II (HPII) [22]. Their activities and expression are dependent on the phase of growth. Whereas HPI had already been produced in the presence of trace amounts of hydrogen peroxide in the exponential phase, the expression of both forms was necessary for higher concentrations of hydrogen peroxide in stationary phase [23]. Jenkins et al. suppose that the detection of a tetrameric typical catalase in the late stationary phase of growth is connected with cross resistance of the cells to starvation that is typical for this growth phase [24]. Some of the stress proteins that are induced during the starvation (without glucose and nitrogen source) overlap the proteins induced by oxidative stress. Only a single form of catalase was detected e.g. in *Helicobacter pylori* [25], *Heamophilus influenzae* [26] and *Bacillus abortus* [27].

Members of *Aspergillus* section *Nigri* are distributed worldwide, growing upon a wide variety of substrates. They have also adapted in soils with high concentration of As, Sb compounds brought by mining industry [28]. Micromycetes are sensitive to the effect of environmental stress mainly in the growth phase of developing hyphae and spore formation [29]. During theses period, a gradual increase in total catalase activity was observed in all tested *A. niger* strains [28]. The production of total catalase was also influenced with nutrients. Enzyme activity was much higher in minimal nitrate medium than that found in minimal ammonium or in ammonium nitrate (Figure 3). Electrophoretic resolution of catalases in cell free extracts of *A. niger* strain grown for 72 h revealed three types of catalases. During this period, only minimal nitrate medium supported production of catalase II with molecular weight of a typical catalase, classified according Witteveen et al. [30].

Three isozymes of catalase were identified during cultivation in the nitrate medium only on the second cultivation day (Figure 4).

Isolates of *A. niger,* selected from coal dust of mine containing As (400 mg/kg) and from river sediment of mine surroundings As (1651 mg/kg), Sb (362 mg/kg), growing in minimal nitrate medium in the phase of developing hyphae and spore formation, exhibited much higher level in total catalase activity than the same species from the collection culture or fungal culture adapted to soil contaminated only with As (5 mg/L). It suggests, that the catalase level reflects the rate of pollution represented by content of metals (Figure 5). However, the production of individual isozymes in crude extracts of isolates grown in nitrate medium in comparison to the collection culture was not significantly influenced and zymogram analysis was similar, as is shown in Figure 4.

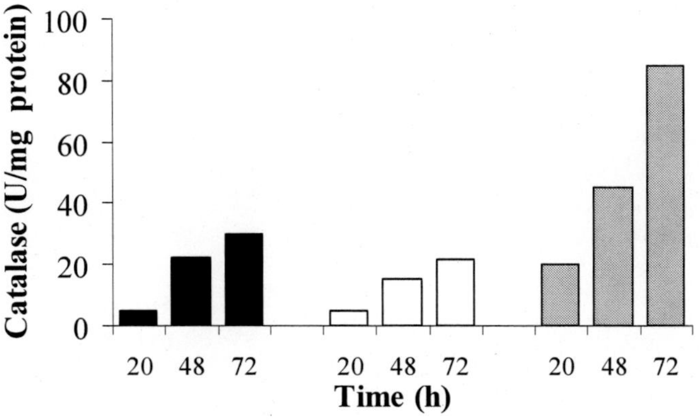

Figure 3. Production of total catalase activity by the collection culture of *Aspergillus niger* during the growth and developing hyphae and spore formation. Grey columns - strain grew in minimal-nitrate medium; white columns – in minimal ammonium medium and black columns – in ammonium nitrate medium.

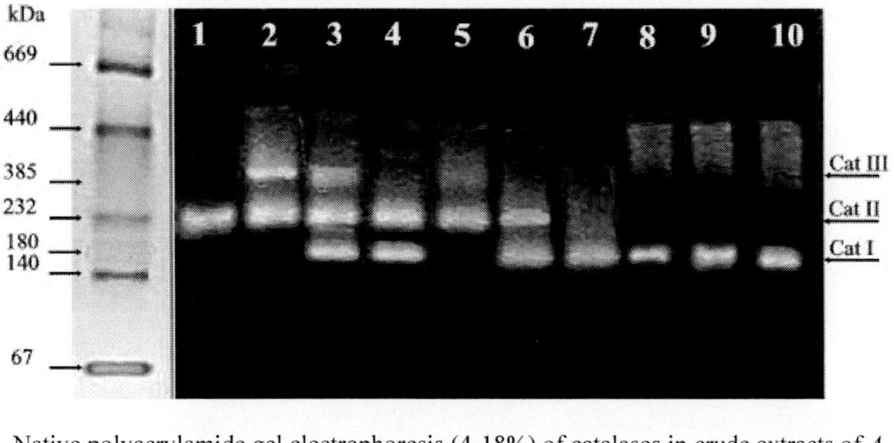

Figure 4. Native polyacrylamide gel electrophoresis (4-18%) of catalases in crude extracts of *A. niger* during growing and developing hyphae and spore formation. Lines from the left: 1 - a typical catalase from bovine liver (marker); 2, 3, 4 - strain grew in minimal nitrate medium for 20, 48 or 72 hours; 5, 6, 7 - growth in nitrate of ammonium medium; 8, 9, 10 – growth in minimal ammonium medium.

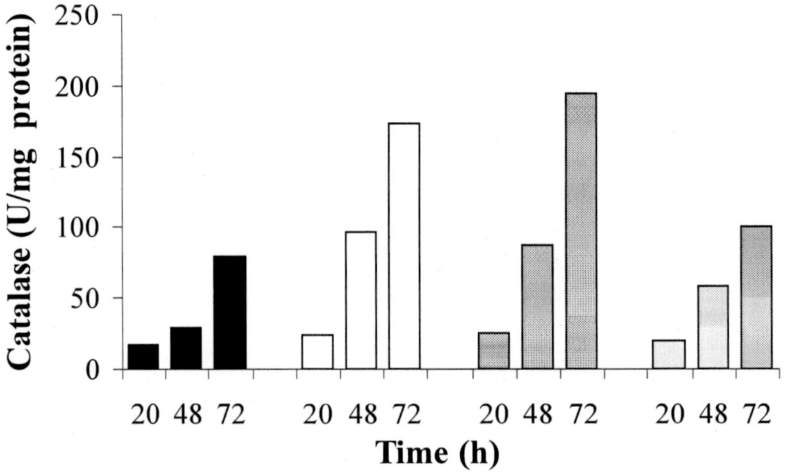

Figure 5. Production of total catalase activity of *A. niger* cultures during the growth and developing hyphae and spore formation in minimal nitrate medium. Black columns - the collection strain of *A. niger*; white columns - isolate selected from river sediment As (400 mg/kg); horizontally hatched columns - isolate from coal dust near mine As (1651mg/kg), Sb (362 mg/kg); vertically hatched columns - *A. niger* - the adapted culture to Sabouraud medium contaminated with As^{5+} (5 mg/kg).

Catalase Expression in Microbial Isolates as a Response to Exogenously Added Stressor

The expression of catales was dependent on individual strain, its growth phase and kind of the stressor in laboratory conditions [31]. For instance the maximum stimulation of total catalase activity (approx. 5 fold) in N3H strain with 0.5 mM peracetic acid (PAA) was in middle of exponential phase observed. Cd^{2+} (25 mg/L) in *C. terrigena* N1C induced highest changes of the activity in the middle growth phase (approx. 2.5 fold), unlike to *C. terrigena* N3H where maximum stimulation (approx. 1.4 fold) in the stationary growth phase was observed (Figure 6). Cumene hydroperoxide significantly increased enzymatic activity in the middle growth phase of *C. testosteroni* isolate (approx. 3 fold). Homogenates of the collection strains *C. testosteroni* 1931^T – ATCC 11996 a *C. terrigena* 2409^T-ATCC 8461, did not exhibit such significant increase of low levels of catalase with the effect of stressors (Figure 6). The presence of added stressors to cultivation medium with *C. terrigena* N3H cells influenced also the production of isozymes cat-1 and cat-2. The expression of the smaller enzyme (cat-1) is supported by 0.5 mM hydrogen peroxide in the medium, and cat-2 (typical catalase) is specifically induced with cadmium ions in 25 mg/L [32].

The effect of metals on expression of catalases of *A. niger* strains was also tested. Higher concentrations of As^{5+}, Cd^{2+}, Cu^{2+} (25 and 50 mg/L) respectively, or 20 and 40 mM H_2O_2 stimulated the total catalase activity mostly only in isolates *A. niger* from surroundings of mines (Figure 7). The control strain and to As^{5+} adapted culture prevailingly exhibited the significant decrease of enzyme activity at these doses of agents [28]. Treatment of *Phanerochaete chrysosporium* mycelium grown under high nitrogen level with 20 mM H_2O_2 caused a slight induction of the catalase activity while concentration 50 mM caused a significant decrease in activity of this enzyme [33].

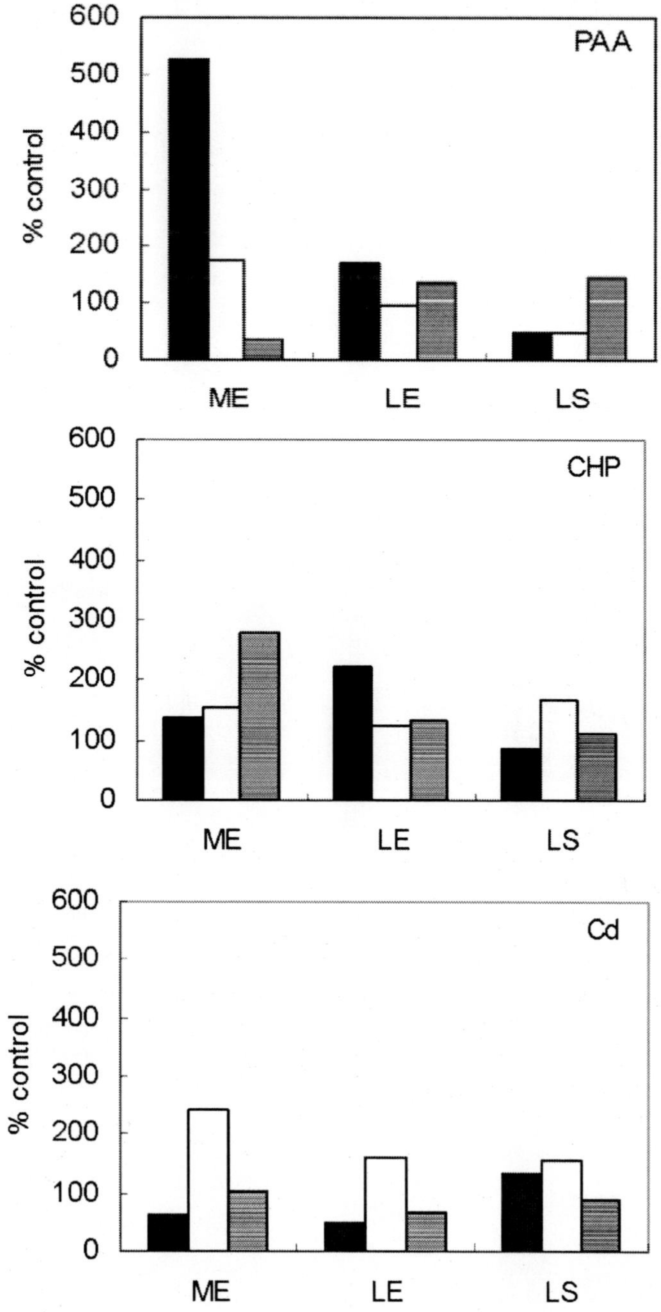

Figure 6. Relative catalase activities (% - compared to bacterial cultures that were grown in the nutrient medium without stressor) in crude extracts of *C. terrigena* N3H (black columns), *C. terrigena* N1C (white columns) and natural isolate *C. testosteroni* (horizontally hatched columns) under various stress conditions. The oxidative agents 0.5 mM peracetic acid (PAA); 0.5 mM cumene hydroperoxide (CHP) or 25 mg/L Cd^{2+} were added at middle exponential (ME), late exponential (LE) and late stationary (LS) phases of growth.

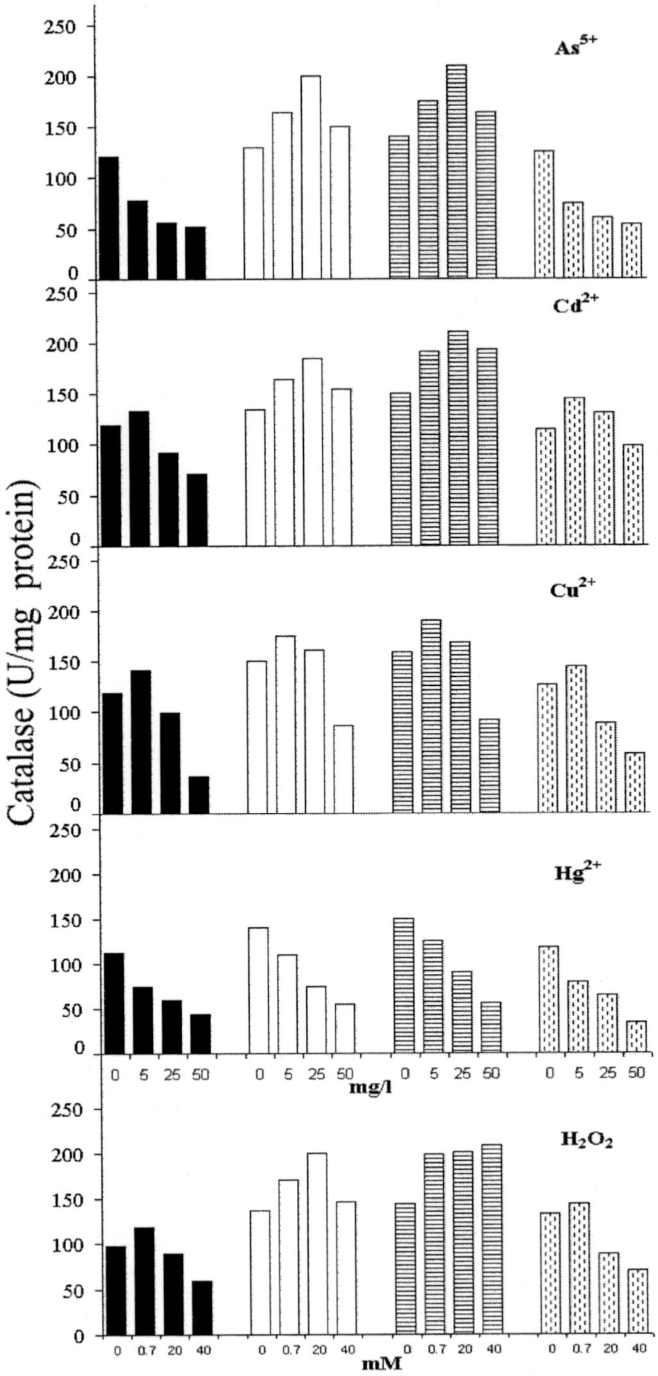

Figure 7. Sensitivity of catalase activity of *A. niger* cultures to exogenous stressor. Cultures grown in minimal-nitrate medium for 48 h were exposed to added stressor during 2 hours. Concentrations of agents: 0.7, 20 or 40 mM H_2O_2 and 5, 25 or 50 mg/L of metal. Black columns - the collection strain of *A. niger*; white columns - isolate selected from river sediment As (400 mg/kg); horizontally hatched columns - isolate from coal dust near mine (As 1651 mg/kg, Sb 362 mg/kg); vertically hatched columns - the adapted culture to Sabouraud medium contaminated with As^{5+} (5 mg/kg).

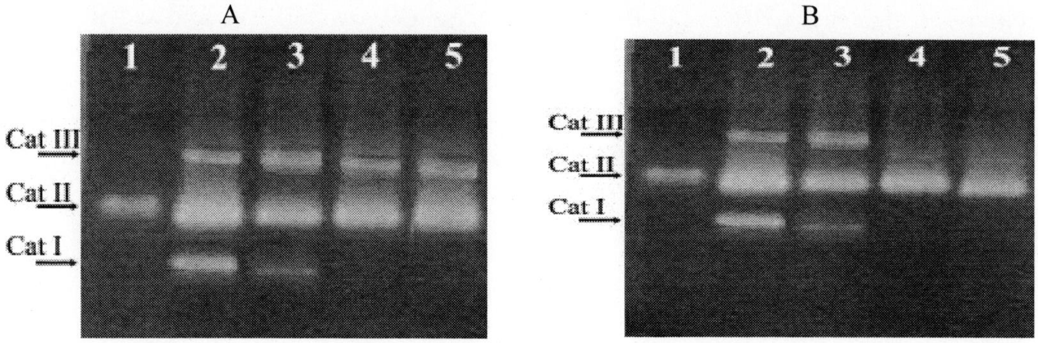

Figure 8. Native polyacrylamide gels electrophoresis (4-18%) of catalases in crude extracts of *A. niger* isolate from coal dust of mine surrounding. Culture grown in nitrate medium for 48 h was exposed to added H_2O_2 (A) and As^{5+} (B) during 2 hours. Lines from the left: 1- a typical catalase from bovine liver (marker); 2 –catalases in crude extract of isolate without effect of stress (control); 3, 4, 5 – the culture stressed with 0.7, 20 or 40 mM H_2O_2 and 5, 25 or 50 mg/L As^{5+}.

Electrophoretic resolution of catalases in cell free extracts of isolates growing in minimal nitrate medium in the phases of hyphal development and spore formation revealed three isozymes of catalases. H_2O_2 (20 or 40 mM) inhibited production the lowest catalase I in all cultures, while As^{5+} (25 or 50 mg/L) catalase I and catalase III as well (Figure 8).

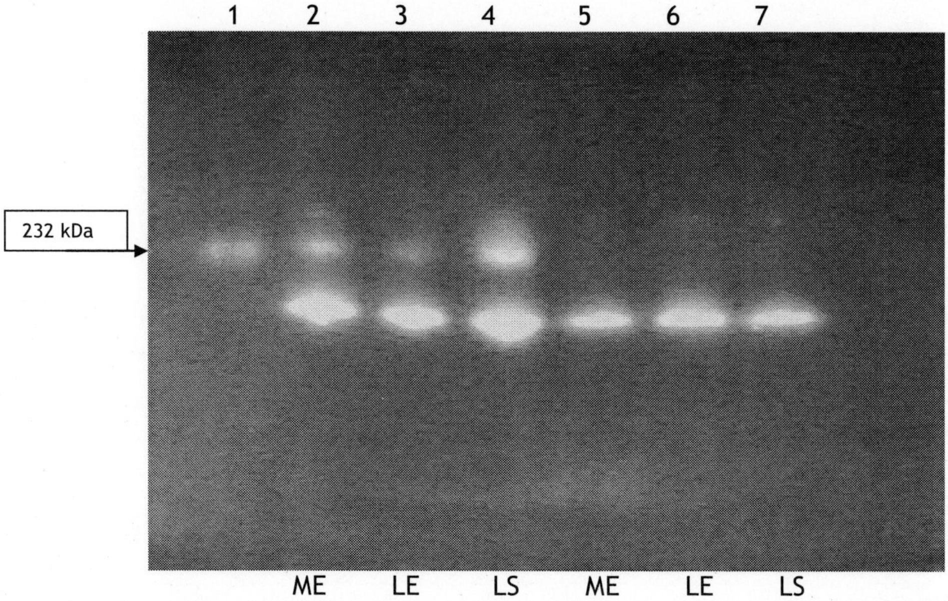

Figure 9. Native polyacrylamide gel electrophoresis (4-18 %) of catalases in cell free extracts of *C. terrigena* N3H (lines 2, 3, 4) and its H_2O_2 - adapted subculture (lines 5, 6, 7). Line 1 - typical catalase from bovine liver (marker). ME – the middle exponential phase, LE -the late exponential phase and LS – the late stationary phase of growth.

Table 1. Catalase activity in cell free extracts from *Comamonas* spp. and sensitivity of the cultures to oxidative stress

Culture	Phase of growth	Catalase activity (mkat/g protein)	Survival after H_2O_2 treatment[a] (%)	
			20 mM	40 mM
C. terrigena N3H	ME	0,9	60	42
isolate	LE		70	58
	LS	6,2	90	83
C. terrigena	ME	0,07	5	3
ATCC 8461	LE	0,1	43	8
	LS	0,2	80	74
C. testosteroni	ME	1,8	15	2
isolate	LE	2,3	76	60
	LS	2,5	89	81
C. testosteroni	ME	0,02	2	2
ATCC 11996	LE	0,03	21	7
	LS	0,08	59	36
C. terrigena N3H	ME	21,7	68	59
H_2O_2 adapted	LE	25	97	94
	LS	37,5	93	90

[a]Expressed as reference to the control (culture without treatment of stressor).

Cultures grown on nutrient medium was exposed to H_2O_2 during one hour. ME – the middle exponential phase, LE – the late exponential phase and LS – the late stationary phase of growth. The measurements were repeated five times and the results represent the mean values.

The Role of Catalase-1 in C. Terrigena N3H

We attempted to identify the role of particular isozyme encoded by *cat-1* gene from *C. terrigena* N3H in the process of adaptation to stress. The H_2O_2 – adapted variant of *C. terrigena* N3H, obtained by subculturings with gradually increasing concentration of oxidative stressor, exhibited a much higher level of catalase activity than the original isolate [34]. Enzyme activity of the transferred culture harvested from the stationary phase was approximately six times higher than in the original isolate (Table 1). Electrophoretic resolution of catalases in crude extracts of the adapted culture during the growth cycle revealed the disappearance of band for larger tetrameric catalase-2 (Figure 9). This suggested that a significant rise in the total hydroperoxidase activity was represented mainly by catalase-1. In addition, H_2O_2-adapted culture exhibited a rise in the peroxidatic activity of the catalases (approximately from 1.4 to 4.3 pcat/g) in comparison to the original isolate, primarily in the 2-electron peroxidations of 1-propanol.

Catalase-1 exhibited significant peroxidase activity towards to phenol, aromatic 1-electron donor substrate. In many prokaryotes phenol is primarily oxidized by phenol hydroxylase, followed by cleavage of the aromatic ring by different catechol dioxygenases. However *C. terrigena* N3H lacks this key enzyme for complete degradation of phenol. We analyzed the possible removal of phenol by the two catalases of this bacterium. Applying oxidative stressor to the culture medium led to increased expression of the catalase-1. Peroxidatic activity and additionally the formation of the secondary product 2, 2'-biphenyl, followed fluorimetricallly were recorded [7]. Our results indicate that this constitutively expressed catalase-1 with smaller molecular weight, its peroxidic activity can also play an important role in the degradation of toxic phenolic compounds.

Diversity of Cat-1 Gene in Isolates of Genus Comamonas

This oxidative enzyme, heme-containing catalase-1, was isolated from exponential phase cells grown in the presence of peracetic acid. The purified sample of catalase exhibited a specific catalase activity of 55 900 U/mg, allosteric behavior in peroxidic reaction, a broad pH optimum, and a rather atypical electronic spectrum. The molecular weight of the subunit of this homodimeric protein was determined as 55 417 Da. N-terminal sequence analysis and internal fragments by the Qq-TOF mass analysis method were obtained. Five such fragments with a total length of 57 amino acids together with several enzymatic properties allowed the classification of this hydroperoxidase as belonging to clade III of monofunctional catalases. The highest sequence similarity is with the catalase from *Vibrio fischeri* [21].

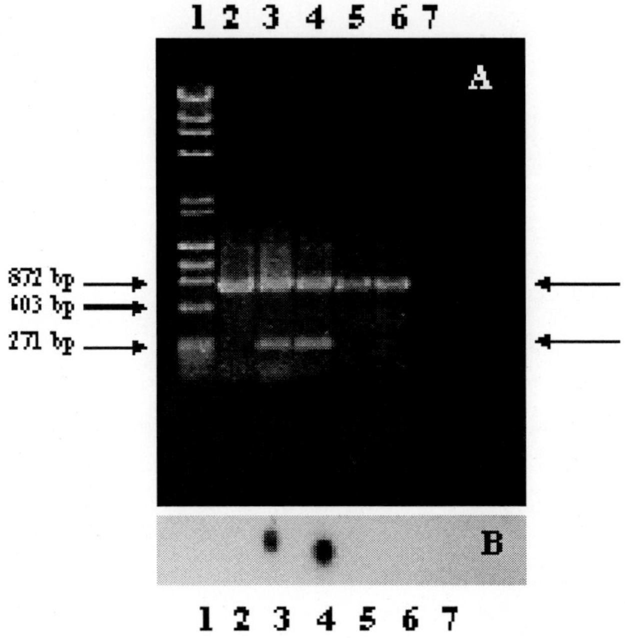

Figure 11. PCR fragments (designated with lower arrow on the right side of the Figure A) obtained by PCR methods with specific primers SCat1 and SCat2 and the hybridization analyses of chromosomal DNAs (B) of the strains *C. terrigena* 2409^T - ATCC 8461 (line 2), *C. terrigena* N3H (line 3), *C. terrigena* N1C (line 4), *C. testosteroni* (line 5) and *C. testosteroni* 1931T – ATCC 11996 (line 6). Line 1 - mix λDNA/ *Hind*III Digest a ϕX174/*Hae*III Digest. Line 7 - *Enterococcus phaecium* as representative of the group catalase-negative microorganisms. The bands (designated with upper arrow on the right side of the Figure A) are the fragments DNA obtained with degenerated primers ComanContrFwd a ComanContr2fo for 16S rRNA that served as inner control of the PCR methods.

```
kat-1    INASNERHWVKSHWVCQQGIENFTDSEAASVVASDRESSQRDLLEAIDRGEFPRWKLCVQ  60
kat-N1C  ----------------------FTDSEAASVVASDRESSQRDLLEAIDRGEFPRWKLCVQ  38
                               **************************************

kat-1    IMPEHEAETYRFHPFDLTKVWSKKDYPLIEVGVMELNRNPRNYHADVEQASFTPANLVPG 120
kat-N1C  IMPEHEAETYRFHPFDLTKVWSKKDYPLIEVGVMELNRNPRNYHADV-------------  85
         ***********************************************

kat-1    IGPSPDRMLQ 130
kat-N1C  ----------
```

Figure 10. Sequence aligment of the amino acid sequence of PCR fragment from *C. terrigena* N1C together with amino acid sequence of catalase-1 from *C. terrigena* N3H. The stars designate an identical amino acid residua.

Consequently, applying sequencing and PCR methods, about 390 bp of *cat-1* gene were obtained [31]. This PCR product was also identified in *C. terrigena* N1C and it was identical with the corresponding sequence of *cat-1* gene (Figure 10, 11A). However, in other strains (not exposed to pollution) this sequence motif could not be found. These results were confirmed by the hybridization analysis, as well (Figure 11B). This fact can indicate a selective pressure of environment on the gene coding for catalase-1. The results suggest that despite the fact that the tested strains belong taxonomically to the same genus or species, they exhibit significant diversity of *cat*-1 gene and respond distinctly to the oxidative stress. They underline the importance of catalase expression in the defense mechanism against oxidative stress in bacterial cells.

The novel-founded internet database PeroxiBase at (http://peroxidase.isb-sib.ch) with over 5000 known peroxidase and catalase sequences was presented in recent papers [35, 36]. It is helpful on the way of gathering information about new attractive practically useful genes and in search for similar genes in contaminated environment. It will allow us to predict how the bacteria can adapt for future stress stimuli in the environment and gives the information on evolutionary potential of the organism.

Tolerance of Microbial Isolates to Toxic Effects

Resistance to H_2O_2 toxicity in *Salmonella* typhimurium can be partially attributed to higher levels of catalase which effectively decreases the H_2O_2 concentration [37]. The psychrophilic bacterium *Vibrio rumoiensis* S-1^T exhibited an extraordinarily high catalase activity that allowed the cells to acquire resistance against two kind of stress; e.g. cold and oxidative environment [10]. It was observed that cells of the bacterium *Pseudomonas putida* in liquid suspension were more sensitive to killing by peracetic acid when they lacked a major isozyme, catalase A [38]. On the other hand, results of Murphy et al. indicated that the protection of *Bacillus subtilis* JH642 cells against killing by lethal levels of H_2O_2 was not mediated by the induction of catalase or superoxide dismutase [39]. Similarly, analysis of enzymes involved in oxidative stress protection in the oxyR2 (Xp) mutant of Xanthomonas *campestris* revealed no significant increase in the levels of catalase and superoxid dismutase [40]. The resistance H_2O_2 toxicity in *B. subtilis* JH642 cells can be attributed to induction of stress proteins [39].

C. terrigena N3H isolated from crude oil contaminated soil and *C. testosteroni* selected from wastewater treatment sludge, exhibited much higher levels of total catalase activity than the same species from the collection cultures. In standard nutrient MPB medium were tested for their sensitivity to exogenously added 20 and 40 mM H_2O_2. The environmental isolates were more resistant against these doses of stressor than collection cultures mainly at the middle or the late exponential phases. The rise of low levels of catalase activity in the collection cultures during the growth cycle was negligible; in spite of this the cultures displayed a relatively high resistance against H_2O_2, added in the stationary phase. The results demonstrate the significance of catalases activity in the resistance to H_2O_2 toxicity mainly in the earlier phases of the growth cycle. In the late stationary phase can be partially attributed to induction of stress proteins, as well (Table 1) [39].

The H_2O_2 – adapted variant of *C. terrigena* N3H, obtained by subculturings with gradually increasing concentration of oxidative stressor, exhibited a much higher level of catalase activity than the original strain. The high level of catalase in H_2O_2-adapted variant was reflected in an increase in the resistance to oxidative stress. Survival of this culture after treatment with 20 and 40 mM H_2O_2 was already at the middle exponential, but more significantly at the late exponential phase the survival rate was higher in comparison to the original isolate (Table 1) [34].

Tolerance of *A. niger* cultures to toxic effect of H_2O_2 and heavy metals was monitored by growth on liquid Czapek-Dox medium, as well [28]. Isolates from river sediment and coal dust were more tolerant to As^{5+}, Cd^{2+}, Cu^{2+} mainly at higher concentrations 25 or 50 mg/L or 20 and 40 mM H_2O_2 in comparison to the control culture. The decrease of growth by the effect of agents at these concentrations was smaller than in the remaining cultures. No strain was tolerant to toxic effect of mercury, even at the lowest tested dose (Figure 12).

Protection Experiments

Altered environmental stress conditions often lead to a high tolerant microbial community. Stress responses are characterized by the adaptation of cells to very toxic or normally lethal level of stressor by pretreatment with a low, nonlethal dose. Hydrogen peroxide in concentrations of 20 and 40 mM significantly reduced the survival of *C. terrigena* N3H cells in the middle exponential phase of growth cycle. This culture was pretreated with 0.5 mM H_2O_2 for 1 h, followed by treatment with 20 and 40 mM for a second hour. This pretreatment reduced the degree of cell mortality in the higher concentrations of oxidative agent. Different results were obtained when an H_2O_2 –adapted culture was used. Pretreatment of these bacteria with a low dose of H_2O_2 did not significantly influence the rate of cell survival (Table 2).

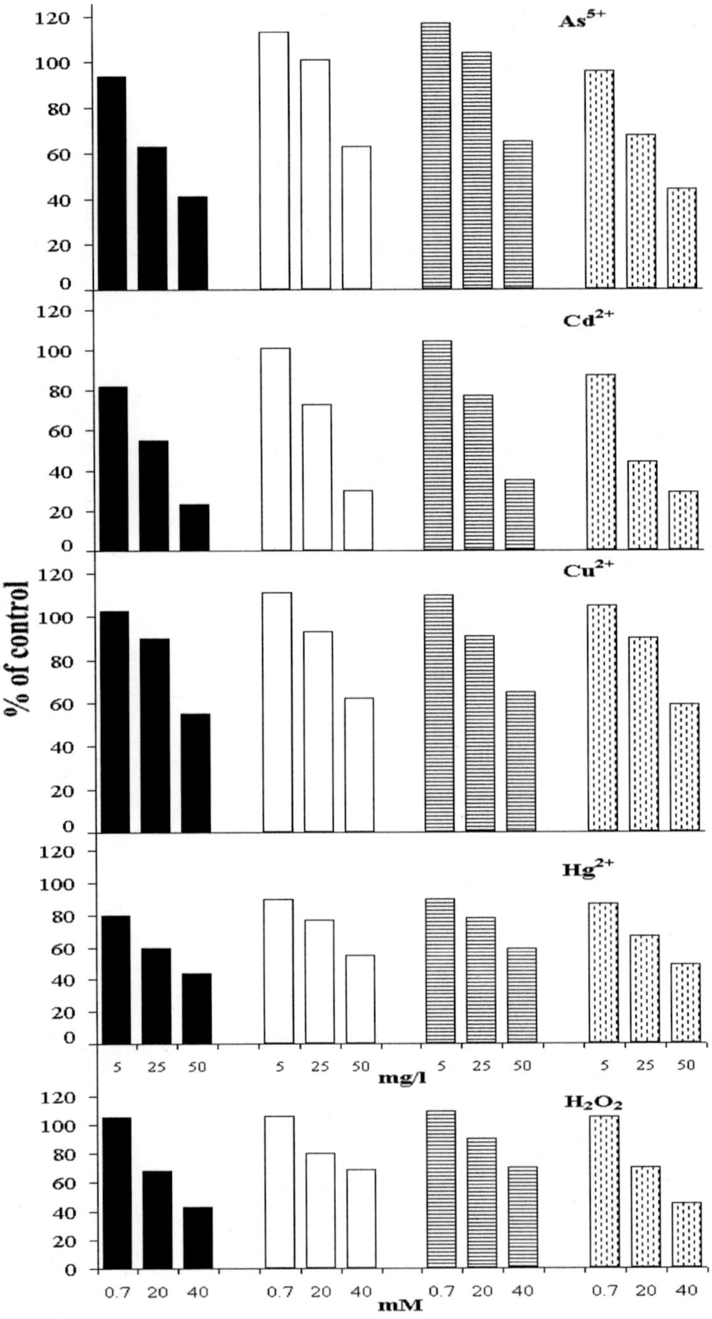

Figure 12. The relative growth of *A. niger* cultures on liqiud Czapek-Dox medium at presence heavy metals after 10 days indicated as dry biomass. Concentrations of agents: 5, 25 or 50 mg/L of metal and 0.7, 20, 40 mM H_2O_2. The results are expressed as reference to the control (dry biomass of individual culture without treatment of stressor). Black columns - the collection strain of *A. niger*, white columns - isolate selected from river sediment As (400 mg/kg); horizontally hatched columns - isolate from coal dust near mine (As 1651 mg/kg, Sb 362 mg/kg); vertically hatched columns - the adapted culture to Sabouraud medium contaminated with As^{5+} (5 mg/kg).

Protection of these bacteria by the pretreatment corresponded with the stimulation of catalase activity in the pretreated culture. It was observed only in the original isolate, while the level of enzyme activity in the adapted culture was not affected (Table 3).

The protection effect of pretreatment has also been reported in other microorganisms. The exposure of *Vibrio harveyi* cells to 250 μM H_2O_2 induced an adaptive protection against a subsequent exposure to a lethal concentrations of hydrogen peroxide [41]. The induced protection against H_2O_2 of *Salmonella typhimurium* was proportional to the amount of induced catalase activity, which accelerated the breakdown of H_2O_2 [37]. Dowds suggested that there is a number of stress genes, which are expressed when the microbial cells are exposed to protective concentrations of H_2O_2 [42]. The products of these genes, including scavenging enzymes (catalase and alkyl-hydroperoxide reductase) protect the cells against the subsequent lethal effects of high concentrations of H_2O_2.

Environmental Factors and Microbial Response

The adaptive changes in microbial communities, their metabolic activities, morphological responses and biomass production, lead them to survive and function under stress. Environmental stress besides the pollutant involves also the other stresses (low or high pH, starvation, osmotic stress, etc.) [43].

Isolates of *Aspergillus niger*, selected from coal dust sediment of a mine containing As (400 mg/kg), pH 3.3-2.8, and from river sediment found near the mine (As, 363 mg/kg, Sb, 93 mg/kg), pH 5.2-4.8, growing on Czapek-Dox agar responded to stress differently [44]. The radial mycelial growth of the isolate from the coal dust in comparison to the control strain from an environment without pollution was reduced approximately to one-half. It formed black, very small compact colonies, with dense sporulation. The opposite, the strain from the river sediment, grew better in Czapek-Dox agar like the control. It formed larger colonies with dense centre and strong sporulation. Also, the culture from river sediment developed faster than the coal dust isolate and control strain from an environment without pollution (Table 4). Differences were also recorded in size and thickness of conidia heads, phialide, metulae, and conidiophores. They correlated with values of the colony growth (Figure 13). These two localities exhibited different pH values. Acidity of the coal dust ranged from 3.3 to 2.8 was higher than that in the river sediment (5.2 - 4.8). Although fungi are known to tolerate more acid conditions than the majority of soil bacteria, in general, the alteration of the soil or sediment pH may affect the solubility, bioavailability of soil macro- and micronutrients and the chemical form of the pollutants [45]. An increase in soil pH tends to decrease the availability of calcium, magnesium, sodium, potassium, ammonia, nitrogen, and phosphorus, whereas a decrease in soil pH results in decreasing availability of nitrate and chloride [46]. This suggests that different pH values of the environments may affect the solubility and bioavailability of As compounds.

Stimulation of fungal growth, which was observed in isolate selected from the river sediment, is very often associated with lower level of metals. Finally, we confirmed the stimulation effect of low concentration of As^{5+}, Cd^{2+} and Cu^{2+} (5 mg/L) on the growth of isolates from surroundings of mines in laboratory conditions (Figure 12).

Babich and Stotzky found, that at concentrations required for the stimulation of growth, both cadmium and zinc may be inhibitors. Sensitivity of fungi to metals is affected by pH of soil, as well [47].

In spite of this, that radial growth of the strain from coal dust in laboratory conditions was weaker than of control isolate, it exhibited higher tolerance against exogenic toxic effects of metals monitored by the growth of biomass [28]. The results support the opinion that tolerance is dependent upon the prevailing environmental conditions [48].

Table 2. The effect of pretreatment of *Comamonas terigena* N3H and its H_2O_2-adapted subculture cells with 0.5 mM H_2O_2 on the survival of cells after the following H_2O_2 treatments. Cultures were used from the midle exponential phase of growth

Culture	Pretreatment[a]	H_2O_2 (mM)	Survival (%)
C. terrigena N3H	-	20	60
	+	20	81
	-	40	42
	+	40	61
C. terrigena N3H	-	20	68
H_2O_2 adapted	+	20	72
	-	40	59
	+	40	63

[a] Pretreatment with 0.5 mM H_2O_2 for 1h, followed by treatment for another hour.

Table 3. Activity of total catalase in cells of *Comamonas terrigena* N3H and its H_2O_2-adapted variant after H_2O_2 treatment during one hour

Culture	H_2O_2 (mM)				
	0,2	0,7	5	20	40
C. terrigena N3H	280	340	210	190	180
C. terrigena N3H H_2O_2 - adapted subculture	100	105	102	100	95

Values are expressed as % of catalase activity of untreated cells.

Table 4. Macroscopic characteristics of *A. niger* isolates and *A. tubingensis* collection culture examined after 6 days of growth on Czapek-Dox agar at 25°C

STRAINS	MACROSCOPIC CHARACTERISTICS	
	COLONY DIAMETER (MM)	COLONY COLOUR
A. NIGER (SOIL	33	DARK BROWN WITHOUT POLLUTION)
A. NIGER (COAL DUST AS 400 MG/KG)	19	DARK BLACK
A. NIGER (RIVER SEDIMENT AS 363, SB 93 MG/KG)	48	BLACK
A. TUBINGENSIS	28	DARK BROWN - BLACK (COLLECTION CULTURE)

The measurements were repeated six times and the results represent the mean values.

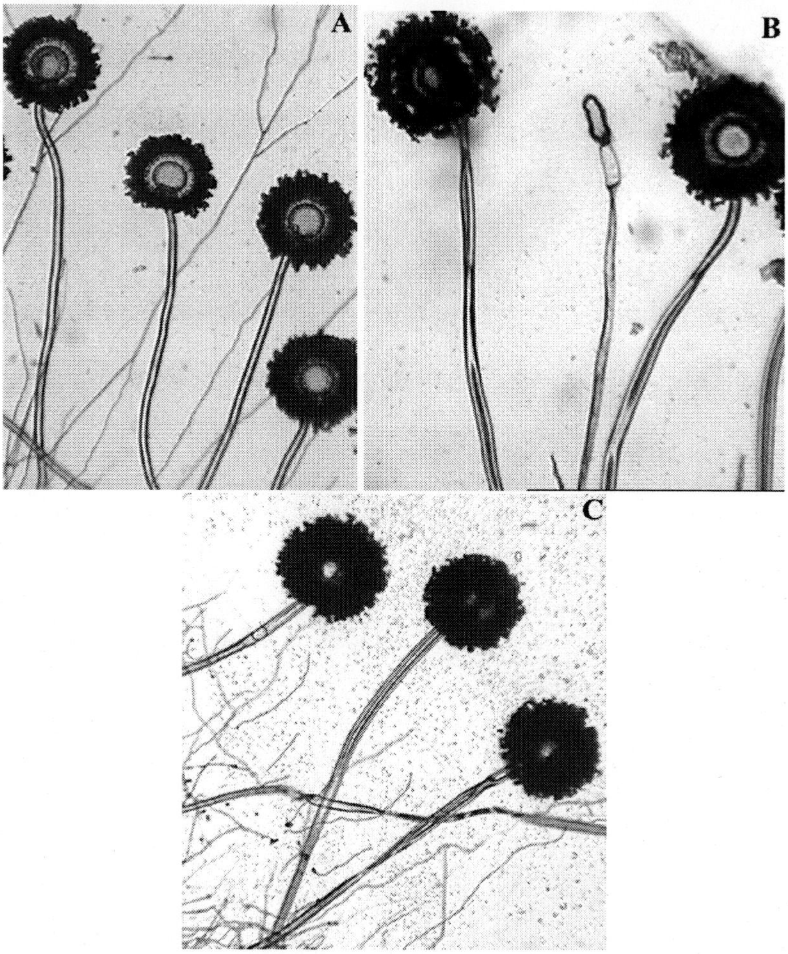

Figure 13. The mycelial morphology of *A. niger* (Tiegh.) isolates as a responses to polluted environments (x 750). A. *A. niger* – isolate from soil without pollution (control sample) B. *A. niger* – selected from the coal dust (As 400 mg/kg) C. *A. niger* - selected from river sediment (As 363 mg/kg, Sb 93 mg/kg).

Nutrients of environment can affect the growth of microorganisms and degradation of contaminants as well. Naphthalene degrading mixed culture (*Pseudomonas putida*, *Klebsiella planticola*, *Xanthomonas maltophilia*) was isolated from a soil contaminated by crude oil [49]. The biodegradation properties of microbial consortium were tested in six media differing in components and concentrations of important nutrients at two different levels of naphthalene (0.1 and 3.0 g/L). The best growth of microbial consortium was observed in Tris-phosphate medium, and then follows acetate- and naphthalene- media in the presence of naphthalene in concentration 0.1 g/L (Tables 5, 6). The naphthalene medium was the most suitable for growth of the mixture culture when cultivation media contained 3.0 g/L. The results confirmed direct relation between biodegradation rate and the amount of total nitrogen in media only at a concentration of 0.1 g/L of naphthalene. They did not confirm direct relation between the amounts of important nutrient elements (P, S), and the growth of the microbial consortium. Adding of nutrients in soil stimulate growth of all microorganisms present and does not select for those with special degradative abilities with complementary metabolic diversity. This suggests that in order to develop an efficient bioremediation, the laboratory tests with stable degraders must be performed in concentration range of contaminant.

The decay of the whole chemical compound of dioctyl sulpho[2,3-^{14}C]succinate required the specific nutrient conditions. The primary degradation of dioctyl sulphosuccinate with *C. terrigena* N3H, the cleavage of ester bonds between octyl groups and sulphosuccinate, lasted significantly shorter time than the subsequent breakdown of the sulphosuccinate moiety of dioctyl sulpho[2,3-^{14}C]succinate. The cleavage of the strongly resistance sulphonate bond, which indicates $^{14}CO_2$ evolution, had a significant shorter lag period with cells in Tris-phosphate medium without inorganic sulphate. The acceleration of the primary degradation by adaptated cells also suggests that some enzymes involved in surfactant degradation are inducible [18].

Table 5. Nutrient composition of the media. BH- Bushnell-Haas medium (g/L), Acetate medium (g/L), MN- mineral nutrients medium (mg/L), TP- Tris phosphate medium (g/L), P- phosphate medium (g/L)

Medium	BH	MN	P	M_{naph}	TP	Acetate
NO_3^- (mM)	12.5	28.6	19.8	50.0	-	-
NH_4^+ (mM)	12.5	-	22.7	50.0	101.0	101.0
Tris $C_4H_{11}O_3N$ (mM)	-	-	-	-	49.5	49.5
PO_4^{3-} (mM)	13.0	-	154.1	9.4	2.9	2.9
$P_3O_{10}^{5-}$ (mM)	-	0.8	-	-	-	-
SO_4^{2-} (mM)	0.8	0.201	22.7	2.0	0.8	1.6
Cl^- (mM)	0.3	0.37	-	1.6	101.0	101.0
CO_3^{2-} (mM)	-	-	-	20.0	-	-
CH_3COO^- (mM)	-	-	-	-	-	121.9
Mg^{2+} (mM)	0.8	0.2	-	2.0	0.8	1.6
Ca^{2+} (mM)	0.1	0.37	-	20.3	-	-
Na^+ (mM)	-	0.8	18.9	-	-	121.9
K^+ (mM)	13.0	28.6	155.0	11.4	2.9	2.9
Yeast extract	-	-	+	+	+	-
Elements (micro-amounts)	+	+	-	+	-	-

Table 6. Growth of microbial consortium (*Pseudomonas putida*, *Klebsiella planticola*, *Xanthomonas maltophilia*) in the stacionary phase in relation to concentration of naphthalene in the individual media[a]

Naphthalene g/L	M_{naph}	BH	TP	MN	Acetate	P
0.1	0.64	0.09	1.64	0.08	1.02	0.14
3.0	2.72	1.65	1.12	1.18	0.75	0.89

[a] Growth was monitored by measuring the optical density of cell suspension at 650 nm. M_{naph} – naphthalene medium, BH - Bushnell-Haas medium, TP - Tris phosphate medium, MN - mineral nutrients medium, P – phosphate medium.

Phenol and substituted phenols are widely used organic compounds and ubiquitous contaminants. *Alcaligenes* spp. belong to bacterial degraders of these pollutants. The highest growth of mixed cultures (*Alcaligenes xylosoxidans* and *Klebsiella pneumoniae*) was observed in phenol concentration 0.5 g/L. Unlike to consortium, *A. xylosoxidans* utilized phenol as a sole source of carbon only very weakly. The growth of *K. pneumoniae* was not observed in concentration range 0.15 – 0.5 g/L. The results indicated that biodegradation of phenol substances demands complementary diversity and special properties of individual cultures in consortium [50].

The taxonomy of black aspergilli (*Aspergillus* section *Nigri*) is not clear because it is primarily based on morphological criteria and in some cases the differences between the species are very subtle. Obvious abnormalities observed in the morphological features of the microorganisms in presence of high levels of pollution very often complicate identification of species (Figure 13A, B, C) [51]. The PCR method which differentiates two strains of *A. niger* and *A. tubingensis*, major species in the *A. niger* aggregate described by Accensi et al. was used [52]. The method uses small differences in chromosomal DNA for 5.8S rRNA of two strains of *A. niger* and *A. tubingensis*, major species in the *A. niger* aggregate. The PCR products of *A. niger* isolates selected from coal dust sediment of a mine containing As (400 mg/kg and from river sediment found near the mine (As, 363 mg/kg, Sb, 93 mg/kg) were digested into two fragments of 519 and 76 base pairs, while the PCR amplicon of *Aspergillus tubingensis* CCF 2818 remained as undigested fragment of 595 base pairs. DNA sequence of the PCR products obtained with *A. niger* isolates confirmed the presence of the recognition site for endonuclease RsaI, respectively taxonomy of *A. niger* (Figure 14).

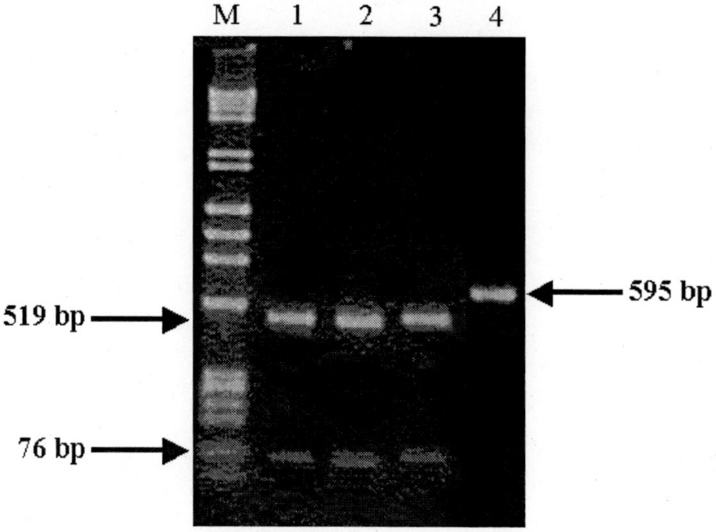

Figure 14. ITS1-5.8S rDNA gene-ITS2 PCR products cleaved by RsaI and separated on a 2% agarose gel. Lane M, molecular marker 100 bp DNA ladder (BioLabs, New England); lane 1, *A. niger* – isolate from the soil (without pollution); lane 2, *A. niger* - selected from the coal dust (As, 400 mg/kg); lane 3, *A. niger* - selected from river sediment (As, 363 mg/kg, Sb, 93 mg/kg); lane 4, A. *tubingensis* – the collection culture.

CONCLUSION

Pollutants in soils are subjected to oxidation processes, which lead to the increasing of toxicity of environment, resulting as the source of potential hazard for terrestrial ecosystems. Regulation of catalases activity, including alteration in expression of isozymes, may be important in the microorganisms surviving in such environment. Pollution affects the diversity of catalases genes that also yields information about the evolutionary potential of the microorganisms. The new, unusual forms of enzymes with convenient the physical and chemical properties (high temperature stability, wide pH optimum, etc.) make them potentially useful in biotechnology. The responses of microbial isolates to pollutant stress refer to complicated relations between biotic and abiotic factors that may directly affect the processes *in situ* environment.

REFERENCES

[1] Stohs, S. J. & Bagchi, D. (1995). Oxidative mechanisms in the toxicity of metal ions. *Free Radical Biol. Med.,* 18, 321-336.

[2] Kappus, H. (1987). Oxidative stress in chemical toxicity. *Arch. of Toxicol.,* 60, 144-149.

[3] Gille, G. & Sigler, K. (1995). Oxidative stress and living cells. *Folia Microbiol.,* 40, 131-152.

[4] Halliwell, B., & Gutteridge M. C. (1999). *Free radicals in biology and medicine.* 2nd edn.Clarendon press, Oxford University Press.

[5] Loprasert, S., Vattanaviboon, P., Praituan, W., Chamnongpol, S. & Mongkolsuk, S. (1996). Regulation of the oxidative stress protective enzymes, catalase and superoxide dismutase in *Xanthomonas* – a review. *Gene,* 179, 33-37.

[6] Xu, F. (2005). Applications of oxidoreductases: Recent progress. *Industrial Biotechnol.,* 1, 38 -50.

[7] Zámocký, M., Godočíková, J., Koller, F. & Polek, B. (2001). Potential application of catalase-peroxidase from *Comamonas terrigena* N3H in the biodegradation of phenolic compounds. *Antonie van Leeuwenhoek,* 79, 109-117.

[8] Zámocký, M. & Dunand, C. (2006). Divergent evolutionary lines of fungal cytochrome c peroxidases belonging to the superfamily of bacterial, fungal and plant heme peroxidases. *FEBS Lett.,* 580, 6655-6664.

[9] Wang, R.F., Wennerstrom, D., Cao, W.W., Khan, A.A. & Cerniglia, C. F. (2000). Cloning, expression, and characterization of the katG gene, encoding catalase-peroxidase, from the polycyclic aromatic hydrocarbon-degrading bacterium *Mycobacterium* sp. strain PYR-1. *Appl. Environ. Microbiol.,* 66, 4300-4304.

[10] Yumoto, I., Iwata, H., Sawabe, T., Veno, K., Ichise, N., Matsujama, H., Okuyama, H. & Kawasaki, K. (1999). Characterization of facultatively psychrophilic bacterium *Vibrio rumoiensis* sp. nov., that exhibits high catalase activity. *Appl. Environ. Microbiol.,* 65, 67-72.

[11] Yeates, Ch., Holmes, A. J. & Gillings, M. R. (2000). Novel forms of ring-hydroxylating dioxygenases are widespread in pristine and contaminated soils. *Environ. Microbiol.,* 2, 644-653.

[12] Brunder, W., Schmidt, H. & Karch, H. (1996). KatP, a novel catalase –peroxidase encoded by the large plasmid of enterohaemorrhagic *Escherichia coli* O157:H7. *Microbiology,* 142, 3305-3315.

[13] Bandyopadhyay, P. & Steinman, H. (2000). Catalase-peroxidases of *Legionella pneumophila*: cloning of the *kat*A gene and studies of KatA function. *J. Bacteriol.,* 182, 6679-6686.

[14] Rafii, F., Lunsford, P., Hehman, G. & Cerniglia, C. E. (1999). Detection and purification of a catalase-peroxidase from *Mycobacterium* sp. Pyr-1. *FEMS Microbiol. Lett.,* 173, 285-290.

[15] Levy, E., Eyal, Z. & Hochman A. (1992). Purification and characterization of a catalase-peroxidase from the fungus *Septoria tritici. Arch. Biochem. Biophys.,* 296, 321-327.

[16] Zámocký, M. (1998). Spectroscopic and enzymic features of yeast peroxisomal catalase. *Chem. Papers,* 92, 875-882.

[17] Bae, H. S., Lee, J. M., Kim, Y. B. & Lee, S. T. (1996). Biodegradation of the mixture of 4-chlorophenol and phenol by *Comamonas testosteroni* CPW301. *Biodegradation,* 7, 463-469.

[18] Godočíková, J., Ferianc, P. & Polek, B. (2004). Lag period of $^{14}CO_2$ evolution from dioctyl sulpho[2,3-^{14}C]succinate in relation to adaptation of bacterium, *Comamonas terrigena,* to dialkyl esters of sulphosuccinate. *Biotechnol. Lett.,* 26, 1497-1500.

[19] Godočíková, J., Polek, B., White, G. F., Ferianc, P. & Toth, D. (1997). Dioctyl sulphosuccinate utilization and mineralization by bacteria *Comamonas terrigena* N3H. *Biologia,* 52, 741-745.

[20] Prokšová, M., Augustín, J. & Vrbanová, A. (1997). Enrichment, isolation and characterization of dialkyl sulfosuccinate degrading bacteria *Comamonas terrigena* N3H and *Comamonas terrigena* N1C. *Folia Microbiol.,* 42, 635-639.

[21] Zámocký, M., Godočíková, J., Gašperík, J., Koller, F. & Polek, B. (2004). Expression, purification, and sequence analysis of catalase-1 from the soil bacterium *Comamonas terrigena* N3H. *Protein Expression Purif.,* 36, 115-123.

[22] Schellhorn, H. E. (1995). Regulation of hydroperoxidase (catalase) expression in *Escherichia coli. FEMS Microbiol. Lett.,* 131, 113-119.

[23] Loewen, P. C., Svitala, J. & Triggs-Rainl, B. L. (1985). Catalase HPI and HPII in *Escherichia coli* are induced independently. *Arch. Biochem. Biophys.,* 243, 144-149.

[24] Jenkins, D. E., Schultz, J. E. & Matin, A. (1988). Starvation-induced cross-protection against heat and H_2O_2 challenge in *Escherichia coli. J. Bacteriol.,* 170, 3910-3914.

[25] Odenbreit, S., Wieland, B. & Haas, R. (1996). Cloning and genetic characterization of *Helicobacter pylori* catalase and construction of a catalase-deficient mutant strain. *J. Bacteriol.,* 178, 6960-6967.

[26] Bishai, W. R., Howard, N. S., Winkelstein, J. A. & Smith, H. O. (1994). Characterization and virulence analysis of catalase mutants of *Haemophilus influenzae. Infect Immun.,* 62, 4855-4860.

[27] Sha, Z., Stabel, T. J. & Mayfield, J. E. (1994). *Brucella abortus* catalase is a periplasmic protein lacking a standard signal sequence. *J. Bacteriol.,* 176, 7375-7377.

[28] Bučková, M., Godočíková, J., Šimonovičová, A. & Polek, B. (2005). Production of catalases by *Aspergillus niger* isolates as a response to pollutant stress by heavy metals. *Curr. Microbiol.,* 50, 175-179.

[29] Kawasaki, L. & Aquirre, J. (2001). Multiple catalase genes are differentially regulated in *Aspergillus nidulans. J. Bacteriol.,* 183, 1434-1440.

[30] Witteveen, C. F. B., Veenhuis, M. & Visser, J. (1992). Localization of glucose and catalase activities in *Aspergillus niger. Appl. Environ. Microbiol.,* 58, 1190-1194.

[31] Boháčová, V., Zámocký, M., Godočíková, J., Bučková, M. & Polek, B. (2006). The expression and diversity of catalases in isolates of genus *Comamonas* in response to the oxidative stress of a polluted environment. *Current Microbiol.,* 53, 430-434.

[32] Zámocký, M., Polek, B., Godočíková, J. & Koller, F. (2002). Oxidative stress-induced expression of catalases in *Comamonas terrigena. Folia Microbiol.*, 47, 235-240.

[33] Kwon, S.I. & Anderson, A.J. (2001). Catalase activities of *Phanerochaete chrysosporium* are not coordinately produced with ligninolytic metabolism. Catalase from a white-rot fungus. *Curr. Microbiol.,* 42, 8-11.

[34] Godočíková, J., Boháčová, V., Zámocký, M. & Polek, B. (2005). Production of catalases by *Comamonas* spp. and resistance to oxidative stress. *Folia Microbiol.,* 50, 113-118.

[35] Passardi, F., Zamocky, M., Favet, J., Jakopitsch, C., Penel, C., Obinger, C. & Dunand, C. (2007). Phylogenetic distribution of catalase-peroxidases: Are there patches of order in chaos? *Gene,* 397, 101-113.

[36] Passardi, F., Theiler, G., Zámocký, M., Cosio, C., Rouhier, N., Teixera, F., Margis-Pinheiro, M., Ioannidis, V., Penel, C. & Falquet, L. (2007). PeroxiBase: The peroxidase database. *Phytochemistry,* 68, 1605-1611.

[37] Winquist, L., Rannung, U., Rannung, A. & Ramel, C. (1984). Protection from toxic and mutagenic effects of hydrogen peroxide by catalase induction in *Salmonella typhimurium. Mutat. Res.,* 141, 145-147.

[38] Anderson, A. J. & Miller, C. D. (2001). Catalase activity and the survival of P*seudomonas putida,* a root colonizer, upon treatment with peracetic acid. *Can. J. Microbiol.,* 47, 222-228.

[39] Murphy, P., Dowds, B. C. A., McConnel, D. J. & Devine, K. M. (1987). Oxidative stress and growth temperature in *Bacillus subtilis. J. Bacteriol.,* 169, 5766-5770.

[40] Vattanaviboon, P., Whangsuk, W. & Mongkolsuk, S. (2003). A suppressor of the menadione-hypersensitive phenotype of a *Xanthomonas campestris* pv. Phaseoli oxyR mutant reveals a novel mechanism of toxicity and the protective role of alkyl hydroperoxide reductase. *J. Bacteriol.,* 185, 1743-1748.

[41] Vattanaviboon, P. & Mongkolsuk, S. (2001). Unusual adaptive, cross protection responses and growth phase resistance against peroxide killing in a bacterial shrimp pathogen, *Vibrio harveyi. FEMS Microbiol. Lett.,* 200, 111-116.

[42] Dowds, B.C. (1994). The oxidative stress response in *Bacillus subtilis. FEMS Microbiol. Lett.,* 124, 255-263.

[43] Fiedurek, J. (2000). Production of *Aspergillus niger* catalase under various stress condition. *Acta Microbiol. Pol.,* 1, 43-49.

[44] Bučková, M., Godočíková, J. & Polek, B. (2007). Responses in the mycelial growth of *Aspergillus niger* isolates to arsenic contaminated environments and their resistance to exogenic metal stress. *J Basic Microbiol.,* 47, 295-300.

[45] Baker, K. H. & Herson, D. S. (1994). Microbiology and biodegradation. In: Bioremediation (KH Baker and DS Herson Ed.), pp. 9-60, McGraw-Hill, New York.

[46] Sharpley, AN. (1991). Effect of soil pH on cation and anion solubility. *Commun. Soil. Sci. Plant. Anal.,* 22, 1391-1399.

[47] Babich, H. & Stotzky, G. (1977). Sensitivity of various bacteria, including actinomycetes, and fungi to cadmium and the influence of pH on sensitivity. *Appl. Environ. Microbiol.,* 33, 681-695.

[48] Cánovas, D., Durán, C., Rodríguez, N., Amils, R. & de Lorenzo, V. (2003). Testing the limits of biological tolerance to arsenic in a fungus isolated from the River Tinto. *Environ. Microbiol.* 5, 133–138.

[49] Polek, B., Dragúň, M., Godočíková, J. & Ferianc, P. (2002). Naphthalene biodegradation by microbial consortium as a contribution to bioremediation treatments. *Biologia,* 57, 305-311.

[50] Boháčová, V., Godočíková, J. & Polek, B. (2001). Biodegradation of phenol substances by pure and mixed culture. *Biologia,* 56, 263-269.

[51] Abarca, M. L., Accensi, F., Cano, J. & Cabanes, F. J. (2004). Taxonomy and significance of black aspergilli. *Antonie van Leeuwenhoek,* 86, 33-49.

[52] Acensi, F., Cano, J., Figuera, L., Abarca, M. L. & Cabanes, F. J. (1999). New PCR method to differentiate species in the *Aspergillus niger. FEMS Microbiol. Lett.,* 180, 191-196.

In: Soil Contamination Research Trends
Editor: Javier B. Dominguez, pp. 143-160
ISBN 978-1-60456-319-1

Chapter 6

CONSTRUCTION OF AN EFFICIENT BIOREMEDIATION SYSTEM FOR PETROLEUM HYDROCARBON-CONTAMINATED SOILS USING SPECIFIC HYDROCARBON-DEGRADING BACTERIA AND BACTERIAL BIOMASS MONITORING

Kouta Hatayama, Yuri Sakihama, Yoshiki Matsumiya and Motoki Kubo*

Department of Bioscience and Technology, Faculty of Science and Engineering, Ritsumeikan University, 1-1-1 Nojihigashi, Kusatsu, Shiga 525-8577, Japan

ABSTRACT

Petroleum hydrocarbons are one of the major energy sources used by humans. Environmental contamination by petroleum hydrocarbons has, however, become a serious problem all over the world, and laws about contaminated soils are enforced in the United States, Germany, the Netherlands and Japan. Incineration and bioremediation systems for the purification of soils contaminated by petroleum hydrocarbons have been investigated and constructed. However, the incineration systems require large-scale equipment and petroleum, and the bioremediation systems need a long time for purification.

To construct an efficient bioremediation system, microorganisms having the ability to degrade long-chain hydrocarbons, which are difficult to degrade in nature, were isolated and characterized (36 isolates belonging to six genera plus one family). The isolates belonging to the genera *Acinetobacter*, *Rhodococcus*, and *Gordonia* degraded long-chain normal alkanes (*n*-alkanes) efficiently. Long-chain cyclic alkanes (*c*-alkanes)

* Tel/Fax: +81-77-561-3901; E-mail: kubo@se.ritsumei.ac.jp

were degraded by the isolates from the genera *Rhodococcus* and *Gordonia* without co-oxidation, but the degradation pathway of *c*-alkanes differed between these two genera.

The bacterial biomass in soil influences biodegradation. To enhance the efficiency of the biodegradation, a bacterial biomass measurement method using environmental DNA (eDNA) was constructed. Moreover, a detection method for the hydrocarbon-degrading genus *Rhodococcus* using alkane hydroxylase gene (*alkB*) was developed. The biodegradation of hydrocarbons in soil was accelerated by controlling and maintaining the hydrocarbon-degrading microorganisms using these methods.

Bioremediation of soil historically contaminated by petroleum hydrocarbons (total petroleum hydrocarbons (TPH): about 4,000 mg/kg-soil, main component: long-chain *c*-alkanes) was carried out. *Rhodococcus* sp. NDKK6 was inoculated into the soil (about 5×10^8 cells/g-soil), and the bacterial biomass was maintained above 1×10^8 cells/g-soil by measurement of eDNA. Moreover, *Gordonia* sp. NDKY76A, having a different *c*-alkane degradation pathway from *Rhodococcus* sp. NDKK6, was added to the soil to enhance the bioremediation. Long-chain *c*-alkanes were efficiently degraded, and about 66% of TPH was removed after 35 days.

To establish a large-scale bioremediation system, a 500 kg scale mixing machine (500 kg soil per batch) was developed. The machine homogenously mixed microorganisms and nutrients into soil (range of bacterial numbers in the mixed soil: 2.44 to 2.80×10^8 CFU/g-soil) in a short time compared with a backhoe method (range of bacterial numbers in the mixed soil: 0.22 to 1.84×10^8 CFU/g-soil). In a large-scale bioremediation project (about 2.5 ton, TPH: about 22,000 mg/kg-soil, main component: aromatic hydrocarbons), enhancement of biodegradation (about 20%) by using this bioremediation system was confirmed.

INTRODUCTION

Petroleum hydrocarbons are one of the major energy sources used by humans. However, environmental pollution by petroleum hydrocarbons is a serious problem all around the world. Oily odors and oil slicks from petroleum hydrocarbon-contaminated environments have adverse effects on both ecosystems and human health.

Laws about the regulation and treatment of petroleum hydrocarbon-contaminated soils are enforced in several countries. The United States has enacted the Hazardous and Solid Waste Amendments (1984) and the Superfund Amendments and Reauthorization Act (1986). The Netherlands and Germany instituted similar laws in 1994 and 1998, respectively. In Japan, the Soil Contamination Countermeasure Law was enforced in 2003, and guidelines for petroleum hydrocarbon contamination were announced in 2006. In the near future, regulation of hydrocarbon pollution will be enacted in even more countries.

Large-scale incineration plants have been developed, and incineration of petroleum hydrocarbon-contaminated soil is carried out to clean up hydrocarbon-contaminated sites. The treatment time is short, but the system requires huge machines and large amounts of heavy oils [1]. Biological treatments for hydrocarbon degradation have also been investigated and developed. Many kinds of hydrocarbon-degrading bacteria have been isolated and characterized for the construction of bioremediation systems [2-7]. Moreover, to enhance the bioremediation efficiency, landfarming and biopile systems have been developed [8-10].

Petroleum hydrocarbons include various types of hydrocarbons, such as normal alkanes (*n*-alkanes), cyclic alkanes (*c*-alkanes) and poly-aromatic hydrocarbons (PAHs), and these different types of hydrocarbons have different microorganism degradation efficiencies [11-14]. Various *n*-alkane degrading microorganisms have been isolated in previous studies, and the metabolic pathways and related enzymes have been investigated in *Pseudomonas putida* GPO1, *Acinetobacter* sp. ADP1, *Rhodococcus* sp. Q15 and *Alcanivorax borkumensis* AP1 [15-18]. In the case of PAH degrading microorganisms, mainly the genera *Pseudomonas*, *Burkholderia*, *Nocardia*, *Rhodococcus*, *Mycobacterium*, and *Sphingomonas* have been isolated from nature [19-21]. Several microorganisms for resin and asphaltene degradation have been isolated, but the degradation efficiencies are very low [22, 23].

Several factors are important for the enhancement of bioremediation. Isolation of high efficiency and/or specific hydrocarbon(s) degrading microorganisms and inoculation of suitable microorganism(s) that respond to hydrocarbon pollutants in soils lead to enhancement of the bioremediation efficiency. Monitoring and control of the microbial biomass in the soil is important for maintaining microbial activity. Moreover, homogeneous addition of bacterial biomass and nutrients is very difficult, therefore a large-scale mixing system is required for large-scale bioremediation.

This chapter describes an investigation of efficient bioremediation based on the use of specific hydrocarbon-degrading microorganisms and monitoring and control of the bacterial biomass by an environmental DNA technique. Construction of a large-scale bioremediation system is also described.

I. Characteristics of Petroleum Hydrocarbon-Contaminated Soils

Petroleum hydrocarbons include various kinds of hydrocarbons such as *n*-alkanes, *c*-alkanes, branched alkanes, aromatics, resins and asphaltenes, and petroleum is refined by distillation into gasoline, kerosene, gas oil, heavy fuel oil, lubricating oil etc. Long-chain hydrocarbons, *c*-alkanes, resins and asphaltenes are generally difficult to degrade by microorganisms [11, 22-24].

To estimate the components and concentration of petroleum hydrocarbons in contaminated soils, analytical methods involving gas chromatography (GC), infrared (IR) detection, gravimetric analysis, and thin layer chromatography (TLC) have been developed. GC [the United States Environmental Protection Agency (EPA) Method 8015M], IR (EPA Method 418.1), and gravimetric methods are used for quantifying total petroleum hydrocarbons (TPH), and the hydrocarbon components present are analyzed by GC and TLC [1, 25-27].

Soils with historical petroleum hydrocarbon contamination were analyzed using these methods. Long-chain *n*-alkanes and *c*-alkanes, derived from heavy crude oil or lubricating oil, tend to remain in historically contaminated soils and can be determined by GC analysis (Figure 1 A and B). Some soils were contaminated by resins and/or asphaltenes (Figure 1 C and D), and the levels of TPH in the historically contaminated soils were in the range 1,000 – 70,000 mg/kg-soil (11 samples). Therefore, microorganisms which degrade long-chain hydrocarbons (including resins and asphaltenes) and *c*-alkanes are required for the bioremediation of soils historically contaminated with petroleum hydrocarbons.

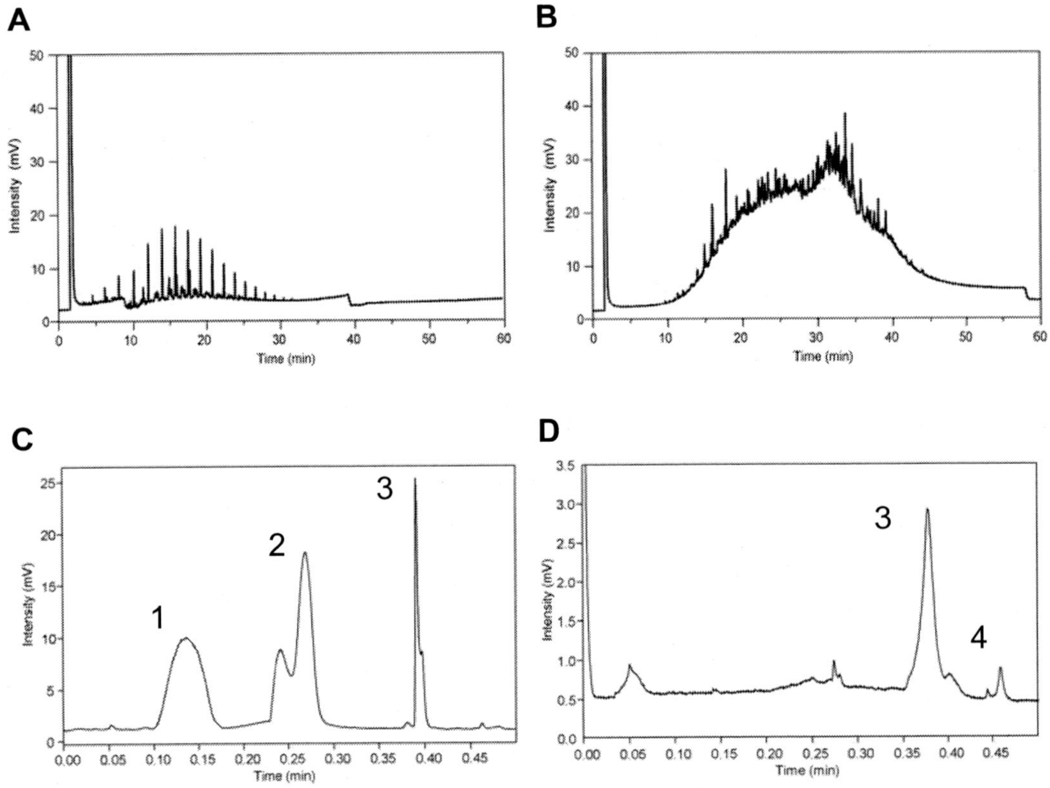

Figure 1. GC (A and B) and TLC (C and D) analysis of soils with historical petroleum hydrocarbon contamination. TLC analysis was carried out by TLC/FID [25, 26], hydrocarbon type: 1; saturated hydrocarbons, 2; aromatic hydrocarbons, 3; resins, 4; asphaltenes.

II. Isolation and Characterization of Hydrocarbon-Degrading Bacteria

Long-chain hydrocarbons and *c*-alkanes are the main residual hydrocarbons in soils historically contaminated with petroleum hydrocarbons, thus long-chain hydrocarbon and *c*-alkane degrading bacteria were isolated and characterized using waste car engine oil, car engine base oil, or the *c*-alkane fraction of car engine base oil as a sole carbon and energy source [24, 28, 29]. Thirty-six strains which showed high cell densities (above 1.0 at OD_{660}) in the hydrocarbon-containing media were isolated, and then the strains were identified and divided into six genera plus one family (*Acinetobacter*, *Rhodococcus*, *Gordonia*, *Pseudomonas*, *Ralstonia*, *Bacillus*, and *Alcaligenaceae*, respectively) by 16S rRNA gene sequencing (Figure 2). In this screening, 3/4 of the isolates were identified as the genus *Acinetobactor*, *Rhodococcus*, or *Gordonia*. When waste car engine oil or car engine base oil was used as a sole carbon source, mainly the genus *Acinetobacter* was isolated. The genera

Rhodococcus and *Gordonia* were isolated using a media containing the *c*-alkane fraction of car engine base oil as a sole carbon source.

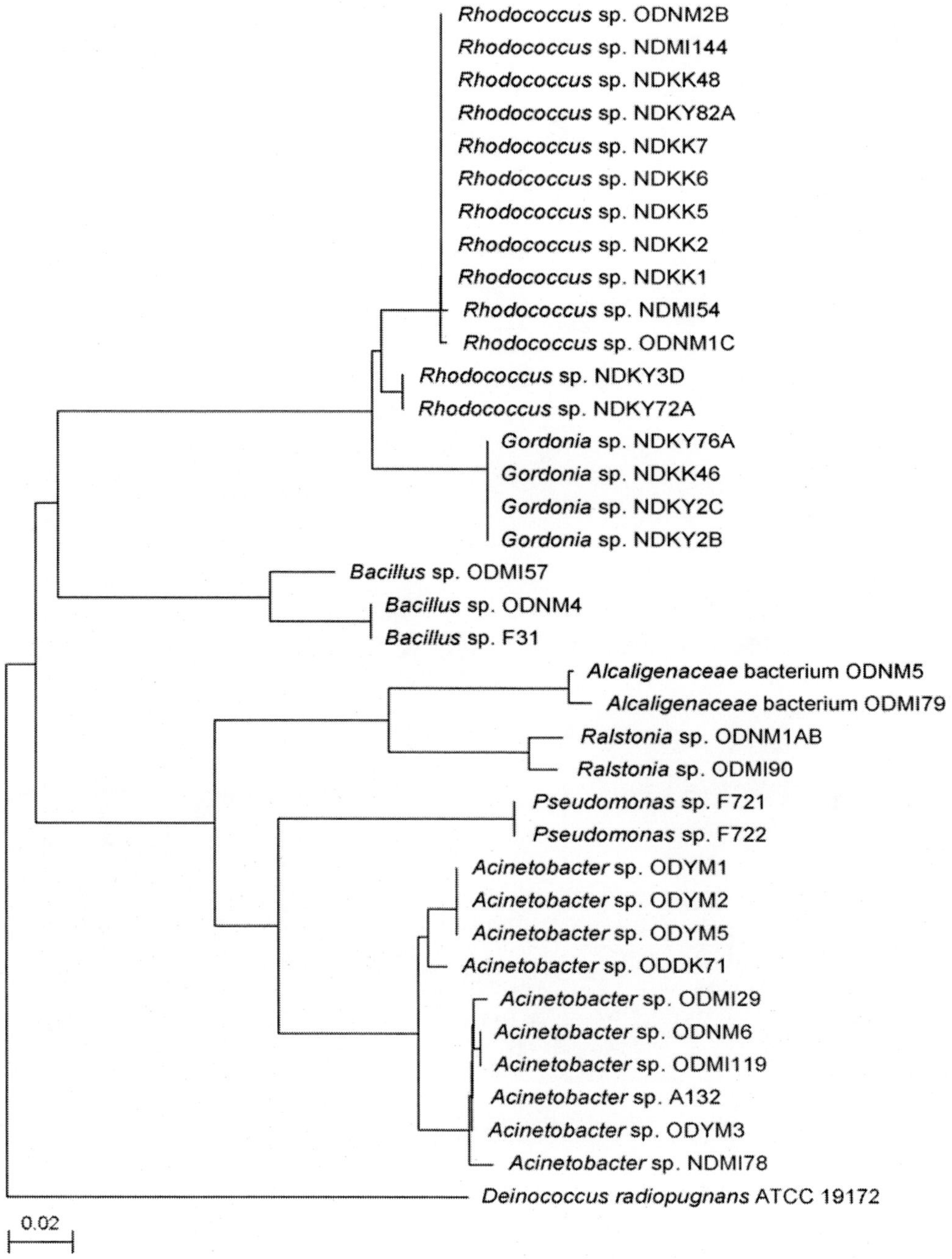

Figure 2. Phylogenetic tree based on the partial 16S rRNA gene sequences (500 bp) of the 36 isolates and related authentic strains. The bar represents 0.02 substitutions per nucleotide position.

Table 1. Evaluation of hydrocarbon degradation by long-chain hydrocarbon degrading bacteria by the 2,6-DCPIP assay[a]

Genus or family	Strain	Petroleum[b]			*n*-Alkane[b]					*c*-Alkane[b]				Aromatics[b]		Polycyclic hydrocarbons[b]			
		BO	HO-A	HO-C	C_6	C_8	C_{10}	C_{16}	C_{20}	CH C_6	MC C_7	HC C_{12}	DC C_{18}	BE C_6	DB C_{18}	DE	TE	NP	AN
Alcaligenaceae	ODMI79	+	+	+	−	+	+	+	+	−	−	−	−	−	+	−	−	−	−
Alcaligenaceae	ODNM5	+	+	+	−	−	+	+	+	−	−	−	−	−	−	−	−	+	−
Ralstonia	ODNM1AB	+	+	+	−	−	−	+	+	+	−	−	−	−	+	−	−	−	−
Ralstonia	ODMI90	+	+	+	+	+	+	+	+	−	−	−	−	−	−	−	−	−	−
Pseudomonas	F721	+	+	+	+	+	+	+	+	−	−	−	−	−	+	−	−	−	−
Pseudomonas	F722	+	+	+	+	+	+	+	+	−	−	−	−	−	+	−	−	−	−
Acinetobacter	NDMI78	+	+	+	−	+	+	+	+	−	−	−	−	−	+	−	−	−	−
Acinetobacter	ODMI29	+	+	+	−	−	−	+	+	−	−	−	−	−	−	−	+	−	−
Acinetobacter	A132	+	+	+	−	−	−	+	+	−	−	−	−	−	+	−	−	−	−
Acinetobacter	ODYM3	+	+	+	−	−	−	+	+	−	−	−	−	−	+	−	−	−	−
Acinetobacter	NDMI119	+	+	+	−	−	−	+	+	−	−	−	−	−	+	−	−	−	−
Acinetobacter	ODNM6	+	+	+	−	−	−	+	+	−	−	−	−	−	+	−	−	+	−
Acinetobacter	ODDK71	+	+	+	−	−	−	+	+	−	−	−	−	−	−	−	−	−	−
Acinetobacter	ODYM5	+	+	+	−	−	−	+	+	−	−	−	−	−	+	−	−	−	−
Acinetobacter	ODYM2	+	+	+	−	−	−	+	+	−	−	−	−	−	+	−	−	−	−
Acinetobacter	ODYM1	+	+	+	−	−	−	+	+	−	−	−	−	−	+	−	−	−	−
Rhodococcus	NDMI54	+	+	+	−	−	−	+	+	−	−	−	+	−	−	−	−	+	−
Rhodococcus	ODNM2B	+	+	+	−	−	+	+	+	−	−	−	−	−	+	−	−	−	−
Rhodococcus	NDKY82A	+	+	+	−	−	−	+	+	+	−	−	−	−	+	+	−	+	+
Rhodococcus	NDKK1	+	+	+	−	−	+	+	+	+	−	−	+	−	+	+	−	+	+
Rhodococcus	NDKK2	+	+	+	−	−	+	+	+	+	−	−	+	−	+	+	−	+	+
Rhodococcus	NDKK5	+	+	+	−	−	+	+	+	+	+	−	−	−	+	+	−	+	+
Rhodococcus	NDKK6	+	+	+	−	−	+	+	+	+	−	−	+	−	+	+	−	+	+
Rhodococcus	NDKK7	+	+	+	+	+	+	+	+	−	−	+	+	−	+	−	−	−	−
Rhodococcus	NDKK48	+	+	+	+	+	+	+	+	−	−	+	+	−	+	−	−	−	−
Rhodococcus	NDMI144	+	+	+	−	−	−	+	+	+	−	−	+	−	+	+	+	−	+
Rhodococcus	ODNM1C	+	+	+	+	+	+	+	+	+	+	+	+	−	+	−	−	+	+
Rhodococcus	NDKY3D	+	+	+	−	−	+	+	+	−	−	+	+	−	+	−	−	−	−
Rhodococcus	NDKY72A	+	+	+	+	+	+	+	+	−	−	+	+	−	−	−	−	+	−
Gordonia	NDKK46	+	+	+	+	+	+	+	+	+	−	+	+	−	+	−	−	+	−
Gordonia	NDKY76A	+	+	+	+	+	+	+	+	−	−	+	+	−	+	−	−	−	−
Gordonia	NDKY2C	+	+	+	+	+	+	+	+	−	+	+	+	−	+	−	−	−	−
Gordonia	NDKY2B	+	+	+	+	+	+	+	+	−	−	+	+	−	+	−	−	+	−
Bacillus	ODMI57	+	+	+	+	+	+	+	+	−	−	−	−	−	+	−	−	−	−
Bacillus	F31	+	+	+	−	−	−	+	+	−	−	−	−	−	−	−	−	−	−
Bacillus	ODNM4	+	+	+	−	−	−	+	+	−	−	−	−	−	+	−	−	+	−

[a] The symbols + and − indicate that the solution was colorless (degraded) or blue(not degraded), respectively.

[b] The abbreviations represent the kinds of hydrocarbons or mineral oils as follows: BO; base oil, HO-A; heavy crude oil (type A), HO-C; heavy crude oil (type C), CH; cyclohexane, MC; methylcyclohexane, HC; hexylcyclohexane, DC; dodecylcyclohexane, BE; benzene, DB; dodecylbenzene, DE; decalin, TE; tetralin, NP; naphthalene, AN; anthracene.

In order to characterize the hydrocarbon degradation of the isolates, hydrocarbon degradation of each strain was evaluated by using 2,6-dichlorophenol indophenol (2,6-DCPIP) which is an oxidation-reduction indicator (when microorganisms oxidize hydrocarbons in 2,6-DCPIP solution, the color of the solution changes from blue to colorless) [30]. Table 1 shows the evaluation of hydrocarbon degradation of the isolates by 2,6-DCPIP assay. All isolates belonging to the respective genera degraded car engine base oil and heavy crude oils (types A and C). As regards *n*-alkane degradation, the long-chain *n*-alkanes hexadecane (C_{16}) and eicosane (C_{20}) were also degraded by all isolates. All strains of the genera *Gordonia* and *Pseudomonas* degraded short- and middle-chain *n*-alkanes, namely hexane (C_6), octane (C_8) and decane (C_{10}), while several strains of the genus *Rhodococcus* degraded hexane (4 of 13 strains), octane (4 of 13 strains) and decane (10 of 13 strains). However, almost none of the strains of the genus *Acinetobacter* (9 of 10 strains) degraded short-chain *n*-alkanes.

On the other hand, *c*-alkanes were mainly degraded by strains belonging to the genera *Rhodococcus* and *Gordonia*. Strains belonging to the genera *Rhodococcus* (5 of 13 strains) and *Gordonia* (4 of 4 strains) degraded hexylcyclohexane, and almost all strains belonging to the genera *Rhodococcus* (10 of 13 strains) and *Gordonia* (4 of 4 strains) degraded dodecylcyclohexane. However, strains belonging to the genera *Rhodococcus* (2 of 13 strains) and *Gordonia* (1 of 4 strains) found it difficult to degrade methylcyclohexane. These results indicate that strains belonging to the genera *Rhodococcus* and *Gordonia* can efficiently degrade *c*-alkanes with long alkyl side-chains.

As regards aromatic hydrocarbon degradation, dodecylbenzene (length of alkyl side-chain: 12) was degraded by many strains throughout all the genera, whereas no strains in any of the genera degraded benzene. With respect to polycyclic hydrocarbons, decalin, naphthalene and anthracene were degraded by several strains belonging to 5 genera, although tetralin was difficult to degrade. Strains of the genus *Rhodococcus* degraded more kinds of polycyclic hydrocarbons than the other genera, since decalin, naphthalene and anthracene were degraded by 6, 8 and 7 strains, respectively, of the genus *Rhodococcus*.

n-Alkanes were degraded by all strains of the respective genera, and the degradable substrates were similar for each genus. *c*-Alkanes were mainly degraded by strains of the genera *Rhodococcus* and *Gordonia*, and almost none of the strains from the other genera degraded *c*-alkanes. These results indicate that the class Actinobacteria may have a specific function for the degradation of *c*-alkanes.

III. Hydrocarbon Degradation Pathways of Isolates

c-Alkanes remain in the environment for a long time, thus the degradation of *c*-alkanes by microorganisms is of importance. The genera *Acinetobacter, Rhodococcus* and *Gordonia* degrade *c*-alkanes, however the *c*-alkane degradation pathways of these strains are different [28, 31, 32]. In the case of the genus *Acinetobacter*, *c*-alkanes are degraded by co-oxidation with *n*-alkanes as a growth substrate. The dodecylcyclohexanes are co-oxidized to 4-dodecylcyclohexanol (ring oxidation pathway) or alkyl side chains (alkyl side chain oxidation pathway) by *Acinetobacter* sp. ODDK71 (Figure 3), and the ring oxidation is the main metabolic pathway for *Acinetobacter* sp. ODDK71.

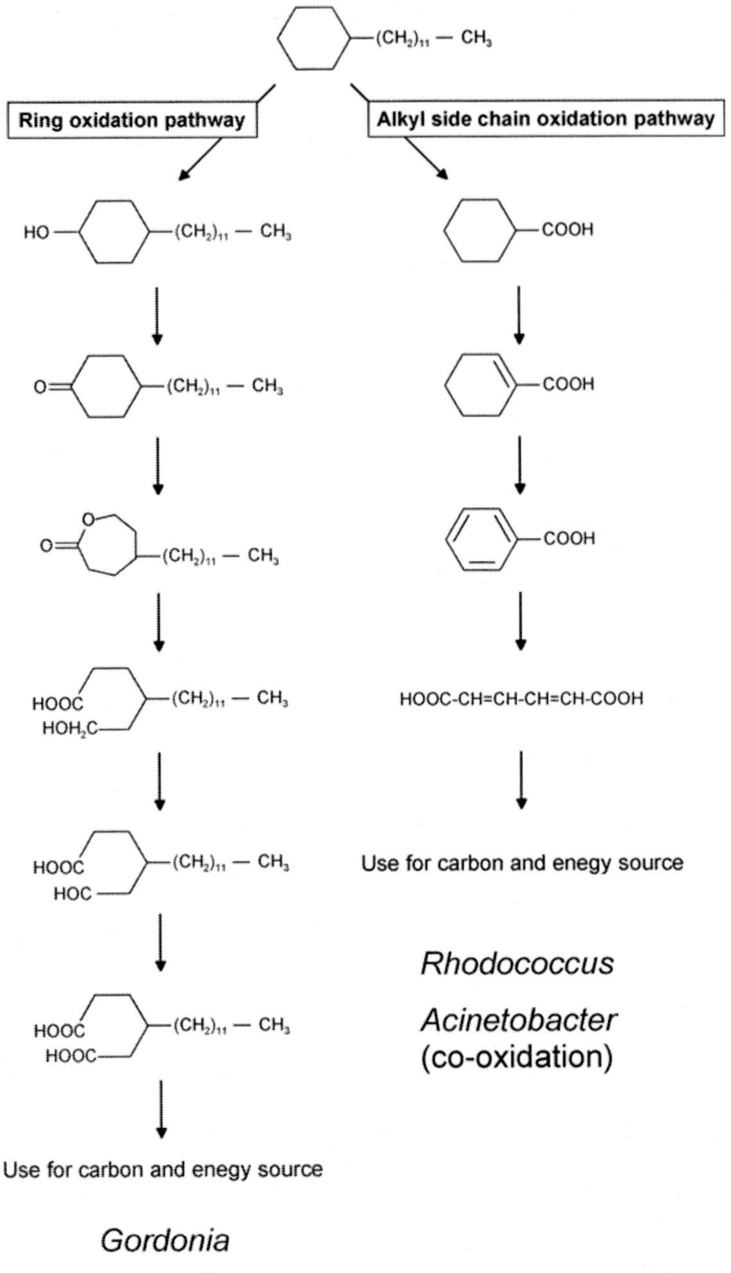

Figure 3. Putative degradation pathway of dodecylcyclohexane in hydrocarbon-degrading bacteria (genera *Acinetobacter*, *Rhodococcus*, and *Gordonia*).

On the other hand, the genera *Rhodococcus* and *Gordonia* degrade *c*-alkanes without co-oxidation. The metabolic pathways of *c*-alkanes by the genera *Rhodococcus* and *Gordonia* are ring oxidation and alkyl side chain pathways (Figure 3). The metabolites of alkyl *c*-hexane using *Rhodococcus* sp. NDKK48 indicate that the genus *Rhodococcus* degrades via both the ring oxidation pathway and the alkyl side chain oxidation pathway (the ratio was estimated as 1:1). In the side chain oxidation pathway of the genus *Rhodococcus*, the cyclohexyl ring is aromatized and then cleaved. In contrast, *Gordonia* sp. NDKY76A degrades *c*-alkanes mainly through the ring oxidation pathway (the ratio of ring oxidation to alkyl side chain oxidation was estimated as 9:1). First, the ring of the *c*-alkane is oxidized, and then the cyclic ketone is oxidized to lactone. The main metabolic pathways of *c*-alkanes by the genera *Acinetobacter*, *Rhodococcus* and *Gordonia* are different, therefore a combination of these strains may lead to enhanced efficient bioremediation of *c*-alkanes.

IV. Bacterial Biomass Monitoring by Environmental DNA in the Environment

Biomass and flora of soil microorganisms are not easy to understand. In order to analyze the flora of soil microorganisms, several methods to quantify and evaluate biomass of soil microorganisms have been constructed. Traditional culture-dependent methods are often used for counting numbers of microorganisms, however 99.7%-99.999% of soil microorganisms are in a viable-but-non-culturable-state (VBNC, [33-35]). Several methods have been developed to evaluate soil microorganisms, such as substrate-induced respiration (SIR), chloroform fumigation extraction (CFE), and a 4',6-diamino-2-phenylindole (DAPI) staining method [36-42].

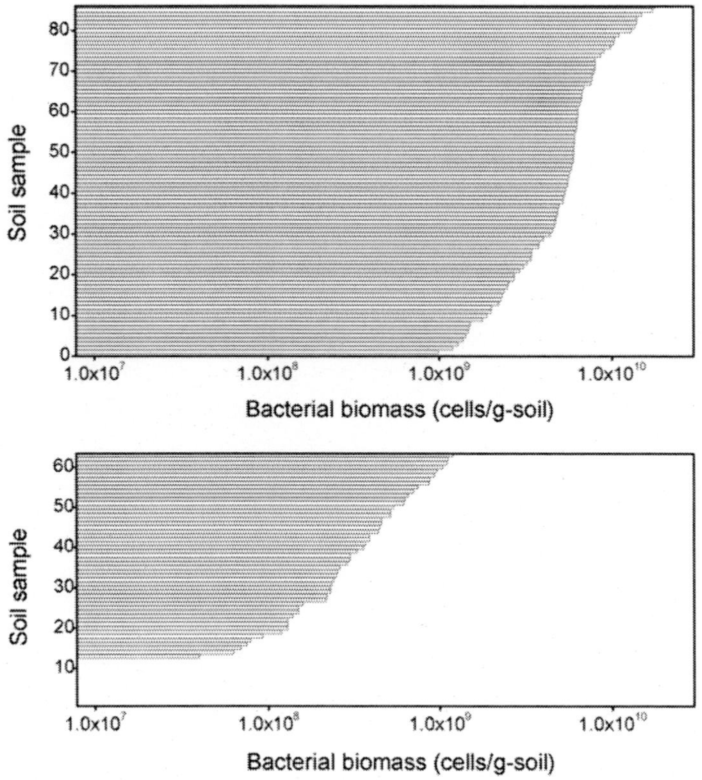

Figure 4. Evaluation of the bacterial biomass in agricultural (A) and oil-polluted (B) soils by the slow-stirring method.

Recently, analysis of environmental DNA (eDNA) has been investigated as a new approach for evaluating soil microorganisms [43-48]. Since eDNA analysis does not require a cultivation procedure, this technique is useful for detecting VBNC [33-35, 44]. Moreover, we have constructed a new method to evaluate bacterial biomass based on the amount of eDNA extracted from soil (slow-stirring method, [49]).

The analysis of bacterial biomass in various soil samples such as petroleum hydrocarbon-contaminated and agricultural soils was carried out by the slow-stirring method (Figure 4). In the agricultural soils, the average of the bacterial biomass was 5.95×10^9 cells/g-soil, but that in the petroleum hydrocarbon-contaminated soils was less than 1×10^9 cells/g-soil. Since the bacteria in the petroleum hydrocarbon-contaminated soils are sensitive to damage by the hydrocarbons, petroleum hydrocarbon pollution might be evaluated by determining the bacterial biomass. Furthermore, monitoring and control of bacterial biomass would contribute to enhancement of bioremediation.

For effective bioremediation, monitoring and control of specific hydrocarbon-degrading bacteria is important. The isolated strains belonging to the genera *Acinetobacter* and *Rhodococcus* carry alkane hydroxylase genes (*alkM* and *alkB,* respectively), however the DNA sequences of the *alk* genes are different [50-52]. Thus, the *alk* genes can be used as a biomarker for monitoring the specific hydrocarbon-degrading bacterial biomass. Specific PCR primers for monitoring *Acinetobacter* (*alkM*) and *Rhodococcus* (*alkB*) were designed, based on differences in the DNA sequences (Table 2), and each primer set increased *alkM* and *alkB* specifically when the specific bacterial biomass was above 3×10^8 CFU/g-soil (an *Acinetobacter* monitoring method and a *Rhodococcus* monitoring method, respectively, unpublished data).

Table 2. Sequences of the specific PCR primers for monitoring of *Acinetobacter* (*alkM*) and *Rhodococcus* (*alkB*)

Target genus (gene)	Primer name	Sequence
Acinetobacter (*alkM*)	Ac81-F	5'-AATAC TGCAC ATGAA TTGAG CCATA A-3'
	Ac81-R	5'-GCATG ATGAT CTGAG TGACG TTGTA-3'
Rhodococcus (*alkB*)	Rh82-F	5'-AACAC CGAAC ACGAG CTCGG TCACA-3'
	Rh82-R	5'-GCGTG ATGAG AGCTC TGACG TTGTA-3'

Bioremediation with bacterial biomass monitoring for efficient hydrocarbon degradation was carried out. When the total and specific bacterial biomass were maintained above around 1×10^8 cells/g-soil with bacterial biomass monitoring and control, removal of petroleum hydrocarbons was enhanced by about 20% (data not shown). The maintenance of bacterial biomass based on total and specific bacterial biomass monitoring lead to enhancement of the bioremediation.

V. Bioremediation of Soil Historically Contaminated with Petroleum Hydrocarbons

Efficient bioremediation was carried out using specific hydrocarbon-degrading bacteria, bacterial biomass monitoring by the slow-stirring method, and control of the bacterial biomass (above 1×10^8 cells/g-soil). Bioremediation of soil with historical petroleum hydrocarbon contamination was carried out using these methods. The contaminated soil contained mainly *c*-alkanes (about 4,000 mg-TPH/kg-soil), and the total bacterial biomass in the soil was less than 3.1×10^7 cells/g-soil.

Rhodococcus sp. NDKK6 was inoculated at about 5×10^8 cells/g-soil, and modified SW medium (W medium [28] supplemented with 1.2 g/l NH_4NO_3 instead of $(NH_4)_2SO_4$, 0.5 g/l polypeptone, and 0.25 g/l yeast extract) was also added into the soil. The total bacterial biomass was monitored by the slow-stirring method, and *Rhodococcus* sp. NDKK6 was maintained (strain NDKK6 was additionally inoculated at the 6th and 26th days) using the *Rhodococcus* monitoring method (Figure 5B and C).

Petroleum hydrocarbons were gradually degraded, and about 47% of TPH was removed by the 28th day (Figure 5B), while about 18% of TPH was removed by the biostimulation method by the 28th day (Figure 5A). To enhance the hydrocarbon degradation, *Gordonia* sp. NDKY76A was also inoculated into the soil on the 30th day. The bacterial biomass was increased, and about 66% of TPH was finally removed by the 35th day. When the order of inoculation with *Rhodococcus* sp. NDKK6 and *Gordonia* sp. NDKY76A was changed, about 58% of TPH was removed by the 35th day.

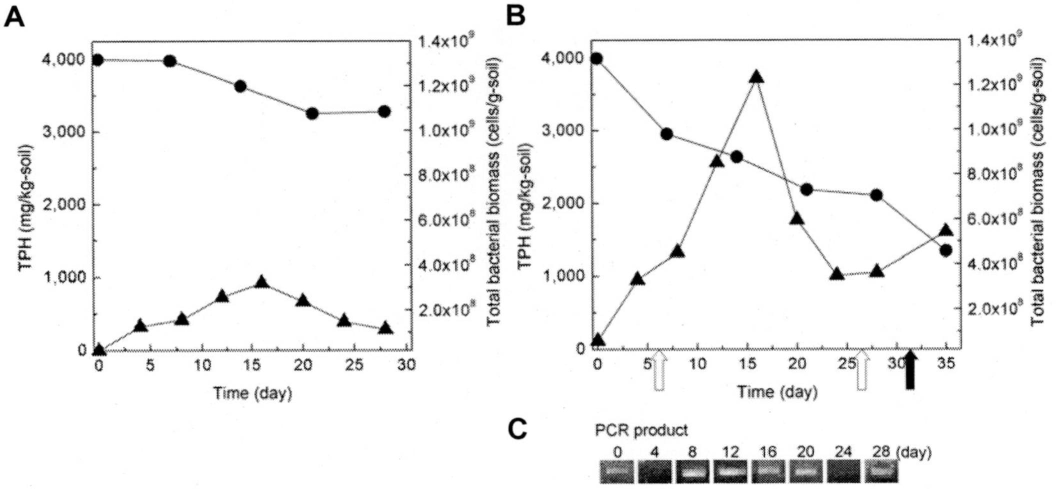

Figure 5. Bioremediation by *Rhodococcus* sp. NDKK6 and *Gordonia* sp. NDKY76A with monitoring and control of the bacterial biomass. A; Biostimulation, B; bioaugmetation with monitoring and control of the bacterial biomass, C; detection of *Rhodococcus* sp. NDKK6 by the *Rhodococcus* monitoring method. Symbols: ●; remaining TPH, ▲; total bacterial biomass. White and black arrows indicate the additional inoculations of *Rhodococcus* sp. NDKK6 and *Gordonia* sp. NDKY76A, respectively.

VI. Construction of Large-scale Bioremediation System

Homogeneous mixing of microorganisms and nutrients into petroleum hydrocarbon-contaminated soil is needed for the construction of a large-scale bioremediation system. A large-scale mixing machine (500 kg soil per batch) was developed (Figure 6A). The 500 kg scale mixing machine was equipped with a double axial blade-type paddle for mixing soil (Figure 6B). This machine can mix microorganisms and nutrients homogenously into various types of soils in 3 minutes, without damage to the microorganisms. This machine can also be transported to a contaminated site by truck.

(A)

(B)

Figure 6. Over (A) and inner (B) views of the 500 kg scale mixing machine.

Bioremediation of historical petroleum hydrocarbon contamination in soil (2.5 ton scale) was carried out using the 500 kg scale mixing machine. The contaminated soil was clay and contained aromatic hydrocarbons as the main hydrocarbon component (about 22,000 mg-TPH/kg-soil), and the total bacterial biomass in the soil was less than 3.1×10^7 cells/g-soil. One % (v/w-soil) culture of *Rhodococcus* sp. NDKK6 (2.9×10^{10} CFU/ml) and modified SW medium were mixed into the contaminated soil (total 2.5 ton) by the 500 kg scale mixing machine. *Rhodococcus* sp. NDKK6 culture was mixed homogenously into the soil (ranging from 2.44 to 2.80×10^8 CFU/g-soil), while the bacterial biomass was varied from 0.22 to 1.84×10^8 CFU/g-soil by the backhoe method. Removal of petroleum hydrocarbons after 6 weeks was about 6,000 (about 27% removal) and 1,700 (about 8% removal) mg-TPH/kg-soil, using the 500 kg scale mixing machine and backhoe method, respectively (Figure 7). These results indicate that the wide-spread distribution of microorganisms and nutrients enhanced efficient removal of the hydrocarbons, due to the increased contact between microorganisms and hydrocarbons. The bioremediation system with specific hydrocarbon-degrading bacteria, monitoring and control of the bacterial biomass, and a 500 kg scale mixing machine is applicable to large-scale contaminated sites.

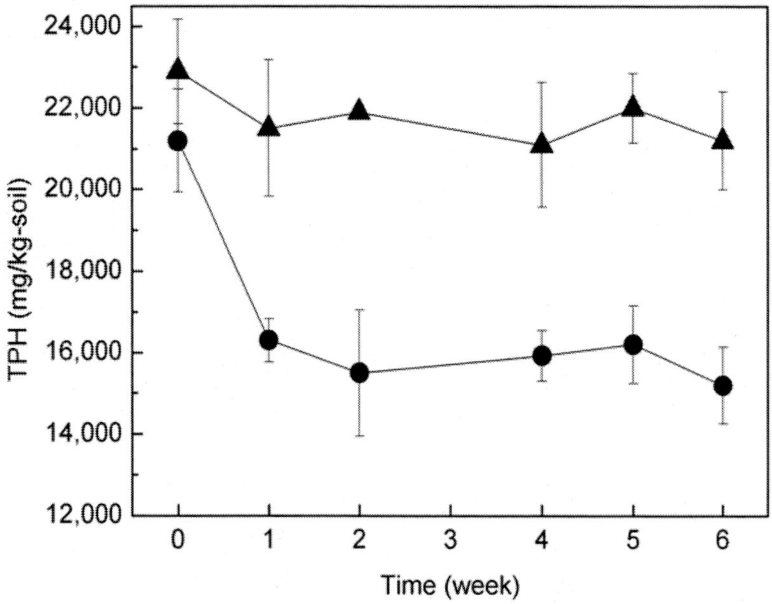

Figure 7. Large-scale bioremediation of soil with historical petroleum hydrocarbon contamination. Symbols: ●; 500 kg scale mixing machine, ▲; backhoe method.

CONCLUSION

In this chapter, we demonstrated the construction of an efficient bioremediation system with specific hydrocarbon-degrading bacteria, monitoring and control of the bacterial biomass, and a 500 kg scale mixing machine. The bioremediation system is illustrated in Figure 8. In the first step, analysis of petroleum hydrocarbons in the soil is important for the selection of specific hydrocarbon-degrading microorganisms. In the second stage, over 1×10^8 cells/g-soil of hydrocarbon-degrading bacteria and nutrients are homogenously mixed into the soil. In the third stage, the total and specific bacterial biomass need to be monitored and maintained above 1×10^8 cells/g-soil during the bioremediation process.

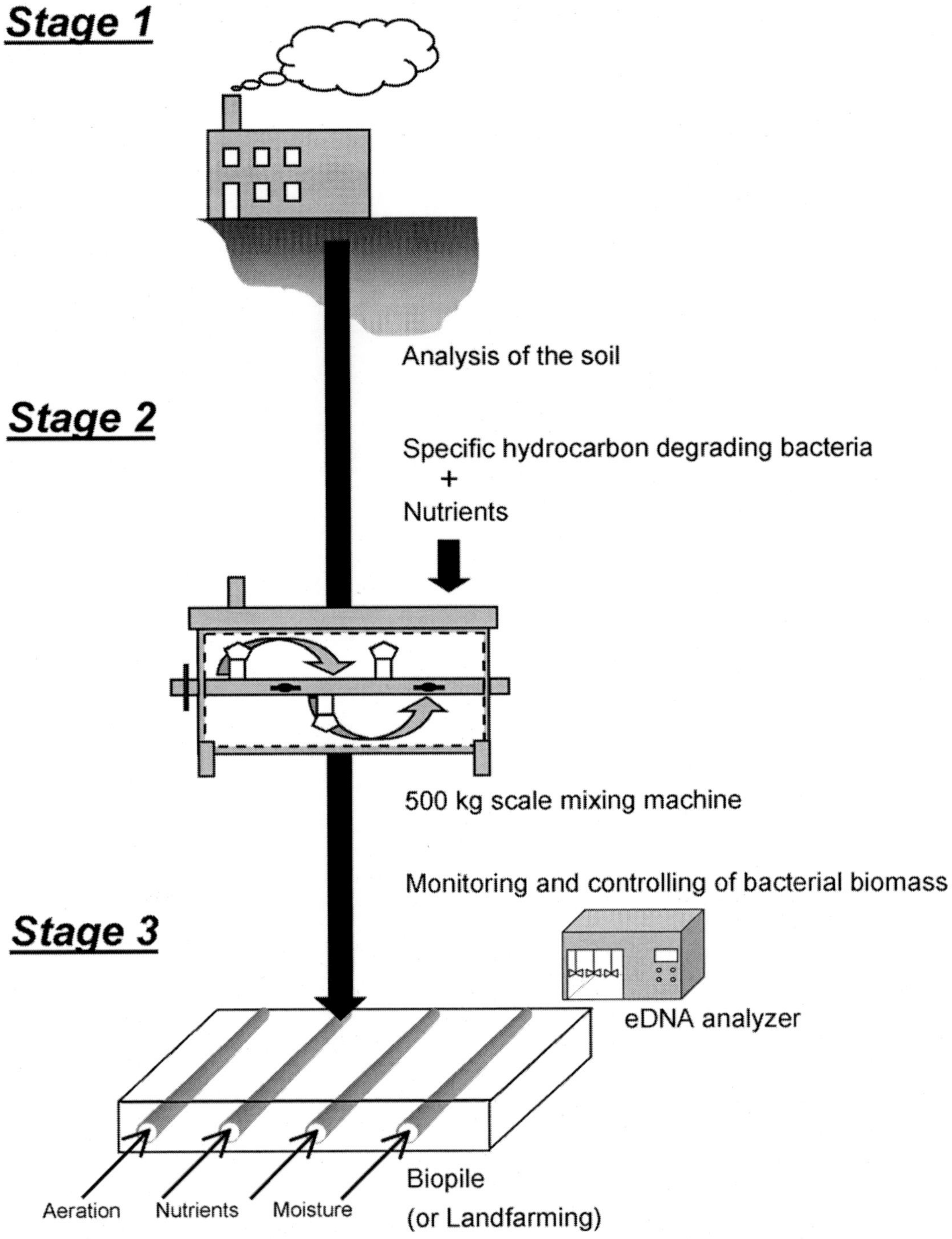

Figure 8. Flowchart of the large-scale bioremediation system using specific hydrocarbon-degrading bacteria and bacterial biomass monitoring.

REFERENCES

[1] Matsumiya, Y., & Kubo, M. (2007). Bioprocess Handbook (in Japanese). Tokyo, *NTS Publishing*, pp. 638-652.

[2] Komukai-Nakamura, S., Sugiura, K., Yamauchi-Inomata, Y., Toki, H., Venkateswaran, K., Yamamoto, S., Tanaka, H., & Harayama, S. (1996). Construction of bacterial consortia that degrade Arabian light crude oil. *J. Ferment. Bioeng.,* 82, 570-574.

[3] Lal, B., & Khanna, S. (1996). Degradation of crude oil by *Acinetobacter calcoaceticus* and *Alcaligenes odorans*. *J. Appl. Bacteriol.,* 81, 355-362.

[4] Vidali, M. (2001). Bioremediation. An overview. *Pure Appl. Chem.,* 73, 1163-1172.

[5] Chaillan, F., Le Flèche, A., Bury, E., Phantavong, Y. H., Grimont, P., Saliot, A., & Oudot, J. (2004). Identification and biodegradation potential of tropical aerobic hydrocarbon-degrading microorganisms. *Res. Microbiol.,* 155, 587-595.

[6] Miller, C. D., Hall, K., Liang, Y. N., Nieman, K., Sorensen, D., Issa, B., Anderson, A. J., & Sims, R. C. (2004). Isolation and characterization of polycyclic aromatic hydrocarbon-degrading *Mycobacterium* isolates from soil. *Microb. Ecol.,* 48, 230-238.

[7] Meintanis, C., Chalkou, K. I., Kormas, K. A., & Karagouni, A. D. (2006). Biodegradation of crude oil by thermophilic bacteria isolated from a volcano island. *Biodegradation,* 17, 105-111.

[8] von Fahnestock, F. M., Wickramanayake, G. B., Kratzke, R. J., & Major, W. R. (1998). *Biopile design, operation, and maintenance handbook for treating hydrocarbon-contaminated soils.* Columbus Ohio, Battelle Press.

[9] Hejazi, R. F., Husain, T., & Khan, F. I. (2003). Landfarming operation of oily sludge in arid region—human health risk assessment. *J. Hazard. Mater.,* 99, 287-302.

[10] Peltola, R., Salkinoja-Salonen, M., Pulkkinen, J., Koivunen, M., Turpeinen, A. R., Aarnio, T., & Romantschuk, M. (2006). Nitrification in polluted soil fertilized with fast- and slow-releasing nitrogen: a case study at a refinery landfarming site. Environ. *Pollut.,* 143, 247-253.

[11] Perry, J. J. (1984). Microbial metabolism of cyclic alkanes. In: Atlas, R. M., editor. Petroleum Microbiology. *New York, Macmillan,* pp. 61-98.

[12] Leahy, J. G., & Colwell, R. R. (1990). Microbial degradation of hydrocarbons in the environment. *Microbiol. Rev.,* 54, 305-315.

[13] Huesemann, M. H., Hausmann, T. S., & Fortman, T. J. (2004). Does bioavaolability limit biodegradation? A comparison of hydrocarbon biodegradation and desorption rates in aged soils. *Biodegradation,* 15, 261-274.

[14] Johnsen, A. R., Schmidt, S., Hybholt, T. K., Henriksen, S., Jacobsen, C. S., & Andersen, O. (2007). Strong impact on the polycyclic aromatic hydrocarbon (PAH)-degrading community of a PAH-polluted soil but marginal effect on PAH degradation when priming with bioremediated soil dominated by mycobacteria. *Appl. Environ. Microbiol.,* 73, 1474-1480.

[15] Whyte, L. G., Hawari, J., Zhou, E., Bourbonnière, L., Inniss, W. E., & Greer, C. W. (1998). Biodegradation of variable-chain-length alkanes at low temperatures by a psychrotrophic *Rhodococcus* sp. *Appl. Environ. Microbiol.,* 64, 2578-2584.

[16] Smits, T. H., Röthlisberger, M., Witholt, B., & van Beilen, J. B. (1999). Molecular screening for alkane hydroxylase genes in Gram-negative and Gram-positive strains. *Environ. Microbiol.,* 1, 307-317.

[17] Dinamarca, M. A., Ruiz-Manzano, A., & Rojo, F. (2002). Inactivation of cytochrome *o* ubiquinol oxidase relieves catabolic repression of the *Pseudomonas putida* GPo1 alkane degradation pathway. *J. Bacteriol.,* 184, 3785-3793.

[18] Sabirova, J. S., Ferrer, M., Regenhardt, D., Timmis, K. N., & Golyshin, P. N. (2006). Proteomic insights into metabolic adaptations in *Alcanivorax borkumensis* induced by alkane utilization. *J. Bacterial.,* 188, 3763-3773.

[19] Bastiaens, L., Springael, D., Wattiau, P., Harms, H., deWachter, R., Verachtert, H., & Diels, L. (2000). Isolation of adherent polycyclic aromatic hydrocarbon (PAH)-degrading bacteria using PAH-sorbing carriers. *Apple. Environ. Microbiol.,* 66, 1834-1843.

[20] Wong, J. W. C., Lai, K. M., Wan, C. K., Ma, K. K., & Fang, M. (2002). Isolation and optimization of PAH-degradative bacteria from contaminated soil for PAHs bioremediation. *Water Air Soil Pollut.,* 139, 1-13.

[21] Van Hamme, J. D., Singh, A., & Ward, O. P. (2003). Recent advances in petroleum microbiology. *Microbiol. Mol. Biol. Rev.,* 67, 503-549.

[22] Venkateswaran, K., Hoaki, T., Kato, M., & Maruyama, T. (1995). Microbial degradation of resins fractionated from Arabian light crude oil. *Can. J. Microbiol.,* 41, 418-428.

[23] Pineda-Flores, G., Boll-Arqüello, G., Lira-Galeana, C., & Mesta-Howard, A. M. (2004). A microbial consortium isolated from a crude oil sample that uses asphaltenes as a carbon and energy source. *Biodegradation,* 15, 145-151.

[24] Koma, D., Hasumi, F., Yamamoto, E., Ohta, T., Chung, S.-Y., & Kubo, M. (2001). Biodegradation of long-chain *n*-paraffins from waste oil of car engine by *Acinetobacter* sp. *J. Biosci. Bioeng.,* 91, 94-96.

[25] Barman, B. N. (1996). Hydrocarbon-type analysis of base oils and other heavy distillates by thin-layer chromatography with flame-ionization detection and by the clay-gel method. *J. Chromatogr. Sci.,* 34, 219-225.

[26] Napolitano, G. E., Richmond, J. E., & Stewart, A. J. (1998). Characterization of petroleum-contaminated soils by thin-layer chromatography with flame ionization detection. *J. Soil Contam,* 7, 709-724.

[27] Nadim, F., Liu. S., Hoag, G. E., Chen, J., Carley, R. J., & Zack, P. (2002). A comparison of spectrophotometric and gas chromatographic measurements of heavy petroleum products in soil samples. *Water Air Soil Pollut.,* 134, 97-109.

[28] Koma, D., Sakashita, Y., Kubota, K., Fujii, Y., Hasumi, F., Chung, S.-Y., & Kubo, M. (2003). Degradation of car engine base oil by *Rhodococcus* sp. NDKK48 and *Gordonia* sp. NDKY76A. *Biosci. Biotechnol. Biochem.,* 67, 1590-1593.

[29] Kubota, K., Koma, D., Matsumiya, Y., Sakihama, Y., Chung, S.-Y., & Kubo, M. (2008) Phylogenetic analysis of long-chain hydrocarbon-degrading bacteria and evaluation their hydrocarbon-degradation by the 2,6-DCPIP assay. *Biodegradation, in press.*

[30] Hanson, K. G., Desai, J. D., & Desai, A. J. (1993). A rapid and simple screening technique for potential crude oil degrading microorganisms. *Biotech. Tech.,* 7, 745-748.

[31] Koma, D., Hasumi, F., Chung, S.-Y., & Kubo, M. (2003). Biodegradation of *n*-alkylcyclohexanes by co-oxidation via multiple pathways in *Acinetobacter* sp. ODDK71. *J. Biosci. Bioeng.,* 95, 641-644.

[32] Koma, D., Sakashita, Y., Kubota, K., Fujii, Y., Hasumi, F., Chung, S.-Y., & Kubo, M. (2004). Degradation pathways of cyclic alkanes in *Rhodococcus* sp. NDKK48. *Appl. Microbiol. Biotechnol.,* 66, 92-99.

[33] Oliver, J. D., Nilsson, L., & Kjelleberg, S. (1991). Formation of nonculturable *Vibrio vulnificus* cells and its relationship to the starvation state. *Appl. Environ. Microbiol.,* 57, 2640-2644.

[34] Rahman, I., Shahamat, M., Kirchman, P. A., Russek-Cohen, E., & Colwell, R. R. (1994). Methionine uptake and cytopathogenicity of viable but nonculturable *Shigella dysenteriae* type 1. *Appl. Environ. Microbiol.,* 60, 3573-3578.

[35] Barer, M. R., & Harwood, C. R. (1999). Bacterial viability and culturability. *Adv. Microb. Physiol.,* 41, 93-137.

[36] Jenkinson, D. S., & Powlson, D. S. (1976). The effects of biocidal treatments on metabolism in soil-V: a method for measuring soil biomass. *Soil Biol. Biochem.,* 8, 209-213.

[37] Schnedl, W., Mikelsaar, A.-V., Breitenbach, M., & Dann, O. (1977). DIPI and DAPI: fluorescence banding with only negligible fading. Hum. Genet., 36, 167-172.

[38] Anderson, J. P. E., & Domsch, K. H. (1978). A physiological method for the quantitative measurement of microbial biomass in soils. *Soil Biol. Biochem.,* 10, 215-221.

[39] Poulin, N., Harrison, A., & Palcic, B. (1994). Quantitative precision of an automated image cytometric system for the measurement of DNA content and distribution in cells labeled with fluorescent nucleic acid stains. *Cytometry,* 16, 227-235.

[40] Kepner, R. L., Jr, & Patt, J. R. (1994). Use of fluorochromes for direct enumeration of total bacteria in environmental samples: past and present. *Microbiol. Rev.,* 58, 603-615.

[41] Yu, W., Dodds, W. K., Banks, M. K., Skalsky, J., & Strauss, E. A. (1995). Optimal staining and sample storage time for direct microscopic enumeration of total and active bacteria in soil with two fluorescent dyes. *Appl. Environ. Microbiol.,* 61, 3367-3372.

[42] Taylor, J. P., Wilson, B., Mills, M. S., & Burns, R. G. (2002). Comparison of microbial numbers and enzymatic activities in surface soils and subsoils using various techniques. *Soil Biol. Biochem.,* 34, 387-401.

[43] Steffan, R. J., Goksøyr, J., Bej, A. K., & Atlas, R. M. (1988). Recovery of DNA from soils and sediments. *Appl. Environ. Microbiol.,* 54, 2908-2915.

[44] Tsai, Y. L., & Olson, B. H. (1992). Detection of low numbers of bacterial cells in soils and sediments by polymerase chain reaction. *Appl. Environ. Microbiol.,* 58, 754-757.

[45] Leff, L. G., Dana, J. R., McArthur, J. V., & Shimkets, L. J. (1995). Comparison of methods of DNA extraction from stream sediments. *Appl. Environ. Microbiol.,* 61, 1141-1143.

[46] Zhou, J., Bruns, M. A., & Tiedje, J. M. (1996). DNA recovery form soils of diverse composition. *Appl. Environ. Microbiol.,* 62, 316-322.

[47] Miller, D. N., Bryant, J. E., Madsen, E. L., & Ghiorse, W. C. (1999). Evaluation and optimization of DNA extraction and purification procedures for soil and sediment samples. *Appl. Environ. Microbiol.,* 65, 4715-4724.

[48] Bergsma-Vlami, M., Prins, M. E., Staats, M., & Raaijmakers, J. M. (2005). Assessment of genotypic diversity of antibiotic-producing *Pseudomonas* species in the rhizosphere by denaturing gradient gel electrophoresis. *Appl. Environ. Microbiol.,* 71, 993-1003.

[49] Aoshima, H., Kimura, A., Shibutani, A., Okada, C., Matsumiya, Y., & Kubo, M. (2006). Evaluation of soil bacterial biomass using environmental DNA extracted by slow-stirring method. *Appl. Microbiol. Biotechnol.,* 71, 875-880.

[50] Tani, A., Ishige, T., Sakai, Y., & Kato, N. (2001). Gene structures and regulation of the alkane hydroxylase complex in *Acinetobacter* sp. strain M-1. *J. Bacteriol.,* 183, 1819-1823.

[51] Whyte, L. G., Smits, T. H., Labbé, D., Witholt, B., Greer, C. W., & van Beilen, J. B. (2002). Gene cloning and characterization of multiple alkane hydroxylase systems in *Rhodococcus* strains Q15 and NRRL B-16531. *Appl. Environ. Microbiol.*, 68, 5933-5942.

[52] van Beilen, J. B., Smits, T. H., Roos, F. F., Brunner, T., Balada, S. B., Röthlisberger, M., & Witholt, B. (2005). Identification of an amino acid position that determines the substrate range of integral membrane alkane hydroxylases. *J. Bacteriol.,* 187, 85-91.

In: Soil Contamination Research Trends
Editor: Javier B. Dominguez, pp. 161-176
ISBN 978-1-60456-319-1

Chapter 7

HEAVY METALS IN MEDITERRANEAN SOILS

J. Navarro-Pedreño, I. Gómez, M. B. Almendro Candel and I. Meléndez-Pastor
Departamento de Agroquímica y Medio Ambiente,
Universidad Miguel Hernández de Elche, Edif. Alcudia, Avd. de la Universidad sn, 03202-Elche, Alicante, Spain

1. HEAVY METALS AND ENVIRONMENTAL RISK

The European Environment Agency (EEA) and the United Nations Environment Program (UNEP) stimulate the study about the role of soil in relation to ecological and economical problems derived from the climatic change, industrial development and trade. Both organisms alert about the pollution problems due to mine and industrial activities and mark several areas affected in the Mediterranean, especially in Spain, Italy and Greece (EEA, 2002). The number of environmental compartments that are related to soil and its quality is great due to the interstitial position of soil among water, atmosphere, rocks and life.

The protection of soils is and should be a principal objective of environmental policy. Soils are finite, increasingly scarce and non-renewable resource with varying biological, chemical and physical properties (European Soil Bureau, 1999). That is why the properties that must be determined in a soil are diverse and soil should be studied from an interdisciplinary point of view.

Heavy metals are considered as contaminants capable of causing environmental problems in soils, water, atmosphere and living organism. The meaning of "heavy metals" is closely related to pollution and the potential toxicity of these elements. The presence of metals in soil is, therefore, not indicative of contamination (McLean and Bledsoe, 1992). However, it is so important the need to measure heavy metal concentrations in soils to assess contamination risks and to identify background values as a basis to establish soil quality standards (van Camp *et al.*, 2004).

The definition of heavy metal is related to those that have a specific gravity greater than about 5.0, but especially those that are poisonous, such as lead or mercury. The last signification makes that arsenic is mostly included in this group although one of its form, yellow As, has a density of 1.97 g/cm^3 and its chemistry is not as a metal.

The elements that can be consider as heavy metals based only in their density would be to long, however, the list of the main heavy metals that are also pollutants and are going to be considered are: cadmium (Cd), chromium (Cr), copper (Cu), mercury (Hg), nickel (Ni), lead (Pb) and zinc (Zn).

Some of them, such Cu and Zn, are essential elements for plants, and others like cadmium have no known biological function and its potentially toxic to both plants and animals (Jackson and Alloway, 1992).

Table 1. Metals and metalloids of primary interest by EPA (2007)

Aluminum	Antimony	Arsenic	Barium
Berylium	Boron	Cadmium	Chromium
Cobalt	Copper	Iron	Lead
Manganese	Mercury (inorg.)	Molybdenum	Nickel
Selenium	Silver	Strontium	Tin
Thallium	Vanadium	Zinc	

Table 2. Some chemical properties of the selected heavy metals (Greenwood and Earnshaw, 1994)

Property	CD	CR	CU	HG	NI	PB	ZN
Atomic number	48	24	29	80	28	82	30
Atomic weight	112.41	51.99	63.55	200.59	58.69	207.2	65.38
Ionic radius$_{\text{most common}}$	(M^{II}) 95	(M^{III}) 61.5	(M^{I}) 77	(M^{I}) 119	(M^{II}) 69	(M^{II}) 119	(M^{II}) 74
(pm)		(M^{VI}) 44	(M^{II}) 73	(M^{II}) 102			
Density$_{20\text{-}25°C}$ (g/cm^3)	8.65	7.14	8.95	13.53	8.91	11.34	7.14

Table 3. Some plant nutrients affected by Cd, Cr and Ni (synergisms +, antagonisms -)

Heavy metal	Macronutrient	Micronutrient
Cd	P (+) Ca (-)	Mn (-) Zn (-)
Cr	P (-)	Fe (-) Mn (-) Zn (-)
Ni	N (+) Mg (-)	Fe (-) Mn (-) Zn (-) Mo (-)

The chemical properties are the key to understand why these heavy metals are so dangerous. Properties like ionic radius in water solution (the soil solution), are the substantial base for their toxicity and availability.

Metals are presented normally at the same time in soil. So, the effects derived from their action can be sometimes due to several metals acting at the same time. It is possible that the effects observed with any one metal may actually have been due to another, or to a combination of metals. In some cases the metal(s) responsible for the adverse effects can be identified even though other metals were presented in the soil (McGrath *et al.*, 1995).

All of them are presented in residues coming from wastewater treatment and also in solid residues from industries. For these residues, soil is most of the times the final destination.

The problem of these chemicals derives from the accumulation in environmental compartments such soils and waters. In this line, the concepts of chemical time bombs should be applied for these contaminants. This concept is focussed on long-term problems of the delayed effects of pollution, due to the accumulation of potentially toxic chemicals and particular attention is being paid to the relationship between toxification and environmental changes (land use/climate), with a view to potential future land use (Smidt, 1990). In this case, the chemical time bomb is mostly associated to the presence of heavy metals in soils and the possible change of soil environments, mobilizing the pollutants to other compartments: water and living organisms.

Many works have demonstrated the negative effects of these pollutants on waters (van Beek *et al.*, 2007) and living organisms (Khan *et al.*, 2007). These metals affect plant nutrition, as an example of Cr and Cd that reduce the nutrient availability and uptake by plants (Moral *et al.*, 1996a; Navarro-Pedreño *et al.*, 1997; Moral *et al.*, 1998; Balaguer *et al.*, 1998).

The environmental importance of metals can be assessment by the metal loadings to media, the metal distribution in media and finally the exposure to metal. One step more is the risk assessments to human that need data of metal distribution in tissues, risk toxicity for individual organisms and populations.

2. Source of Heavy Metals

The source of heavy metals is variable: from anthropogenic sources (for instance, addition of sewage sludge, industrial or mine activities) and natural found. However, the pollution is commonly associated with man actions disturbing the soil and favouring the pollution.

Table 4. Estimated occurrence in soils and earth crust of heavy metals (1) (Navarro-Pedreño *et al.*, 1993) and soils (2) (Lindsay, 1979)

Occurrence (mg/kg)	Cd	Cr	Cu	Hg	Ni	Pb	Zn
Soil (1)	0.01-0.7	10-5000	2-100	0.01-0.3	5-2000	2-200	10-300
Soil mean value (1)	0.06	84	30	0.03	35	10	30
Earth crust (1)	0.2	200	70	0.1	80	16	70
Soil common range (2)	0.01-0.7	1-1000	2-100	0.01-0.03	5-500	2-200	10-300
Soil average (2)	0.06	100	30	0.03	40	10	50

For a determined area, anthropogenic sources (like the addition of sewage sludge in agricultural lands) use to favour the presence of a great variety of metals (not need in high amounts) although natural found should be related to a few kinds of metals.

An important source in industrial countries is the atmospheric deposition, especially for Cd, Hg and Pb (Davis *et al.*, 2001; Slootweg *et al.*, 2007; Norton, 2007). Other important sources are the roads and highways, especially for Cu, Zn, Cd and Pb (Ellis *et al.*, 1987;

Makepeace *et al.*, 1995; Muthanna *et al.*, 2007). However, in Mediterranean environments, organic matter is usually presented at low level in arid and semiarid soils. The scarcity of animal organic manures and the addition of other organic materials may promote the pollution and mobility of heavy metals.

The natural occurrence of these metals are the minerals and rocks that containing amounts of the metal. In the case of Cd, the presence is detected in zinc and phosphorus ores and greenockite is the mineral where it is found. The principal ore for Cr is chromite. For Pb, the main presence is in anglesite, cerussite and galena. The last mineral has representative ores in Mediterranean regions of Spain. Ni is obtained from pentlandite but it's a common element in iron ores, presented in the Mediterranean region. Cu and Zn and widely distributed in rocks, soil, water, air and living organisms.

The main sources in Mediterranean environments are:

- Agriculture activities: fertilizers, manure, use of sewage sludge and other organic residues (is every time more frequent the use of compost from sewage sludge as organic manure)
- Industrial activities: mining waste and abandoned industrial areas, chemical waste dumps (Cartagena in the south-east of Spain, Sagunto in Valencia or Aznalcollar in Seville)
- Urban, domestic and commercial activities: incineration of wastes, landfills and urban wastes dumps (presenting along the Mediterranean coast)
- Energy use: batteries and electrical components

The major pathways of input and presence of heavy metals in soils are:

- Vertical transport through soils to deeper layers after atmospheric deposition
- Lateral surface runoff and subsurface runoff within the soils
- Inputs from groundwater (evaporation, evapotranspiration, waterlogging, raising of water table, sea-level rise)
- Direct anthropogenic inputs, such agriculture, settlements, industrial and mining dumps, and waste dumps
- Natural occurrence.

Major pathways by which heavy metals are dispersed from soils:

- Leaching into water
- Uptake by plants or other biota
- Volatilization to the atmosphere
- Degradation of soil compounds.

The interest is also associated to the presence of these elements in soil amendments, like sewage sludge, as the legislation of Europe and America indicated.

The application of sewage sludge to land in member countries of the European Economic Commission (EEC) is governed by Council Directive No. 86/278/EEC (Council of the European Communities, 1986). This Directive prohibits the sludge from sewage treatment plants for being used in agriculture unless specified requirements are fulfilled, including the testing of the sludge and the soil. Parameters subject to the provisions of the Directive include the following:

- Dry matter (%)
- Organic matter (% dry solids)
- Copper (mg/kg dry solids)
- Nickel (mg/kg dry solids)
- pH
- Nitrogen, total and ammoniacal (% dry solids)
- Phosphorus, total (% dry solids)
- Zinc (mg/kg dry solids)
- Cadmium (mg/kg dry solids)
- Lead (mg/kg dry solids)
- Mercury (mg/kg dry solids)
- Chromium (mg/kg dry solids).

To these parameters, the UK Department of the Environment (1989) has added molybdenum, selenium, arsenic and fluoride in the "Code of Practice for Agricultural Use of Sewage Sludge". Sludge must be analyzed for the Directive parameters at least once every 6 months and every time significant changes occur in the quality of the sewage treated. The frequency of analysis for the additional four parameters may be reduced to not less than once in five years provided that their concentrations in the sludge are consistently no greater than the following reference concentrations: Mo - 3mg/kg dry solids, Se - 2mg/kg dry solids, As - 2mg/kg dry solids and Fl - 200mg/kg dry solids.

In Spain, the RD 1310/1990 is the transposition of European directive. Nowadays, a discussion about introducing new parameters to be analyzed in sewage sludge is in the EC countries, for instance, the determination of several microbial parameters like helminths in wastewater and sewage sludge as several European projects are studying, for instance the WATERREUSE project (http://www.waterreuse.wur.nl).

Table 5. Load and cumulative load of heavy metals from sewage sludge

	Annual Load rate (kg/ha dw)	Cumulative loading rates (kg/ha dw)	Annual load (mg/kg dw) mean of ten years
Cd	1.9	39	0.15
Cr	150	3000	3
Cu	75	1500	12
Hg	0.85	17	0.1
Ni	21	420	3
Pb	15	300	15
Zn	140	2800	30
	USEPA 1993	USEPA 1993	86/278/EEC

Table 6. Concentration of heavy metals in composted organic matter

	Concentration limits (mg/kg dw)	Total concentration in organic farming compost	Sludge content (mg/kg dw) for agricultural purposes (depending soil pH under or over 7)
Cd	85	0.7	20-40
Cr	3000	70 (Cr(VI) below detection)	1000-1500
Cu	4300	70	1000-1750
Hg	57	0.4	16-25
Ni	420	25	300-400
Pb	840	45	750-1200
Zn	7500	200	2500-4000
	USEPA 1994	Organic farming regulation (EEC) N° 2092/91	86/278/EEC

Another source of heavy metals can be the composted or fermented household waste, provided that it comes from source separated waste of vegetable or animal origin in a closed and monitored collection system. The maximum concentrations in mg/kg of dry matter of the metals are also determined.

3. SOIL PROPERTIES AND AVAILABILITY OF HEAVY METALS

The bioavailability of metals is associated to the environmental properties and the kind of organisms exposed to the metal and their interactions. Mostly, the plants are the main source of metals to animal diet. The importance of this fact produce that the main works related to the availability of heavy metals have been done using plants.

The availability of heavy metals to plants and the mobility along the soil profile (source of pollution of waters) depend mainly on the type of metal and the soil properties. These properties favour the presence or not of the metal in soil solution and after that, metal can be available for living organisms or can produce pollution of waters.

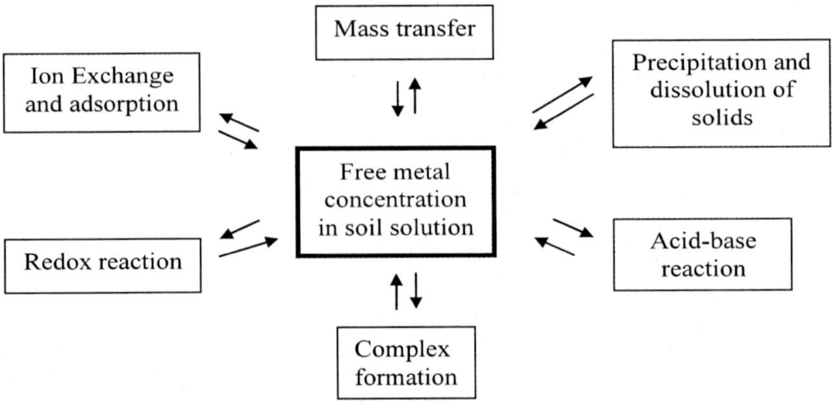

Figure 1. Principal controls on free metal presence in soil solution (Mattigod *et al.*, 1981).

Mattigod *et al.* provide and scheme to explain the principal controls on free trace metal concentrations in soil solution (1981) that it is still useful to understand the behaviour of heavy metals in soils.

The main soil properties affecting the availability of the metals in the Mediterranean environments are:

a. Surface charges and sites. The sorption of heavy metals in soil surfaces is one of the most important mechanisms to immobilize heavy metals in soils. However, the redox state of soil surfaces is affected by other parameters like pH and the sources of organic matter. In both cases, the pH should be 7 or more and organic matter is most of the times in low rates. The surface charges are usually negative and positive ions are easily retained more than negative ions such chloride, nitrate or sulphate. The adsorption, the accumulation of ions at the interface between a solid phase and an aqueous phase, has the key control of metals. Obviously, the sorption is proportional to surface area and surface site density, so the presence of clay compounds and organic matter in soils could improve the capacity of sorption of heavy metals, but pH is still a key factor in surface charge.
b. Water/dry soil. The size of the ion and the interaction with other metals define the behaviour of the element in water solution (Navarro-Pedreño *et al.*, 1997). In this line, the movement up or down of the metal and the hydrated sphere around the ion are determined by the presence of water in the soil. The metals can go down into the soil profile and may be considered as pollutants of subsurface waters due to the presence of an excess of water in soil and the permeability. In the same way and like the rest of the soluble compounds in water, a dry soil makes easy the movement to soil surface due to evaporation and evapotranspiration (in the case of the presence of plants). Iron and zinc are found in all the water samples and lead, aluminium and nickel were the second most frequently (Yilmaz *et al.*, 2003) found in the Mediterranean Turkey and this result may be extended to the rest of the Mediterranean area. The redox state of metals is also important to determine the presence in aqueous or solid phase. The changes in the water content of the soil alter the redox state, especially for iron and manganese.
c. pH. Soil pH controls all aspects of soil environment. Increasing soil pH also results in an increase in the number of negatively charged soils sites, the sorption and removal from pore water (Bohn *et al.*, 1985). Many metals may precipitate to form a three dimensional solid, depending on the pH of the soil solution. The substitution of Cd for Ca in calcium carbonate is an example, due the existence of a competitive Cd-Ca interaction (Moral *et al.*, 1996b). The stability diagrams of solubility of chemical compounds containing the heavy metals depending on pH may help to predict the possible formation of precipitates of metals in soils. The pH regulates the activity of the metal in the solution. A basic pH usually promotes the retention and/or precipitation of the heavy metals in form of carbonates, phosphates and oxides. Harter (1983) indicated that retention did not increase significantly over pH 7 for Cu, Ni, Pb and Zn, so precipitation may be the most important process in high pH values.
d. Soil organic matter. The organic matter of soils is composed by several fractions: fresh organic matter, recent decomposed organic matter, humic substances and recalcitrant organic matter. Metal bioavailability often is not related to the differing

composition of organic matter (Walker *et al.*, 2003). But the most important fraction related to heavy metals and their availability is the dissolved organic matter and the humic substances. These substances provide sites for the sorption of metals (mainly due to the presence of acid functional groups). For instance, Ashworth and Alloway (2007) found a good correlation between the dissolved organic matter from sewage sludge and the complexation of Cu in pH close to neutrality, which means that Cu may be more available to plant absorption due to the addition of organic matter, and may be a possible pollutant source of heavy metals. It could be expected that metals with chemistry close to Cu, may have the same behaviour. In this way, Linde *et al.* (2007) showed that the concentration of dissolved organic matter was the most important governing factor for dissolution of Cu, Cr, Hg and Pb.

e. Humic substances. In the case of humic substances, low ratios of heavy metal/humic substances are found under natural environmental conditions, but after the addition of exogenous materials the ratio could change. The presence of humic substances could favour the adsorption of heavy metals and a competition among them could be established. Jeong *et al.* (2007) studied the competition for the sorption sites and type of ligand model for Cd, Cu and Zn, and the results show that Cu has a greater binding affinity than the other two metals. Cadmium is a soft cation, larger and more polarizable than zinc, which is considered to be a borderline case (Ramalho and Figueroa-Villar, 2002). Thus, the complex fulvic acids and Cd should interact better with water molecules than Zn complex, leading to a more stable complex (Ramalho *et al.*, 2007) which means that Cd presence in soil solution could be favoured by the presence of fulvic acids. These results suggest that is so important the kind of metals presented in soils and the opposite effects of dissolved organic matter and humic substances. When sewage sludge is added to soil, it is desiderable that organic matter of the residue may be stabilized previously and poor dissolved organic matter should be added to avoid heavy metal solubility. Processes such as composting can reduce the toxicity (Roldan *et al.*, 1996).

f. Salt concentrations. The presence of soluble salts in soil solution can modify the solubility of the heavy metals. But not of them are affected by this property in the same way. Those that have a higher ionic radius are less strongly bounded to the soil surfaces and consequently, the metals are more easily solubilized. Linde *et al.* (2007) demonstrated that the increment of salt concentration caused considerably increased dissolution of Cd, Ni and Zn, but Cu, Cr, Hg and Pb were less affected. In the other hand, salt concentration affects microbial activity of soils, and can be inhibitory to compost decomposition (Tester and Parr, 1983).

g. Carbonates. The presence of carbonates, especially calcium carbonate, is very important in Mediterranean environments. As it has been shown, there are interactions between metals and carbonates depending on pH and most of the metals can be precipitated as carbonates (Navarro-Pedreño *et al.*, 2003). The combination of these properties determines the heavy metal behaviour and the availability. Wong *et al.* (2007), found that dissolved organic matter from sewage sludge, in calcareous soils, with a pH over 7, affected the sorption capacity and facilitated the presence of complexed organic forms of Zn and Cd in the soil solution. Although the work was done in Chine, these factors are frequent in Mediterranean soils and attention may be paid to these results.

h. Biological activity and vegetation. Most of the metals affect negatively the living organisms, excepting those that are especially adapted to the metal. It has been demonstrated in several works that Cd affected negatively the biological activity of soil (Moral *et al.*, 1999). The use of plants adapted to the metal for soil pollution restoration is a technique that involves the activity of plants and microorganisms in the soils. The phytoextraction of heavy metals may be a friendly technique to remediate these soils, and research is mainly focused on plants with high heavy metal uptake, so called "hyperacumulating" plant species (Koopmans *et al.*, 2007). Also, these plants can produce a great amount of biomass living in extreme conditions. *Thlaspi caerulescens* is a plant that can resist in mine soils and accumulate important amounts of Pb and Zn (Martínez *et al.*, 2006) and are used in semiarid Mediterranean environments. However, most of these plants are not adequate for Mediterranean environments because they are out of their common environments. So, a good strategy to look for a phytoextraction method in Mediterranean environments is to find which kind of plants naturally presented in this areas are able to accumulate heavy metals and can be used for soil remediation. For instance, Chopin and Alloway (2007) studied the distribution and mobility of trace metals in Huelva (Spain). The mining area has acid soils under semiarid climate and they studied the *cistacea* family represented by *Cistus ladanifer*, with strongly aromatic and sticky leaves, and *Plantago coronopus* L. that constitutes the main ground-covering vegetation. They found that the latest specie has accumulation ratios higher than the first one for As, Cu and Pb. But, the problems are not that plants adapted to Mediterranean environments were or not able to absorbed heavy metals, the main problem is if the species are able to live in soils with high heavy metal concentrations. But, the existence of a natural vegetable cover, in areas presenting an atmospheric source of contamination by heavy metals, would thus have, in an indirect way, a protective effect not only on the environment but also on population health and thus on quality life (Kadem *et al.*, 2004).

4. Analytical Methods to Estimate the Availability of Metals

The total concentration can determine the vertical or horizontal concentration of the metal in the soil. But this may not indicate the possible toxicity of the element. The potential migration of metals from soil systems promotes the necessity of applying sequential extraction methods and in fact, the development or toxicity test, especially for human exposure.

In the case of total concentration, a digestion with nitric acid, perchloric acid and hydrochloric acid or a combination and the use of aqua regia are the main analytical procedures to determine heavy metal content.

Examples of methods to determine the total concentration are the following:

a. Total digestion. Digestion of 2.5 g of soil with 50 mL of 7 M HNO_3 with recourse to heating for 2 hours in a heating block with a temperature of 120 ºC (Fitamo *et al.*, 2007).
b. USEPA method 3051A (1988): 0.500 ± 0.001 g of soil sample weighed in a tetrafluoromethoxyl (TFM) vessel (100 mL capacity), then 9 mL of HNO_3 is added and 3 mL of HCl too. Sample and reagents were digested for 5 min at 700 W, and 10 min at 500 W in a microwave oven. After cooling, separation of the extracts from the solid residue is achieved by centrifugation and then diluted to 25 mL with ultrapure water. Really, the methods provides more reliable results if it is used closed vessels and assures a quick and safe extraction process (Micó *et al.*, 2007a).

The most common method used is the sequential extraction. The use of one extracting agent in a sequential method is the most common for soil characterization and availability.

The common fractions determined by several methods are: exchangeable, fraction associated to carbonates, fraction associated to oxides, fraction associated to organic matter and the residual fraction (recalcitrant components of the metal).

The extractants used differ from one author to other. However, it is possible to reduce them to the following scheme:

- Exchangeable: chloride, nitrate and acetate salts at concentrations of 0.2 to 1 M.
- Fraction associate to carbonates: Acetic and acetate salts at pH from 4.5 to 6.5.
- Fraction associated to oxides: acetic, acetate salts and chelant agents like EDTA or citrate at pH from 2 to 6.5.
- Fraction associated to organic matter: phosphate, acetate salts and also H2O2.
- Residual fraction: hard acids like HF, nitric acid or perchloric acid.

Abollino *et al.* (2006) applied the sequential extraction to assess the metal availability in Piedmont (Italy) using the Telssier's scheme in five fractions as follow:

- 8 mL of 1M $MgCl_2$, for 1 h, at room temperature;
- 8 mL of 1 M CH_3COONa, adjusted to pH 5.0 with CH_3COOH, for 5 h, at room temperature;
- 20 mL of 0.04 M $NH_2OH{\cdot}HCl$ in 25% CH_3COOH, for 6 h at 96 ± 3 ºC;
- 5 mL of 30% H_2O_2 and 3 ml of 0.02 M HNO_3, for 5 h at 85 ± 2 ºC, followed by addition (after cooling) of 5 mL of 3.2 M CH_3COONH_4 in 20% HNO_3, dilution to 20 mL, and further extraction for 30 min at room temperature.

Cresser (2007) used the following sequential extraction method:

i) 50 mL of 1 M NH_4NO_3 pH 7.1, 1 hour shaking, 20:1;
ii) 50 mL of 0.5 M CH_3COONH_4 + 0.02 M EDTA, pH 4.65, 1 hour shaking, 10:1;
iii) 100 mL of 0.175 M $(NH_4)_2C_2O_4$/0.1 M $H_2C_2O_4$ pH 3.25, boiling or 15 min occasional agitation, 40:1;
iv) 50 mL 0.1 M $Na_4P_2O_7$, shaking overnight, 20:1; total minus sum of the extractable (1-4) 50 mL of 7 HNO_3, 20:1.

In both extraction procedures, the expected chemical forms of the five phases are the following: exchangeable, bound to carbonates, bound to iron and manganese oxides (reducible), bound to organic matter (oxidizable), and residual phase.

Other methods are designated to study the availability for plants more than exchangeable metals and the fraction where the metal is bounded. The most used are the extraction of Soltanpour *et al.* (1976, 1979) for Fe, Zn. Mn, Cu (and also P and K) and Lindsay-Norwell (1978) developed for Zn, Fe, Mn and Cu, based in the use of DTPA. For extension, these methods are also used to study the bioavailability of heavy metals. But the procedure is sometime questioned because the metals differ in their chemistry and behaviour in soil.

5. Heavy Metal Content in Mediterranean Environments

Along the past thirty years, many researches have been done in Mediterranean soils, but mainly related to fertilization parameters and the use of organic manures. However, works related to soil pollution and especially related to heavy metals have been increased in the last years. Some of then are revised and discussed focussed in the relation among heavy metals and soil characteristics.

The Mediterranean agricultural soils use to be basic, carbonated, having organic matter content and loam textures (Peris *et al.*, 2007). Although natural soils are very variable, most of these properties of the agricultural soils are also presented, but salinity or low organic matter content are characteristic factors of natural soils in arid and semiarid Mediterranean regions.

In Agricultural soils, it has been demonstrated that the use of organic manures and biosolids under good management, can be used to keep important levels of soil organic matter. Relatively low application rates of sewage sludge could be used for several years to maintain crop production in Mediterranean-types climates (Antolín *et al.*, 2005) and they improve soil physical properties in mechanized vineyards (Ramos, 2006) and horticultural crops (Navarro-Pedreño *et al.*, 1997; Casado-Vela *et al.*, 2006).

In general, low bioavailability of metals and leaching may be expected in this type of soils. The levels of organic matter, high percentages of clay and the presence of carbonate seem to suggest an important retention of heavy metals by these components (Micó *et al.*, 2006a).

In natural soils, the factors controlling the availability of metals are the carbonate content, soil organic matter and salinity. The input of heavy metals in natural soils is mainly due to atmospheric deposition and the use of organic residues for soil restoration (Larcheveque *et al.*, 2006), and also the uncontrolled waste dumps. For instance, in the case of restoration of soils in quarries activities, poor displacement of heavy metals have been found due to the majority of quarries are carbonate ores and the presence of carbonates is a factor that can control the movement of the metals, even though in the case of using sewage sludge for soil restoration and rehabilitation (Almendro-Candel *et al.*, 2007).

In Spain and as a part of the EC countries, the limit value of heavy metals in soils in dry matter, is closely related to the soil pH, as the following table shows.

In general, there are not so many works in Mediterranean regions studying the content of heavy metals to establish the baseline of them in oils. Baseline values for heavy metals were proposed in Alicante (Spain), a representative agricultural area of European Mediterranean region. Micó *et al.* (2007b) analyzed fifty-four agricultural plots and proposed a baseline. It could be compare with the mean values of the Segura River Valley (Micó *et al.*, 2006b), Serpis River Valley (Almendro-Candel *et al*, 2002). and Ramos (2006) in area of Penedès (Catalonia, Spain).

It is important to remark that in many Mediterranean countries, the scarcity of water and the important increment of wastewater treatment produce amounts of sewage sludge and their destination is frequently the soil (agriculture land and others). The addition of low metal content sewage sludge does not increase the leaching of metals in basic or acid Mediterranean soils. Extractable metals in calcareous Mediterranean soils are not great increased due to addition of low pollutant sewage sludge, as an example of Cu and Zn; the increment was less than 10% in DTPA extractable fraction (Casado-Vela *et al.*, 2006) or Cd, Cu, Ni and Zn (Almendro-Candel *et al.*, 2002).

In calcareous soils, not important amounts of heavy metals (Ni and Cd) were found in experimental soil columns (Navarro-Pedreño *et al.*, 2004). The same results were observed by Toribio and Romanyà (2005) for Cu and Zn and suggested that the low mobility of metals in the Mediterranean basic soils may be related to high $CaCO_3$ and organic matter content.

Table 7. Limit value of heavy metals content in soils (mg/kg dw) in Spanish law

Heavy metal	Limit value Soil pH<7	Limit value Soil pH>7
Cd	1	3
Cr	100	150
Cu	50	210
Hg	1	1.5
Ni	30	112
Pb	50	300
Zn	150	450

Table 8. Heavy metal content in Mediterranean coastal area of Spain

Metal	Baseline proposed (mg/kg d.w.)	Segura River Valley mean values (mg/kg d.w.)	Serpis River Valley (mg/kg d.w.)	Total content in Penedès (NE Spain) (mg/kg dw)
Cd	0.7	0.38	0.39	
Cr	36	28.3	21.2	
Cu	28	21.6	9.7	33-77
Ni	31	23.7	18.2	
Pb	28	19.6		
Zn	83	57.8	36.2	44-90

In general, Mediterranean soils are capable of assimilating low amounts of heavy metals and the natural occurrence of the pollutants is limited to mine sites. The organic matter content and the carbonate presence are key factors that control the heavy metals in these environments associated to the pH. The studies done in the Southeast of Spain can be useful to understand the problem of the metals in arid and semi-arid environments of other countries of the Mediterranean basin.

6. References

Abollino, O.; Giacomino, A.; Malandrino, M.; Mentaste, E.; Aceto, M.; Berberis, R. (2006). Assessment of metal availability in a contaminated soil by sequential extraction. *Water Air Soil Pollut.* 137:315-338.

Almendro-Candel, M.B.; Navarro-Pedreño, J.; Gómez, I.; Jordán, M.M.; García, E.; Mataix-Solera, J. (2002). Movement of Fe, Mn, Cu and Zn in a sewage sludge-treated soil. *Waste Management and the Environment* 2002, 311-320. Ed. WitPress, Southampton.

Almendro-Candel, M.B.; Jordán, M.M.; Navarro-Pedreño, J.; Mataix-Solera, J.; Gómez, I. (2007). Environmental evaluation of sewage sludge application to reclaim limestone quarries wastes as soil amendments. *Soil Biol. Biochem.* 39(6):1328-1332.

Antolín, M.C.; Pascual, P.; García, C.; Polo, A.; Sánchez-Diaz, M.; (2005). *Field Crops Research* 94(2-3):224-237.

Ashworth, D.J.; Alloway, B.J. (2007). Complexation of copper by sewage sludge-derived dissolved organic matter: effects on soil sorption behaviour and plant uptake. *Water Air Soil Pollut.:* 182:187-196.

Balaguer, J.; Almendro, M.B.; Gómez, I.; Navarro-Pedreño, J.; Mataix, J. (1998). Tomato growth and yield affected by nickel presented in the nutrient solution. *Act. Hort.* 458:269-272.

Bohn, H.L.; McNeal, B.L.; O'Connor, G.A. (1985). *Soil chemistry,* 2nd ed. Ed. John Wiley & Sons, New York.

Casado-Vela, J.; Sellés, S.; Navarro-Pedreño, J.; Bustamante, M.A.; Mataix, J.; Guerrero, C.; Gómez, I. (2006). Evaluation of composted sewage sludge as nutricional source for horticultural crops. *Waste Management* 26:946-952.

Chopin, E.J.B.; Alloway, B.J. (2007). Distribution and mobility of trace elements in soils and vegetation around the mining and smelting areas of Tharsis, Riotinto and Huelva, Iberian Pyrite Belt, SW Spain. *Water Air Soil Pollut.* 182:245-261.

Cresser, M.S. (2007). Why critical loads of acidity and N for soils should be based on pollutant affective concentrations rather than deposition fluxes. Water Air Soil Pollut: Focus 7:407-412.

Davis, A.P.; Sholouhian, M.; Ni, S. (2001). Loadings of lead, copper, cadmium, and zinc in urban runoff from specific sources. *Chemosphere* 44(5):997-1009.

Ellis, J.B.; Revitt, D.M.; Harrop, D.O.; Beckwith, P.R. (1987). The contribution of highway surfaces to urban stormwater sediments and metal loadings. *Sci. Total Environ.* 59:339-349.

EPA (2007). *Framework for metals risk assessment.* EPA 120/R-07/001 March, 2007.

European Environment Agency (2002). Down to earth: Soil degradation and sustainable development in Europe - *A challenge for the 21st century.* ISBN: 92-9167-398-6. Edited EEA, Copenhagen (Denmark).

European Soil Bureau – Scientific Committee (1999). Heavy metals (trace elements) and organic matter content of european soils. Ed. Joint Research Center, European Commission. *European Communities,* Luxembourg.

Fitamo, D.; Itana, F.; Olsson, M. (2007). Total contents and sequential extraction of heavy metals in soils irrigated with wastewater, Akaki, Etiopía. *Environ Manage* 39:178-193.

Greenwood, N.N.; Earnshaw, A. (1994). Chemistry of the elements. Ed. Pergamon Press Ltd, Oxford (UK).

Harter, R.D. (1983). Effect of soil pH on adsorption of lead, copper, zinc, and nickel. *Soil Sci. Soc. Am. J.* 47:47-51.

Jackson, A.P.; Alloway, B.J. (1992). The transfer of cadmium from agricultural soils to the human food chain. In *"Biogeochemistry of trace metals",* chapter 5, pages 109-158, edited by D.C. Adriano. Ed. CRC Press, Boca Raton (USA).

Jeong, C.Y.; Young, S.D.; Marshall, S.J. (2007). Competitive adsorption of heavy metals in humic substances by simple ligand model. *Soil Sci. Soc. Am. J.* 71:515-528.

Kadem, D.E.D.; Rached, O.; Krika, A.; Gheribi-Aoulmi, Z. (2004). Statistical analysis of vegetation incidence on contamination of soils by heavy metals (Pb, Ni and Zn) in the vicinity of an iron steel industrial plant in Algeria. *Environmetrics* 15(5):447-462.

Khan, K.S.; Chander, K.; Hartmann, G.; Lamersdorf, N.; Joergensen, R.G. (2007). Sources of heavy metals and their long-term effects on microbial C, N and P relationships in soil. *Water Air Soil Pollut.* 181:255-234.

Koopmans, G.F.; Römkens, P.F.A.M.; Song, J.; Temminghoff, E.J.M.; Japenga, J. (2007). Predicting the phytoextraction duration to remediate heavy metal contaminated soils. *Water Air Soil Pollut.* 181:355-371.

Larcheveque, M.; Baldy, V.; Montès, N.; Fernandez, C.; Bonin, G.; Ballini, C. (2006). Short-term effects of sewage sludge compost on a degraded Mediterranean soil. Soil Sci. *Soc. Am. J.* 70:1178-1188.

Linde, M.; Öborn, I.; Gustafsson, J.P. (2007). Effects of changed soil conditions on the mobility of trace metals in moderately contaminated urban soils. *Water Air Soil Pollut.* 183:69-83.

Lindsay, W.L.; Norwell, W.A. (1978). Development of a DTPA soil test for zinc, iron, manganese, and copper. *Soil Sci. Soc. Am. J.* 42:421-428.

Lindsay, W.L. (1979). *Chemical equilibria in soils.* Ed. John Willey & Sons, New York.

Makepeace, D.K., Smith, D.W.; Stanley, S.J. (1995). Urban stormwater quality: summary of contaminat data. *Environmental Science & technology* 25(2):93-139.

Mattigod, S.V.; Sposito, G.; Page, A.L. (1981). Factors affecting solubilities of trace metals in soils. In D.E. Baker (ed.) Chemistry in the soil environment. ASA Special Publication N° 40. *Amer. Soc. Agronomy,* Madison, WI.

McGrath, S.P.; Chaudri, A.M.; Giller, K.E. (1995). Long-term effects of metals in sewage sludge on soils, microorganisms and plants. *J. Industrial Microbiology* 14: 94-104.

McLean, J.E.; Bledsoe, B.E. (1992). *Behavior of metals in soils.* EPA/540/S-92/018 October-1992.

Martínez, M.; Bernal, P.; Almela, C.; Vélez, D.; García-Agustín, P.; Serrano, R.; Navarro-Aviño, J. (2006). An engineered plant that accumulates higher levels of heavy metals than *Thlaspi caerulescens*, with yields of 100 times more biomass in mine soils. *Chemosphere* 64(3): 478-485.

Micó, C.; Peris, M.; Sánchez, J.; Recatalá, L. (2006a). Heavy metal content of agricultural soils in a Mediterranean semiarid area: The Segura River Valley (Alicante, Spain). *Spanish Journal of Agricultural Research* 4(4):363-372.

Micó, C.; Recatalá, L.; Peris, M.; Sánchez, J. (2006b). Assessing heavy metal sources in agricultural soils of an European Mediterranean area by multivariate analisys. *Chemosphere* 65(5):863-872.

Micó, C.; Peris, M.; Recatalá, L.; Sánchez, J. (2007a). Baseline values for heavy metals in agricultural soils in an European Mediterranean region. Sci. Total Environ. 378(1-2):13-17.

Micó, C.; Recatalá, L.; Peris, M.; Sánchez, J. (2007b). A comparison of two digestión methods for the análisis of heavy metals by flame atomic absorption spectroscopy. *Spectroscopy Europe* 19(1):23-26.

Moral, R.; Gómez, I.; Navarro-Pedreño, J.; Mataix, J. (1996a). Absorption of Cr and effects on micronutrient content in tomato plant (*Lycopersicum esculentum* M.). *Agrochimica* 40(2-3):132-138.

Moral, R.; Gómez, I.; Navarro-Pedreño, J.; Mataix, J. (1996b). Cadmium adsorption in different substrata: influence exerted by Ca. *Agrochimica* 40(4):191-196.

Moral, R.; Navarro-Pedreño, J.; Moreno, J.; Gómez, I.; Mataix-Solera, J. (1998). Nutrients in a calcareous soil affected by cadmium. *J. Plant Nutr.* 21(9):1933-1941.

Moral, R.; Navarro-Pedreño, J.; Moreno, J.; Gómez, I.; Mataix-Solera, J. (1999). Effects of fertilisation on physico-chemical properties and nitrogen of a calcareous soil under Cd addition. *Arch. Acker-Pfl. Boden.* 44:123-130.

Muthanna, T.M.; Viklander, M.; Gjesdahl, N.; Thorolfsson, S.T. (2007). Heavy metal removal in cold climate bioretention. *Water Air Soil Pollut.* 183:391-402.

Navarro-Pedreño, J.; Gómez, I.; Mataix, J. (1993). Ciclos Biogeoquímicos de oligoelementos metálicos. In *"Metales en sistemas biológicos"* (Mas, A. and Azcue, J.M. editors). Ed. PPU, Barcelona (Spain).

Navarro-Pedreño, J.; Gómez, I.; Moral, R.; Palacios, G.; Mataix, J. (1997). Heavy metals and plant nutrition and development. *Recent. Res. Devel.* in Phytochem. 1:173-179.

Navarro-Pedreño, J.; Almendro-Candel, M.B.; Jordán-Vidal, M.M.; Mataix-Solera, J.; García-Sánchez, E. (2003). Mobility of cadmium, chromium, and níkel through the profiel of a calcisol treated with sewage sludge in the southeast of Spain. *Environment. Geology* 44:545-553.

Navarro-Pedreño, J.; Almendro, M.B.; Jordán, M.M.; Mataix-Solera, J.; García-Sánchez, E. (2004). Risk areas in the application of sewage sludge on degraded soils in the province of Alicante (Spain). Geo-Environment: Monitoring and remediation of the geological environment 2004:293-302.

Norton, S.A. (2007). Atmospheric metal pollutants-archives, methods and history. *Water Air Soil Pollut.:* Focus 7:93-98.

Peris, M.; Micó, C.; Recatalá, L.; Sánchez, R.; Sánchez, J. (2007). Heavy metal contents in horticultural crops of a representative area of the European Mediterranean region. *Sci. Total Environ.* 378(1-2):42-48.

Ramalho T.C.; da Cunha, E.F.F.; de Alencastro, R. B.; Espínola, A. (2007). Differential complexation between Zn^{2+} and Cd^{2+} with fulvic acid: a computacional chemistry study. *Water Air Soil Pollut.* 183:467-472.

Ramalho, T.C.; Figueroa-Villar, J.D. (2002). Thermodynamic evaluation of complexes of zinc and cadmium that mimetize metallic centres in transcription factors. *J. Molecular Structure (Theochem)* 580:217-226.

Ramos, M.C. (2006). Metals in vineyard soils of the Penedès area (NE Spain) after compost application. *J. Environment. Management* 78(3):209-215.

Roldan, A.; Albadalejo, J.; Thornes, J.B. (1996). Agrégate stability changes in a semiarid soil alter treatment with different organic amendments. *Arid Soil Res. Rehabil.* 10:139-148.

Slootweg, J.; Hettelingh, J.P.; Posch, M.; Schütze, G.; Spranger, T.; de Vries, W.; Reinds, G.J.; van t'Zelfde, M.; Dutchak, S.; Hyin, I. (2007). European critical loads of cadmium, lead and mercury and their exceedances. *Water Air Soil Pollut.: Focus* 7:371-377.

Smidt, G.R.B. (1990). Report of a European workshop on chemical time bombs. *Ed. Ministry of VROM,* Utrecht (The Netherlands).

Soltanpour, P.N.; Khan, A.; Lindsay, W.L. (1976). Factors affecting DTPA-extractable Zn, Fe, Mn, and Cu from soils. *Commun. Soil Sci. Plant Anal.* 7:797-821.

Soltanpour, P.N.; Khan, A.; Schwab, A.P. (1979). Effect of grinding variables on the NH_4HCO_3-DTPA soil test values for Fe, Zn, Mn, Cu, P and K. *Commun. Soil Sci. Plant Anal.* 10:903-909.

Tester, C.F.; Parr, J.F. (1983). Decomposition of sewage sludge compost in soil: IV. Effect of indigenous salinity. *J. Environ. Qual.* 12(1): 123-126.

Toribio, M.; Romanyà, J. (2005). Leaching of heavy metals (Cu, Ni and Zn) and organic matter after sewage sludge application to Mediterranean forest soils. *Sci. Total Environ.* 363(1-3):11-21.

Van Beek, C.L.; Droogers, P.; van Hardeveld, H.A.; van den Eertwegh, G.A.P.H.; Velthof, G.L.; Oenema, O. (2007). Leaching of solutes from an intensively managed peat soil to surface water. *Water Air Soil Pollut.* 182:291-301.

Van Camp, L.; Bujarrabal, B.; Gentile, A.-R; Jones, R.J.A.; Montanarella, L.; Olazábal, C.; Selvaradjou, S.-K (2004). *Report of the Technical Working Groups established Ander the thematic Strategy for Soil Protection,* vol. IV. Contamination and Land Management EUR 21319 EN/4, European Communities, Luxembourg.

Walker, D.J.; Clemente, R.; Roig, A.; Bernal, P. (2003). The effects of soil amendments on heavy metal bioavailiability in two contaminated Mediterranean soils. *Environmental Pollut.* 122(2):303-312.

Wong, J.W.C.; Li, K.L.; Zhou, L.X.; Selvam, A. (2007). The sorption of Cd and Zn by different soils in the presence of dissolved organic matter from sludge. *Geoderma 137:310-317.*

Yilmaz, H.; Erbatur, O.; Erbatur, G. (2003). Monitoring of pesticida and heavy metal discharge of the major rivers and drainage chanels into the Mediterranean sea in the north-east coastal zone, Turkey. *Fresenius Environ. Bull.* 12(11):1407-1411.

In: Soil Contamination Research Trends
Editor: Javier B. Dominguez, pp. 177-201

ISBN 978-1-60456-319-1

Chapter 8

THE ROLE OF NATURAL SELF-PURIFICATION PROCESSES IN IN-STREAM ON SITE CLEAN-UP ACTIVITIES

Anatoliy I. Fisenko[*]
ONCFEC, Inc., 625 Evans Avenue, Suite 1108,
Toronto, Ontario M8W 2W5, Canada

ABSTRACT

An in-stream on site clean-up approach for stream polluted sites, including the benthic soil and the entire water is proposed. The given approach is based on the proper complex of natural self-purification processes in a stream, leading to formation of froth on a surface with a large amount of all kind of polluting particles - organic and inorganic. Once formed, the natural froth is collected using a boom system and can then be removed and treated appropriately. As a result, we can intervene directly on site in the clean-up of the stream sites that are heavily polluted from various polluting sources, including the non-point ones - runoffs from agricultural and mining operations as well as discharges from sewage treatment plants and others. One important self-purification process in entire water and in/on benthic soil is the decomposition of total organics, including dead fauna and flora and also industrial organic pollution, by fungi, anaerobic and aerobic bacteria and other microorganisms. During the decomposition, the entire water body as well as the benthic soil is periodically enriched with biological surfactants, such as amino, pyruvic, fatty and other acids and generated dissolved biogases – oxygen, ammonia, carbon dioxide, methane, nitrogen and others. Further, stream polluted sites also contain dissolved air and various polluting particles, including man-made surfactants. Therefore, the next step in river self-purification – the generation of biogas and air bubbles, and the attachment of the polluting particles to the bubbles in the presence of surfactants should be considered. For the latter, the proper level of turbulence must exist. Water cascading over weirs, waterfalls and other obstacles, creating the shallow-turbulent character of

[*] E-mail: afisenko@oncfec.com

water current, is the suitable condition. The particle-bubble aggregates as well as decomposed masses from the benthic soil enriched with biogas bubbles, will rise to the water surface and concentrate in the froth and the surrounding thin top layer of surface water upstream from the installed boom system. The resulting froth as well as the thin top layer of surface water comprises a high concentration of all kinds of polluting particles.

Keywords: self-purification processes, pollution, non-point sources, stream clean-up, spill, natural froth formation, activated sludge foaming, flocculation

1. Introduction

It is well-known that rivers and creeks have a natural environmental capacity to restore their polluted sites through a complex of biological, physical, and chemical self-purification processes [1-31].

The biological self-purification processes play an essential role in the river's self-restoration. One important process of the latter is aerobic and anaerobic decomposition of total organic matter in entire water as well as in/on benthic soil by fungi, bacteria, and other microorganisms to the stable inorganic solids. During these processes, biological surfactants are produced and dissolved biogases and microbubbles of biogases are generated.

Several functional biological filters [7-14] are also important for purifying aquatic systems. These include the aquatic plants, organisms and microorganisms that intercept pollution. Their natural functions are: a) preventing the aquatic system from rapid eutrophication [7-9] by filter-feeding organisms for removing phytoplankton and algae in the stream water body. A typical example of the organisms are: rotifers, bryozoans, crustaceans, and others; b) preventing the entry of pollution and the biogenic elements, such as nitrogen and phosphorus from the surrounding land into the stream [10, 11] by communities of aquatic plants; c) preventing the entry of the polluting particles and the biogenic elements into the stream from the benthic soil [12] by benthic organisms such as *Tubifex, Chironomus, Asellus,* green macroalgae *Dictyosphaeria cavernosa*, and others; d) filtering water by attaching the microorganisms to the particles suspended in the stream [13, 14].

In [32-34], the selected species of metal-reducing microorganisms such as *Geobacter* species for bioremediation of uranium-contaminated groundwater was studied. *Geobacter* are best known as iron-eaters. It is shown that these species allow to reduction of a soluble, oxidized form of uranium, U(VI), to a harmless insoluble U(IV) one.

Typical chemical and physical self-purification processes are: a) chemical-oxidation, hydrolysis, photochemical reactions, free radical-dependent destruction, and others; b) physical- sedimentation, adsorption, evaporation, aeration, flocculation, and others.

The complex of the natural self-purification processes is essential for maintaining the quality of the entire water and the benthic soil in rivers and creeks, and for providing habitats for a number of aquatic species with a unique set of advantages such as: 1) they act fast; 2) they act on the entire body of water and the benthic soil, and all the categories of pollutants; and 3) they work in all the seasons of the year.

In [35], a new stream self-purification process - natural froth formation has been discovered and described. It has been shown that a combination of the biological self-purification process, such as the decomposition of total organics, together with the physicochemical ones, stimulated by the proper level of turbulence, leads to the natural formation of the froth on the stream surface. The naturally formed froth collects a wider spectrum of polluting particles, including organic, inorganic as well as pathogenic bacteria, from the entire water and the benthic soil.

Based on new insights into stream self-purification, an in-stream on site clean-up method dealing with the *non-point* (unregulated) sources of pollution (NPS) has been proposed [35, 36]. The pollution from non-point sources originates mainly from stormwater, carrying pollutants from numerous origins such as farms, building and road construction, housing developments, towns, mining, and untreated sewage on and through the soil into the river water.

The *point* (regulated) sources of pollution are smoke stacks and discharge pipes. A general idea of the latter is to control the "ends of pipes" in order to reduce the pollutants going to rivers and creeks from discharge pipes, and thus to give them the possibility to restore themselves through the natural self-purification processes. Through pollution prevention, including stormwater management and spill control, the proposed *point* source approach provides benefits. This has been a typical approach around the world.

In fact, the problem of stream pollution originated from the *non-point* sources is more intractable. Authorities anywhere are now increasingly aware of this issue.

Photo 1. A large amount of the froth on creek surface.

The present paper is devoted to the subsequent development of the *non-point* sources approach for cleaning-up the polluted sites of streams, which was previously proposed in [35, 36]. By way of example, several sites of creeks in Greater Toronto Area, Ontario, Canada where the naturally formation of froth takes place have been investigated. It is shown that the froth collects a wider spectrum of polluting particles.

In Section 2, a brief description of how the polluted sites of the creek self-restored after a toxic spill are considered. After understanding that the natural froth formation process plays an essential role in river and creek natural self-restoration, we must answer the question: What are the natural processes involved? The nature of froth formation and the contribution of biological foam in it are considered in Section 3. The total picture of how a stream self-purifies through froth formation process is described in Section 4. Here, similar processes involved in the foam formation in different aquatic environments are considered. Materials and methods used in research are presented in Section 5. In Section 6, the chemical analysis of the removed froth contents as compared to the stream water quality both downstream and upstream from the installed boom system are presented in detail. As a result, the froth as well as the surrounding thin layer of surface water upstream contains a much higher amount of pollutants. Conclusion and practical recommendations are given in the final section. In appendix, additional evidence from entire water body is considered.

2. Natural Self-restoration of the Polluted Sites of the Creek after Toxic Spill

In [35], all the stages of the extensive natural self-purification processes in the polluted sites of the creek after the toxic spill during the self-restoration time were investigated in detail. Let us briefly describe these processes. One of the important processes in the creek self-restoration was the periodical appearance of froth on the creek surface below the weir and at the places downstream from the shallow-turbulent area. See Photo 1.

The proposed mechanism of natural froth formation process was the following. In the decomposing process of the large amount of the dead organic materials by fungi, bacteria and other microorganisms, the affected polluted sites of the creek, including the entire water and on/in the benthic soil were enriched with the biological surfactants such as humic, amino and other acids. Meanwhile, the dissolved biogases and the micro-bubbles of biogases (ammonia, methane and others) are generated. This system also contains the dissolved air, the man-made surfactants and the polluting particles.

For the new biogas and air bubbles to be generated, and for the polluting particles to be attached to the air and the biogas bubbles in the presence of the biological and the man-made surfactants, the proper level of turbulence must exist. Water cascading over the weirs, the waterfalls and other obstacles, creating the shallow-turbulent character of the stream current, is the suitable condition for the bubble-particle attachments. The particle-bubble aggregates will rise to the water surface of the river downstream from the turbulent areas and concentrate in the froth. The resulting froth comprises a high concentration of polluting agents.

Biogas bubbles, resulting from the decomposition of dead matter in the affected area of the benthic soil, are now rising to the surface when the water and the benthic soil were shaken up. See Photo 2.

Photo 2. Many biogas bubbles are coming up.

Having been enriched, during the decomposition of a large amount of dead organic matter, with a large number of the micro-bubbles of biogases, the black decomposed masses rose to the water surface. See Photo 3. These contained pollutants such as cyanide and heavy metals removed from the benthic soil.

General conclusions were:

- A period of three months was necessary for the selected polluted sites of the creek to restore themselves through the extensive natural self-purification processes after the toxic spill.
- The appearance of both the black masses from the benthic soil and the large amount of froth on the water surface below the weir and at the places downstream from the shallow-turbulent area was the important facts in the self-restoration of the creek.
- Important in the natural froth formation process are the activities of benthic and planctonic bacteria, including Pseudomonas, Bacillus, Chlorella, Carshesium, Oscillatoria [37] and others, which are the most resistant to cyanide and heavy metals, and other microorganisms, breaking up dead organic matter in the entire water and in/on benthic soil.
- The natural froth formation process in rivers is similar to the well-known froth flotation. As a result, the froth should collect cyanide and heavy metals from the entire stream, including the benthic soil.

Photo 3. The creek bottom becomes dark, with black masses rising to the surface.

With the knowledge of the natural stream purification through the froth formation process, our next step was to understand the mechanisms involved in this process.

3. THE NATURE OF FROTH FORMATION

What do you immediately notice when you observe the Photos of Niagara Falls? Of course, you are first struck by the majesty of the view. Then you are impressed by the power and volume of water. Look more closely. You will also see a great deal of froth on the water surface. See Photo 4. The froth gathers along the banks in large amount. Did you consider the question, why does the froth accumulate? This is the manner in which the river purifies itself. The qualitative explanation of removal process is the following: the water upstream from Niagara Falls (and any other) appears homogenous, but if you examine the water downstream, you will find it in two phases: the froth with heavy pollution, and the less polluted water, from which pollutants of all kinds were removed by air and biogas bubbles into the froth.

Photo 4. A view of Niagara Falls, with a great deal of the froth on the water surface.

Photo 5. The froth appears downstream from the weir in the creek.

Water cascading over weirs, waterfalls and other obstacles, creating the shallow-turbulent character of water current, is also the suitable condition for removal mechanism. See Photos 5-7.

When walking along any river and creek, you often notice the froth floating on the water surface, particularly downstream from waterfalls (Photo 4), weirs (Photo 5), rocks in water (Photos 6) and other places where the water current is shallow-turbulent (Photo 7).

The froth on rivers and creeks can be caused by both natural processes and man-made pollution. It is usually a combination of both. The simplest definition of froth is that it's a gas dispersed in a liquid.

The froth on river and creek surfaces results from a combination of the following:

- The presence of natural organic matter (e.g., dead fauna and flora), and/or industrial organic pollution (e.g., sewage and domestic waste, textile and paper manufacturing waste, runoff from inadequately stored animal waste, soaps, etc.).
- The activity of benthic and planctonic bacteria, fungi and other microorganisms decomposing the entire organic matter present in the polluted sites of rivers and creeks.

Photo 6. The froth downstream from the rocks.

Photo 7. The froth also occurs downstream from the shallow-turbulent area.

Due to the decomposition, the polluted site of stream contains: a) biological surfactants such as acids: amino, pyruvic, fatty, acetic and others; and b) dissolved biogases and the bubbles of biogases such as methane, oxygen, ammonia, carbon dioxide, nitrogen, hydrogen sulfide and others;

- The inorganic pollution (e.g., detergents, oil and chemical contamination).
- The proper level of natural or artificially created turbulent water flow or pumping air into a stream.

To summarize, every river possesses the own natural capacity to purify itself. The more organic matter (natural and organic pollution) is present in the water, the more biological and/or man-made surfactants the water contains. A river also includes the generated bubbles of air and biogases. Surfactants simply lessen the surface tension of stream water, making it more susceptible to foaming. When the bubbles are mixed with the natural or man-made surfactants in the water, the surfactants can cause a stable protective film around each bubble, which results in the generation of froth on the surface. The amount of froth formed on the surfaces of streams is strongly dependant upon the concentration of surfactants, the generated biogas and air bubbles and the proper level of turbulence.

Photo 8. The froth on Mimico Creek surface water.

Photo 9. The froth on Don River surface water.

Photos 8-11, depict examples of the froth formation on several streams (Ontario, Canada). The amount of froth differs in each case because of varying concentrations of surfactants and, air and biogas bubbles, and also due to the different turbulence levels.

As seen in Photos 8-11, froth arises from the same causes in all the streams.

Photo 10. The froth on Humber River surface water.

Photo 11. The froth on Etobicoke Creek surface water.

In conclusion, it is important to note the following. The naturally formed froth on river surface was usually considered the result of man-made pollution only, but the contribution of the biological froth in it, created by fungi, bacteria and other microorganisms, which play an essential role in river natural clean-up, was omitted. A typical example of the latter is the biological froth in sewer treatment plants. Therefore, we should answer the question: Does the naturally formed froth contain some biological froth? To prove that it does, we tested the froth and creek water samples in order to determine the concentrations of the groups of polyamine such as *putrescine* and *cadaverine*, which are two products of the breakdown of protein by fungi, bacteria and other microorganisms. Polyamines are biogenic amines and can come into stream water from living phytoplankton and zooplankton, and microorganisms as well as and their decaying cells.

The structural formulas of these compounds are the following:

$N_2H - (CH_2)_4 - NH_2$ *putrescine*
$N_2H - (CH_2)_5 - NH_2$ *cadaverine*

As seen, the *putrescine* and *cadaverine* contain two amino groups. A particular feature of the polyamine structure is the presence of a carbonic chain with amino group at the end of molecules.

The concentrations of tested polyamines were the following: a) *putrescine*: creek water - $0.03\,\mu g / L$; froth – $74.82\,\mu g / L$; b) *cadaverine*: creek water - $0.04\,\mu g / L$; froth – $105.1\,\mu g / L$. As seen, the test results are supported our concept that the naturally formed froth contain the contribution of biological froth.

4. Natural Self-Purification through Froth Formation Process

In Section 2, one of the conclusions was that the nature of froth formation resembles the well-known flotation. In this Section, we will explain this fact in more detail. It is known that flotation means a separation or removal process [38, 39]. As air is pumped into a mixture, it makes certain particles rise to the liquid surface, where they can be skimmed off. It is the tiny air bubbles which move the particles to the top.

What do we learn from the flotation? How can we apply the principle of flotation to river self-purification? Perhaps we can best approach this topic by determining the fundamental requirements for flotation in general. These are the following [40]:

- *"A gas bubble and a particle must come in contact with each other".*

To achieve this requirement, air or gas must be introduced into a mixture under pressure. This is, in fact, what nature does to streams. As you have already noticed earlier in the text, the air captured from the atmosphere enters the water. Besides, the bubbles of biogases are present there, due to the action of microorganisms in the decomposition of total organic matter.

The proper level of turbulence is a necessary condition for generating air and biogas bubbles, and also puts the particles in contact with the bubbles. The streams with a shallow-turbulent character of water current illustrate this phenomenon because of good mixing. Other examples are water cascading over the weirs, waterfalls, and others.

- *"The particle should have an affinity for attaching itself to the bubble".*

To achieve this requirement, a number of chemical reagents, which fall into the following five basic categories: a) collection; b) conditioning; c) levitation, d) frothing, and f) depressant should be added to the water. Some of the above reagents are:

- Frothers, surface active chemicals. They form an adsorbing film on bubble surfaces. As a result, bubbles could obtain an electric charge. Example: amino acids, detergents, cresylic acids, alcohols, pine oils, and others.
- Collectors, surface active organic chemicals. They form an adsorbing film on particle surface and could, like the frothers, get an electric charge too. Example: fatty acids, soaps, amine salts, and others.

Now let us consider what happens in a river.

The natural and/or man-made surfactants are already present because of the degradation process and the organic pollution. As a result of it, the river water (including the benthic soil) contains:

- Frothers such as amino acids, detergents and others.
- Collectors: fatty acids, soaps and others.

Air and biogas bubbles, which are generated by the proper level of turbulence, may adsorb the available surface active chemicals and obtain an electric charge. Besides, when the polluting particles adsorb the surface active organic chemicals, they obtain an electric charge, too. Upon being charged, the polluting particles are attracted to the bubbles and form the bubble-particle aggregates. Then their aggregates will be rising to the river water surface, and will concentrate in the froth and the surrounding thin top layer of surface water.

To summarize, natural self-purification through froth formation, similar to flotation, will occur in any site of river and creek with the proper level of turbulence. These sites are downstream from: a) an area where the river is shallow-turbulent; b) weirs; d) rapids and c) rocky ramp.

Here it is important to note that if the polluted site of a stream has deeper water, the air should be pumped into a site for separating polluting particles from stream water.

In brief, the following Chart 1 concisely present the material discussed above.

The natural self-purification process through froth formation described in Chart 1 relates to any polluting sources, including the non-point ones. Therefore, this process applies to pollution removal in general.

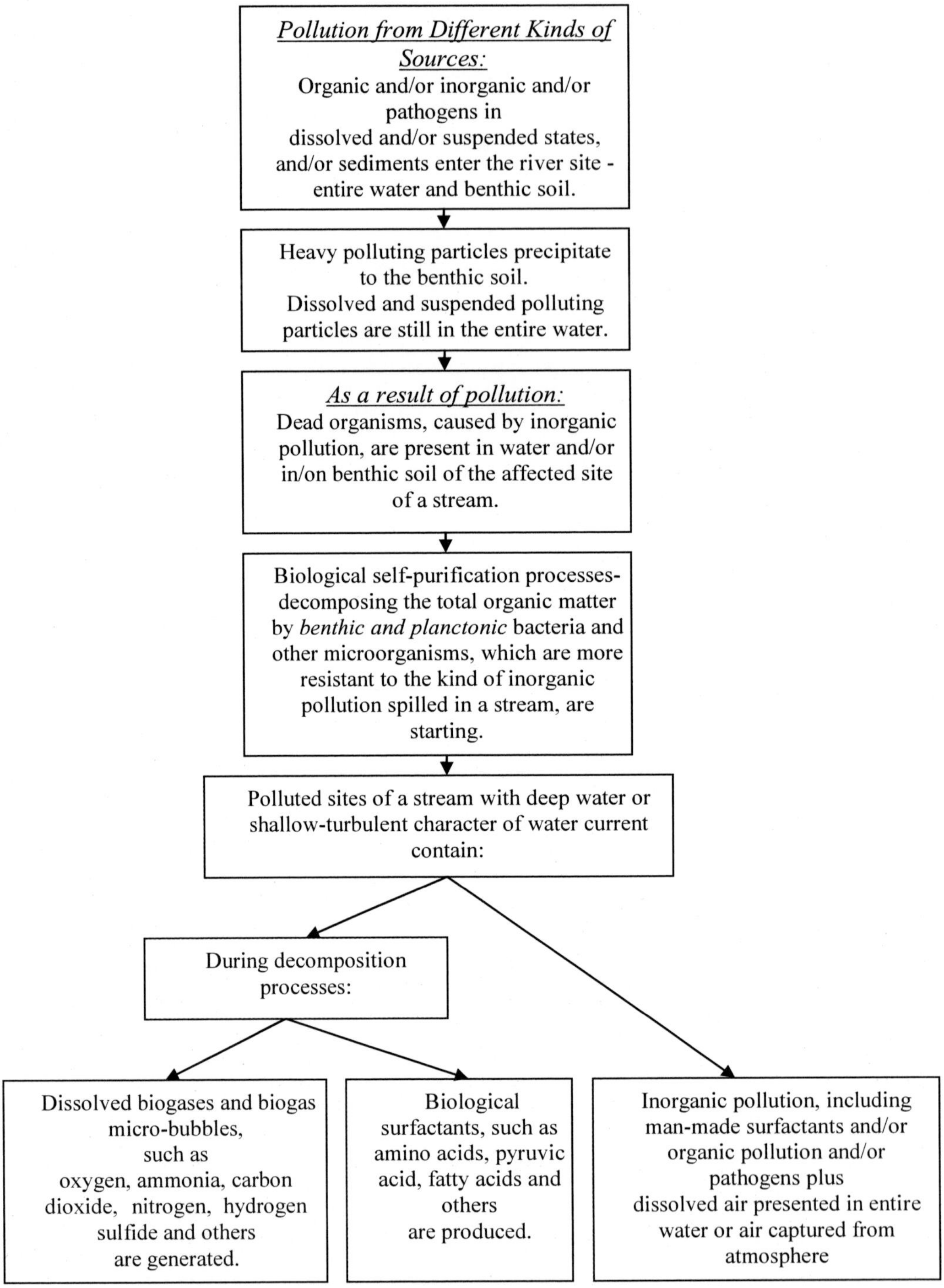

a.

Chart 1. Continued on next page.

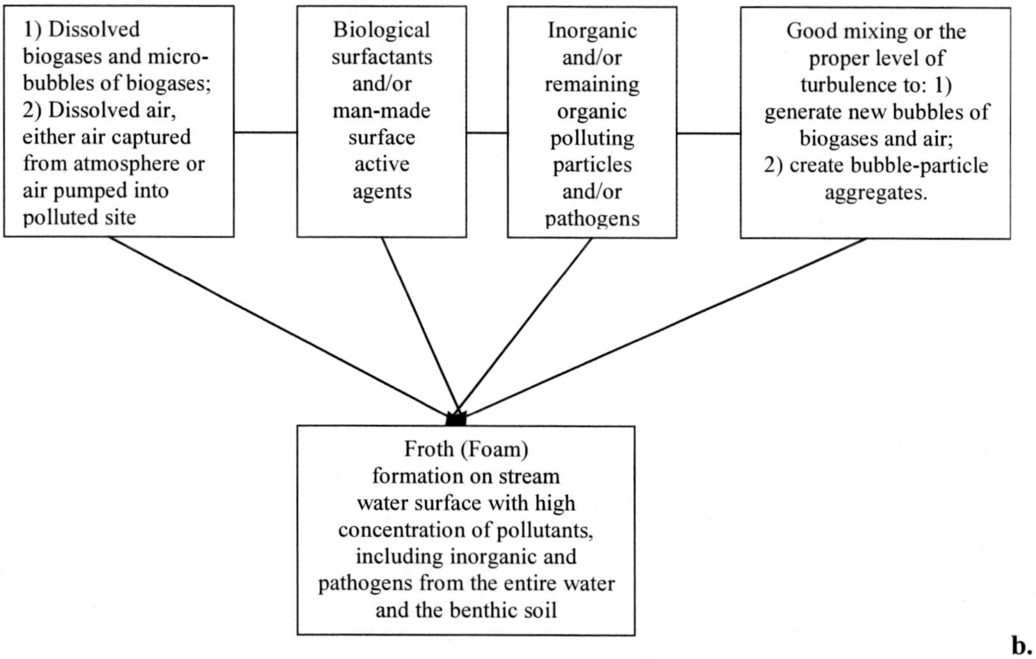

Chart 1. Natural Stream Self-Purification Through the Froth Formation Process.
a, Contents of polluted sites, including the result of decomposition.
b, Removal process through froth formation.

The lower concentration of heavy metals downstream from the polluting sources in Ref. [4] can be explained by the proposed natural froth formation process. In this case, the bubbles of biogases and air are attached to heavy metal particles in the presence of natural or man-made surfactants and form the bubble-particle aggregates. Rising to the water surface, the aggregates concentrate in the froth and the surrounding thin top layer of surface water. Then, as the current moves the thinning froth mass downstream, the bubbles creating the froth gradually disappear. As a result, the heavier aggregates precipitate to the river bottom. Consequently, the farther from pollution sources, the lower will be the concentration of the heavy metals.

In conclusion, it is essential to note that the natural self-purification process through froth formation in a stream is similar to the biological foam in activated sludge plants. Let us consider this question in more detail.

Activated sludge foaming is a major problem in biological wastewater treatment plants by causing serious operating problems and leading to poor quality effluence [41]. Several mechanisms of biological foam formation have been proposed but they have not been yet proven [42].

It is well recognized that the formation and stabilization of biological foam utilizes gas bubbles, surfactants and hydrophobic particles [41, 42]. Surfactants are already present in wastewater and also produce by microorganisms during decomposition. Gas bubbles are formed in the aeration tanks either by the aeration system with mechanical aerator or using a submerged diffused air system. Nitrogen gas bubbles are formed in the anoxic tank. The hydrophobic particles are the Mycolata group of bacteria (foam forming bacteria) attached to the bubble and it is considered to play a key role in the formation of the biological foam.

In Table 1, two mechanisms of foam formation in different aquatic environments are presented.

Table 1. Foam formation in different aquatic environments

Froth (Foam) Formation In A Stream	Foam Formation In Activated Sludge Plants
Source Of Organic Matter: Dead Fauna And Flora, And/Or Industrial Organic Pollution (E.G., Sewage And Domestic Waste, Textile And Paper Manufacturing Waste, Runoff From Inadequately Stored Animal Waste, Soaps, Etc.).	**1. Source Of Organic Matter:** Raw Wastewater Enriched With Organic Substances.
Decomposition Of Organic Matter By: Anaerobic: Benthic And Planctonic Bacteria And Other Microorganisms.	**2. Decomposition Of Organic Matter By:** Anaerobic: Mycolata Group Bacteria (Nocardia, Rhododoccus, Tsukamurella And Mycobacteria) And Others.
Products During Decomposition Process: A) Biological Surfactants, Such As Such As Amino Acids, Pyruvic Acid, Fatty Acids And Others; B) Dissolved Biogases And Biogas Micro-Bubbles, Such As Methane, Carbon Dioxide And Others.	**3. Products During Decomposition Process:** A) Biological Surfactants – Organic Acids Such As Acetic, Propionic And Mycolic Asids, Amino Acids, Fatty Acids And Others; B) Dissolved Biogases And Biogas Micro-Bubbles, Such As Methane And Carbon Dioxide.
Good Mixing: Water Cascading Over Weirs, Waterfalls And Other Obstacles, Creating The Shallow-Turbulent Character Of Water Current.	**4. Good Mixing:** Gas (Air Or Nitrogen) Pumped Into Aeration System.

As seen from Table 1, similar processes are involved in foam formation for different aquatic environments. Let us note that our new insights in a stream self-purification through natural froth formation process and the appearance of the decomposed black masses on creek surface in extraordinary circumstances allow us to approach a problem of activated sludge foaming and focus on the modeling physiochemical, and biological aspects of flocculation with stream-water point of view. This and other questions will be a subject of our further research.

5. Materials and Methods

Boom System

One of the ways for collecting samples for researching is to entrap and concentrate the froth in the proper place for removal. For this purpose, the boom system has been designed and constructed.

While the boom system may be made of various materials, you should consider the following conditions:

- The boom systems should be inexpensive, since they are easily subject to damage.
- It should be possible to make collectors in different dimensions for concentrating the pollutants in the proper places for further removal.
- The boom system has to be floating on the river surface.
- The boom system has to be light enough to be removed for maintenance and to be moved to various sites where the natural froth formation process occurs.
- The fish must swim underneath without hindrance.

It is essential to note that the boom system could be constructed in different ways, depending on the nature of the stream and the availability of materials.

Methods

For proving our hypothesis that the selected sites in creeks have its own natural capacity to clean-up itself through the froth formation process, we decide to test the samples from the froth and the top layer of surface water downstream from the froth collector respectively. For this purpose, the Palintest Photometer 5000 Instrument has been used.

The Palintest Photometer 5000 is the water-testing instrument for measuring the color intensity, which is produced when reagents are added to the sample solution. The color intensity is proportional to the concentration of the investigated parameters under test. Reagents are supplied in the form of test tablets. The Photometer shows different chemical elements in the water samples in varying the color intensity. It is simple to use in field research and is accurate.

6. Results and Discussion

Pollution Removal through Natural Froth Formation Process

To prove that the natural froth formation process affects a wider range of polluting particles, the boom system in the proper place at the site in two creeks downstream from the shallow-turbulent area has been installed. See Photos 8, 11.

There are two ways of determining the pollutants in the froth.

Table 2. Test results of the filtered black mixture from the froth taken at the site of the creek

Chemical elements:	Filtered Mixture from Froth mg/L	Ontario Provincial Water Quality Objectives mg/L
1) Zinc (Zn)	0.24	0.02
2) Phosphate LR (P)	0.36	0.01-0.02

First, the froth can be studied directly at the creek site, but in this case, after filtration, the Photometer can read test results on wavelengths of 570 nm and 640 nm only. A filter with pores of one micron has been used. The comparison test results are present in Table 2.

As seen from Table 2, the froth contains much higher concentrations of tested elements than the Ontario Provincial Water Quality Objectives [43] for the safety of aquatic life. For example, consider phosphorus. The objectives indicate the highest level that does not lead to the growth of algae.

Next step was to provide test analysis of filters after black mixture from the froth was passed through set of filters (three filters for the filling of 10 mg of the black mixture from the froth). In this case, particles larger than one micron had remained in the filters.

The test results for chemical elements in the filters are presented in Table 3.

As seen in Table 3, the concentration of the heavy metals in the froth is considerably higher than recommended in the Ontario Provincial Water Quality Objectives. Thus, copper rates sixty times higher. As for iron, its concentration is fourfold.

As seen in Tables 2 and 3, the froth collects the heavy metals and other chemical elements from Etobicoke Creek.

The fact that the Photometer can read the results for 570 nm and 640 nm gives the limitation for investigating a wider range of chemical elements; therefore, we decided to leave the black mixture sample from the froth in the laboratory for several days until the heavier particles had settled to the bottom, and then to analyze the samples from the top layer. Further in the text, the mixture from the froth means the mixture removed from the top layer of the froth sample after the heavy particles had settled.

For performing the test analysis, the mixture from the froth had poured through one filter with pores of one micron. Therefore, the test results for the filtered mixture from the froth are obtained for the particles smaller than one micron in size. This range also includes polluting particles in the colloid and the ionic states.

Table 3. The concentrations of the heavy metals in the filters

Metals:	Filters mg/L	Ontario Provincial Water Quality Objectives mg/L
Copper	0.307	0.005
Chromium	0.172	0.1
Nickel	0.076	0.025
Iron	1.458	0.3

Table 4. The comparison test results for the Etobicoke creek samples

Chemical elements:	The top layer of surface water downstream from boom system, mg/L	Filtered mixture from the froth, mg/L
1) Phenol (detection limit: 0.05mg/L)	0.00	0.98
2) Hexavalent Chromium (Cr^{VI})	0.04	0.17
3) Copper (Cu) (detection limit: 0.03mg/L) a) total	0.00	0.24
4) Turbidity	8	>400
5) Color	10	120
6) Nitrate (N)	0.15	0.70
7) Chloride	152	315
8) Ammonia (N)	0.10	12.0
9) Phosphate (P)	0.020	1.11

It is essential to note that the quality of the water body and that of the top layer of surface water, both originating downstream from the froth collector, are similar. Therefore, further in the text we will use term - the top layer of surface water.

The comparison test results samples taken for the several sites of creeks are presented in Tables 4, 5.

As clearly seen in Tables 4 and 5, the concentrations of all the investigated elements in the filtered mixture from the froth are significantly higher than in the top layer of surface water downstream from the collector. As a result, the naturally formed froth collects pollutants from the investigated sites of creeks.

Here it is essential to note that the naturally formed froth contains contaminants in the suspended state and the heavy particles, too. This means that the test results presented in Tables 4 and 5 do not provide us with the full information regarding the concentration of pollutants in the naturally formed froth. But as clearly seen in Tables 4 and 5, the concentration of the tested pollutants in the froth is significantly higher than that in the top layer of surface water downstream from the froth collector.

Table 5. The comparison test results for the Mimico creek samples

Chemical elements:	The top layer of surface water downstream from boom system, mg/L	Filtered mixture from the froth, mg/L
1) Turbidity	8	>400
2) Color	20	160
3) Iron LR (Fe)	0.02	>>1
4) Ammonia (N)	0.11	11.06
5) Phosphate (P)	0.073	1.056

Table 6. The comparison test results for the Etobicoke creek samples

Elements:	Upstream from boom system, mg/L	Downstream from boom system, mg/L
1) Zinc (Zn)	0.13	0.08
2) Hexavalent Chromium(Cr^{VI})	0.04	0.01
3) Copper (Cu): Total (detection limit: 0.03mg/L)	0.00	0.00
4) Nitrate (N)	0.032	0.015
5) Iron LR (Fe)	0.68	0.14
6) Ammonia (N)	>1.0	0.16
7) Molybdate HR (Mo)	0.96	0.60
8) Nickel (Ni)	0.06	0.00
9) Phosphate LR (P)	0.32	0.059
10) Phenol (C_6H_5OH) (detection limit: 0.05mg/L)	0.05	0.00

Table 7. The comparison test results for the Mimico creek samples

Chemical elements:	Upstream from boom system, mg/L	Downstream from boom system, mg/L
1) Iron LR (Fe) (detection limit: 0.01mg/L)	0.02	0.00
2) Turbidity (FTU)	5	5
3) Ammonia (N)	0.060	0.050
4) Hexavalent Chromium (Cr^{VI})	0.02	0.01

Some comparison test results of the top layers of surface water upstream rather distant from the boom system and downstream from it are presented in Table 6 and Table 7.

In [35, 36] the comparison test results for several sites of creeks throughout the year in order to observe the removal effect at different seasons are presented.

As seen in Tables 4-7, the filtered mixture from the froth in the colloid and the ionic states and the thin top layer of surface water upstream from the boom system contain much greater concentration of the investigated elements than the water body. Therefore, our hypothesis presented in Chart 2, that streams have the environmental capacity to purify themselves from pollutants through the froth formation process has been proved.

In general, based on our researches, we conclude: 1) every river and creek has the environmental capacity to purify itself from all kinds of pollutants through the froth formation process; 2) the natural froth formation process allows us to concentrate in the froth and the thin top layer of surface water upstream from a boom system a large amount of polluting particles, taken from entire water and benthic soil.

In conclusion of this section, it is essential to answer a question: how much of the black mixture from the froth can be removed within a certain period?

In fact, we needed a half hour to obtain manually two liters of the black mixture from the froth (2.2 kg.). Let's assume that we proceed at the same rate of success for a year. We can estimate what might be accomplished even if we remove the froth from only one site of the creek. Because of inclement weather in winter, we will include only 7 months in our calculation. We obtain: 2.2 kg. X 4 hours X 5 days per week X 28 weeks equals *2 tons 464 kg.*

If we install boom systems at many places, the effect will be much more beneficial.

Conclusion

During our research, the natural froth formation process by which the selected sites of creeks purify themselves has been investigated. The biological self-purification processes (such as the decomposing and degrading of the total organic matter) together with the physico-chemical ones, caused by the proper level of turbulence will, in the presence of man-made surfactants, lead to the natural formation of froth on river surface. The naturally formed froth, as well as the surrounding thin top layer of surface water upstream from boom system, comprises a higher concentration of polluting particles than the top layer of the surface water downstream from the installed boom system. The naturally formed froth collects pollutants of all kinds, ranging from the organic, heavy metals to the ammonia and the phosphate, from the polluted sites of the creeks.

Based on the new insights of river self-purification through the froth formation process, a practical long-term clean-up method for streams directly on site is proposed. A principal idea of the latter is to install boom systems in the proper places, where the natural froth formation process takes place, and subsequently to remove the froth and the surrounding thin top layer of surface water. As a result, we can, by utilizing the proper set of biological, chemical and physical natural self-purification processes, intervene directly on site in the clean-up of a creek that is heavily polluted from polluting sources, including the non-point ones.

The benefits of a proposed in-stream on site clean-up method are the following:

1) Reduction of nutrient concentration and concomitant reduction in the algae population of lakes, creeks and rivers.
2) Decrease of toxins (heavy metals and et.) in creeks, rivers and lakes.
3) Elimination of oil products flowing on the water's surface in streams, resulting in an improved habitat for aquatic life.
4) Reduction of *E-coli* bacteria in raw water traveling to intake water supply. As a result, the risk of *E-coli* bacteria in drinking water taken from rivers and creeks can be reduced.

Some applications of an in-stream on site clean-up method are the following:

a) Removal of nitrogen and phosphorus nutrients discharged from sewage treatment plants.
b) Reduction of the pollution originating with pesticides, herbicides, fertilizers, nutrients and bacteria from agricultural runoffs.
c) Runoffs from industries such as mining and others would be pretreated.
d) Boom systems should be used for at least three months after toxic spills.

The practical recommendations that we propose should be of great advantage to urban centers all over the world, where streams and creeks often suffer from severe pollution.

In conclusion, it is important to expand research in the following directions:

1) Conduct research on froth quality in comparison to water qualities both downstream and upstream from Niagara Falls.
2) Conduct research on several creeks and rivers worldwide, where the natural self-purification processes through froth formation takes place.
3) Investigate groups of bacteria and other microorganisms that are responsible for biological foam in a stream.
4) Investigate groups of bacteria and other microorganisms that are responsible for biological foam in a stream under extraordinary circumstances (after toxic spills, etc.).
5) Investigate processes involved in the formation of biological microbubbles (nucleation, grow, etc.) in a stream as well as in biological wastewater treatment plants.

These and other topics will be points of discussion in subsequent publications.

ACKNOWLEDGMENTS

This work was supported by the EJLB Foundation, to which we are grateful. Further, the author cordially thanks Professor N.P. Malomuzh and Professor N.K. Berdinskikh for fruitful discussion.

APPENDIX

If we tested samples from the running water in a stream, we would usually find a turbidity range of mostly 5 FTU, as in our samples. This range covers particles mostly in the dissolved state only. The raw water, for this especial parameter, could be used for drinking water preparation. Therefore, our next step was to find the percentage of removed dissolved particles from the entire water by installing the boom system.

In [35] a schematic of the experiment and a number of samples needed from polluted sites of streams are described in detail. There, it has been assumed that most of the physical, chemical and biological reactions in the shallow-turbulent area lead to the natural formation of froth downstream from it. The pollutants, removed from water body by bubbles of biogas and air, concentrate in the froth and the surrounding thin top layer of surface water.

The water samples originated from points 1, 2, 3 and 4 before the installation of the boom system and from the identical points three hours after we had installed the boom system have been taken for test analysis.

By using the *t*-distributions [44], the true average results for pollutants removed from the entire water with the 95% probability of the results are presented in Table 8.

Table 8. Average removal effect of pollutants by the boom system in creeks

Chemical elements:	Mimico Creek	Etobicoke Creek
	The average removal effect in percent φ_A (%)	
1) Ammonia (N)	25	32
2) Molybdate HR (M)	24	22
3) Phosphate LR (P)	19	19

As seen in Table 8, the polluting particles in the dissolved state could be removed from the water body.

As a result, if we place the boom systems upstream from the intake for drinking water purposes, we will have accomplished a pre-treatment stage. The subsequent procedures for preparing the drinking water will be less costly.

Simultaneously, water to be used for recreational use will also be safeguarded.

In conclusion, it is important to provide more research in this direction.

REFERENCES

[1] Wuhrmann, K. (1972) *Stream Purification*, in Ralph Mitchell, ed. Water Pollution Microbiology. New York: John Wiley and Sons, Inc.

[2] Gehm, H., and Bregman, J. (eds.). (1976). *Handbook of Water Resources and Pollution Control*. New York: Van Nostrand Reinhold Co.

[3] Spellman, F.R., and Drinan, J.E. (2001). *Stream Ecology and Self-Purification.* 2nd Edition. A Technomic Publishing Company book.

[4] Koppe, P. (1973). Investigation of the behavior of waste water constituents from metal-working industries in water cycle and their influence on drinking water supply. *Gas und Wasserfach* (in German), 114, 170-175.

[5] Beyers, R. and Odum, H.T. (1993). *Ecological Microcosms*. New York, Berlin, Heidelberg, London, Paris, Tokyo: Springer.

[6] Grice, G. and Reeve, M. (eds.). (1982). *Marine Mesocosms: Biological and Chemical Research in Experimental Ecosystems*. New York, Berlin, Heidelberg: Springer.

[7] Loo, L. and Rosenberg, R., Bivale. (1989). Suspension-Feeding Dynamics and Benthic-Pelagic Coupling in an Eutrophicated Marine Bay, *J. Exp. Mar. Biol. Ecol.*, 130, 253-276.

[8] Sommer, U. (1998). *Biologische Meereskunde*. Berlin, Heidelberg, New York: Springer.

[9] Hily, C. (1991). Is the Activity of Benthic Suspension Feeders a Factor Controlling Water Quality in the Bay of Brest? *Mar. Ecol. Prog.*, Ser.69, 179-188.

[10] Mandi, L., Houhoum B., Asmama S., and Schwartzbrod J. (1996). Wastewater Treatment by Reed Beds as Experimental Approach, *Water Res.*, 30, 2009-2016).

[11] Robach, F., Eglin, I. and Carbiener, R. (1991). Rhine Hydrosystem: Aquatic Vegetation in Relation with Water Quality (Rhinau Insland), *Bull. Ecol.*, 22, 227-241.

[12] Stimson, J. Larned, S. and McDermid, K. (1996). Seasonal Growth of Nutrient Availability, Temperature and Herbivory on Growth Rate, *J. Exp. Mar. Biol. Ecol.*, 196, 53-77.

[13] Otsuki, S., Fukuyo, Y., Stephens, K., Levings, C., Yamagishi, H., Kotyk, M., and McAllister, C. (1988). Influence of Wastewater Discharge on the Uptake Kinetics of Dissolved Organic Materials by Microorganisms in the Fraser River Estuary, British Columbia, *Arc. Hydrobiol.*, 112, 55-65.

[14] Logan, B., and Hunt, J. (1987). Advantages to Microbes of Growth in Permeable Aggregates in Marine Systems, *Limnol. And Oceanogr.*, 32, 1034-1048.

[15] Konstantinov, A.S. (1979). *General Hydrobiology*. (Vysshaya Shkola Press, (in Russian)).

[16] Nemerov, N.L. (1991) *Stream, Lake, Estuary and Ocean Pollution*. New York: Van Nostrand Reinhold.

[17] Ostroumov, S.A. (1999). *Water Self-purification in Ecosystems and Sustainable Development. In: Aquatic Ecosystems and Organisms*. Moscow: Dialogue-MSU Press, (in Russian).

[18] Vavilin, V.A. (1983). *Non-linear Models for biological purification and self-purification processes in rivers*. Moscow: Nauka Press, (in Russian).

[19] Jackson, G.B. (1993). *Applied Water and Spentwater Chemistry. A Laboratory Manual*. New York: Van Nostrand Reinhold.

[20] Wuhrmann, K. (1954). High-rate activated sludge treatment and its relation to stream sanitation. II Biological river tests of plant effluents. *Sewage Industrial Wastes*, 26, 212-220.

[21] Klein, L. (1971). *River Pollution. II. Causes and Effects*. London: Butterworths Scientific Publications.

[22] Skurlatov, Yu. I. (1988). Fundamentals of the management of quality of water.- *Ecologicheskaja Khimija Vodnoi Sredy (Ecological Chemistry of Aquatic Environment)*, 1, 230-255.

[23] Izrael, Yu. A. and Tsyban, A. V. (1989). *Anthropogenic Ecology of the Ocean*. Leningrad: Gidrometizdat (in Russian)..

[24] Horne, A.J. and Goldman, C.R. (1994) *Limnology*. New York, St. Louis: McGraw-Hill.

[25] Koronelli, T.V. (1996). Principles and methods for raising efficiency of biological degradation of hydrocarbons in the environment. *Prikladnaja Biokhimija I Mikrobiologija (Applied Biochemistry and Microbiology)*, 32. 579-585.

[26] Tilzer, M. and Serruya, C. (eds.). (1990). *Large Lakes: Ecological Structure and Function*. Berlin, Heidelberg, New York: Springer.

[27] Ostroumov, S.A. (1986) *Introduction to Biochemical Ecology*. Moscow: Moscow University Press, (in Russian).

[28] Ostroumov, S.A. and Donkin, P. (1997). Effects of pollution of aquatic environment on biological mechanisms of removal of particles. Physical Problems of Ecology. *Proceedings of the Conference*. Moskva, 2, 23-27 July, 71-72.

[29] Ostroumov, S.A. (1999) Water Self-Purification and Natural Bioremediation in the Biosphere: New Concepts and New Experimental Data. *The Journal of Biospheric Science,* 1, Number 2.

[30] Klein, L., with chapters by Erichsen Jones, J.R. and Hawkes, H.A. (1957). *Aspects of River Pollution*. London: Butterworths Scientific Publications.

[31] Sykes, G., and Skinner, F.A. (1971). *Microbial Aspects of Pollution*. London and New York: Academic Press.

[32] Lovley DR, Phillips EJP, Gorby YA, Landa ER. (1991). Microbial Reduction of Uranium. *Nature*. 350. 413-416.

[33] Lovley D.R. (2003). Cleaning Up With Genomics: Applying Molecular Biology to Bioremediation, *Nature Reviews Microbiology* 1 (October 2003):36-44.

[34] Methé BA et al, (1967). Genome of *Geobacter sulfurreducens*: Metal Reduction in Subsurface Environments, *Science* 302 (5652).

[35] Fisenko, A.I. (2004). A New Long-Term on Site Clean-Up Approach Applied to Non-Point Sources of Pollution, *Water, Air and Soil Pollution*, 156, 1-27.

[36] Fisenko, A.I. (2006). Natural Purification of Rivers and Creeks through Froth Formation Process, *Res. J. Chem. Environ.,* 10(1), 24-29.

[37] Haghighi-Podeh, M., and Siyahati-Ardakani, G. (2000). Fate and toxic effects of cyanide on aerobic treatment systems. *Wat. Sci. Tech.* 42, 125-129.

[38] Gaudin, A.M. (1957). *Flotation*. (second edition). New York: McGraw-Hill.

[39] Leja J. (1982). *Surface Chemistry of Froth Flotation*. New York: Plenum Press.

[40] Considine, D.M., and Cousidine, G.D. (1989). *Van Nostrand's Scientific Encyclopedia.* (seventh edition). Volume 1. New York: Van Nostrand Reinhold.

[41] Andrews J. F. (1992) *Dynamics and Control of the Activated Sludge Process*, Water Quality Management Library - Volume 6, edited by J. F. Andrews, Technomic Publishing Company, Inc., Lancaster, PA, USA.

[42] Stratton, H., Sevior, B., and Brooks, P., (1998). Activated sludge foaming: what causes hydrophobicity and can it be manipulated to control foaming?, *Wat. Sci. Tech.* 37, No 4-5, 503-509.

[43] Ontario Water Quality Objectives. http://www.agatlabs.com/ontario1.htm

[44] Huntsberger, D.V. and Billingsley, P. (1981). *Elements of Statistical Inference.* (Fifth edition). Boston, London, Sydney, Toronto: Allyn and Bacon, Inc.

In: Soil Contamination Research Trends
Editor: Javier B. Dominguez, pp. 203-215
ISBN 978-1-60456-319-1

Chapter 9

USE OF PLANT BIOASSAYS FOR THE DETECTION OF GENOTOXIC COMPOUNDS IN SOILS

Miroslav Mišík[1,2], Armen Nersesyan[1], Katarína Mišíková[2], Siegfried Knasmüller[1,*]

[1]Institute of Cancer Research, Medical University of Vienna, Austria,
[2]Department of Botany, Comenius University in Bratislava, Slovakia

ABSTRACT

Plant bioassays (PBA) are widely used for the detection of genotoxins in soils. It is known that the release of DNA damaging agents into the environment may affect the stability of ecosystems and may cause cancer, age related diseases and infertility in exposed humans (e.g. via consumption of plants grown in contaminated soils or via ground waters). The most widely used PBAs are micronucleus (MN) assays with pollen tetrads of Tradescantia and MN and chromosome aberration analyses with root tip cells of *Vicia faba* and *Allium cepa*. In addition, also other endpoints have been developed for plant experiments (e.g. gene mutation assays with *Zea mays*, *Arabidopsis thaliana*, somatic mutation assays as well as the single cell gel electrophoresis test). One of the advantages of the aforementioned assays is that they are sufficiently sensitive to detect the effects of contaminations in native soils in *in situ* experiments whereas for other genotoxicity tests with soils extraction and concentration procedures are required. The evaluation of the currently available data shows that PBAs are quite insensitive towards polycyclic aromatic hydrocarbons (PAHs) and aromatic amines which are often found in soils and sediments but are highly sensitive to certain pesticides, herbicides, heavy metals such as As^{3+}, Cr^{6+}, Cd^{2+}, Ni^{2+} and radionuclides which can not be easily detected with other bioassays (e.g. with bacterial mutagenicity tests which are among the most abundant genotoxic carcinogens found in soils and sediments. These characteristics make

* Corresponding author: Institute of Cancer Research, Medical University of Vienna, Borschkegasse 8a, Vienna 1090, Austria. Tel.: +43-1-427765142; fax: +43-1-42779651. E-mail address: siegfried.knasmueller@meduniwien.ac.at

PBAs important integral components of test batteries aimed for detection of DNA damaging contamination in soils and sediments.

INTRODUCTION

The deposition of genotoxic compounds in soils may lead to two main adverse effects. First, the contamination of ground water and the uptake of plants used for food production may cause exposure of humans to these toxins. It is well documented that DNA-damage is causally related to a number of diseases such as cancer [1], atherosclerosis [2], age related as well as neurological disorders [3], reduced fertility and heritable diseases [4]. On the other hand, damage of genetic material has also impact on the fertility of wildlife species and may affect the stability of ecosystems [5,6].

In order to monitor and control the extent of pollution with such compounds, numerous investigations have been carried out. Most of the experiments (approximately 37%) were conducted with bacterial assays. In a equally high number of studies, plant bioassays were used [7], which have the advantage that it is not necessary (in contrast to other methods) to use extraction procedures which may lead to loss of potentially active compounds.

The present article gives a short overview on the methodological principles of the most common used plant bioassays (PBAs) and their usefulness for the detection of genotoxins commonly found in soils. In the last paragraph, examples of results from recent investigations are described which are not contained in the comprehensive review of White and Claxton [7].

CURRENTLY USED PLANT BIOASSAYS

In the majority of soil investigations, tetrads of *Tradescantia* (Trad MN assay) and meristematic root tip cells of *Vicia* and *Allium* were used.

The Trad MN assay has been established by Sax [8], Steinitz [9] and Ma and coworkers [10,11]. It is based on scoring of MN in early tetrad cells which are highly synchronized under normal conditions. MN are formed as a consequence of chromosome breaks (structural chromosomal aberration) and aneuploidy (numerical chromosomal aberrations) [12] and can be conducted with a number of different organisms including mammalian cells. A number of *Tradescantia* species and hybrids (e.g. *T. paludosa*, *T. pallida*, on clones #4430, #02 and #03) can be used in the experiments. Under moderate climatic conditions, the plants have to be cultivated in a greenhouse (for details see Ma et al. [11]). In older investigations, soil studies were conducted mainly with aqueous leachates [13-16]; more recently Steinkellner et al. [17] developed a modified protocol for direct exposure of plants and showed that this modification is by far more sensitive for detection of heavy metals.

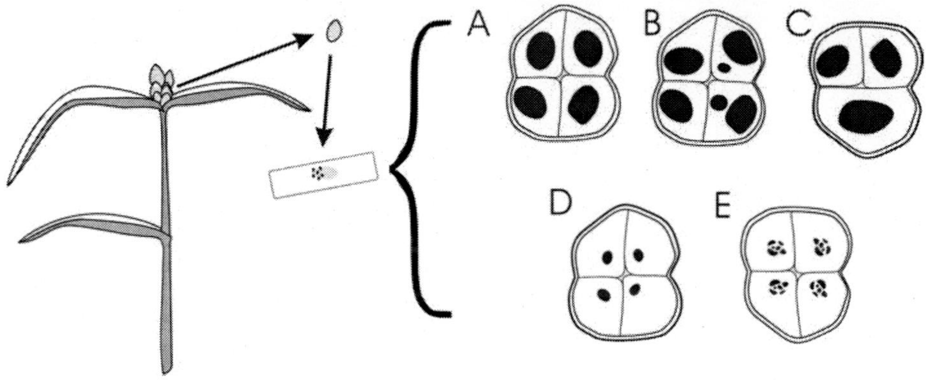

Figure 1. Schematic representation of Trad MN assays. Micronuclei are scored in early pollen tetrads after exposure of stems to soil leachetes or after exposure of whole plants (with roots).
A – Normal synchronized early pollen tetrads ; B – MN (2x) in tetrads; C – asynchronous cell division; D – cells with pycnotic nuclei ; E –cells with nuclear fragmentation

Additional endpoints which can be scored in Tradescantia hybrids (e.g. #4430, #02) are stamen hair mutations (SHM). However, this procedure requires immediate scoring in contrast to MN experiments in which the evaluation can be conducted with fixed material. As an endpoint, the change in the stamen hair cells color from wild-type blue to pink are scored [18]. The plant is heterozygous for genes responsible for the production of delphinidine/cyanidine and loss of the dominant allele leads to a pink cell color [19,20].

MN assays can also be conducted in root tips cells of higher plants such as *Allium cepa* and *Vicia faba* [21-23]. Bulbs [24] or seedlings [25] are kept under wet conditions till primary roots are formed and subsequently directly exposed in soils. It has been shown that mitotic chromosomes are less susceptible to DNA damage than meiotic chromosomes in experiments with ionizing radiation [8,26] and heavy metals [17,27], therefore the Trad MN test is more sensitive than root tip MN assays [17].

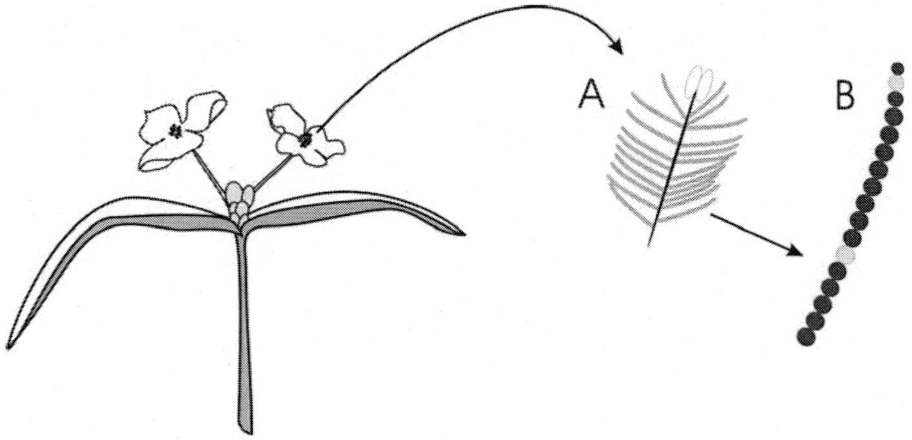

Figure 2. Schematic procedures of the Trad SHM test, based on the scoring of "pink" mutations in stamen hair cells. A – Stamen with hair; B – hair containing blue (black) normal cells and two pink (grey) cells.

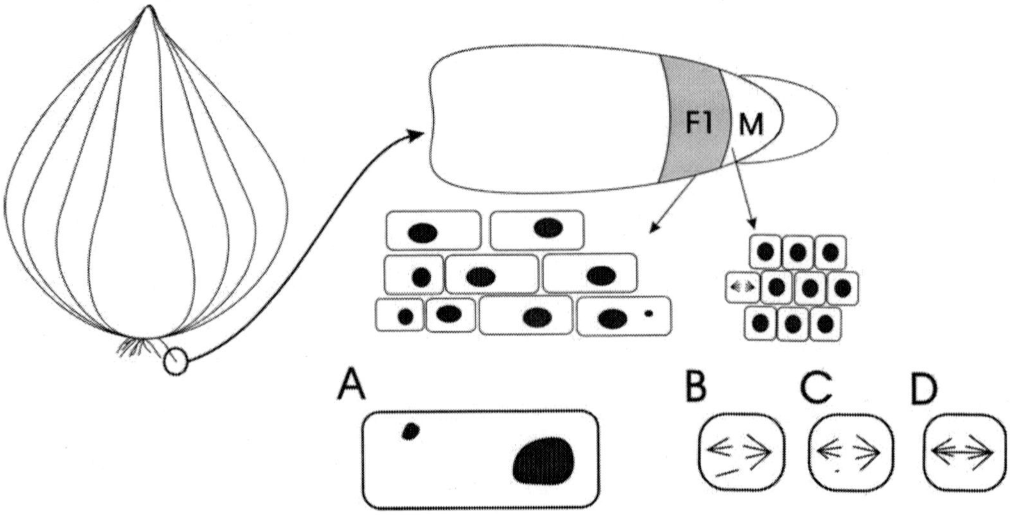

Figure 3. Anaphase analyses and MN test with Allium.
F1 – region recommended by Ma et al. [24] as more sensitive for MN assay; M – meristematic region; A – cells with micronuclei; B - D – anaphase analyses: lagging chromosomes (B), chromosome breaks (C) and chromosomal bridge (D). The target cells are located in meristematic zone of roots.

For all MN assays standardized protocols have been developed and the assays have been validated with variety of compounds [13,28-31].

In anaphase (anatelophase) assays the chromosomal aberrations are scored in late anaphase and early telophase of mitotic division. Most often fragments, bridges and whole chromosomes (laggards) [32,33] were evaluated. Some authors used also mixed categories i. E. c-anaphases, multipolarity and polyploidy [34-37] or evaluated also categories such as sickness and gap [38].

Apart from MN or anaphase assays, also other plant test systems were used occasionally for experiments with soils. The initial test with *Arabidopsis thaliana* was developed by Müller [39] and is based on scoring of lethal, chlorophyll deficient embryos in the siliquae (for recent protocol see Gichner et al. [40]). The method developed by Kriukov et al. [41] is based on the determination of total dominant lethal mutations.

The *Zea mays waxy* locus assay employs iodine staining to detect *wx* revertant pollen grains, which are stained blue – black in contrast to *Wx* pollen which stains black-brown [42-45]. The relative large number of plants, and the time and cost requirements of these assays are the most important limitations of this procedure.

Another type of experiments which are increasingly used in ecogenotoxicity studies with plants is the single cells gel electrophoresis (comet) assay. It is based on the determination of DNA migration in an electric field [46-48]. One of the advantages of this procedure is that a variety of different plant organs can be evaluated, but it should be keep in mind that DNA migration is not necessary reflecting persisting lesions and can be repaired. To date, a number of experiments have been published with heavy metals [49,50] and alkylating agents [51-53], but the test has been used only rarely in experiments with soils [54].

White and Claxton [7] evaluated the data from PBAs (462 observations in 29 publication); in 27.7% the Trad MN assay were used, 24.4% of the data come from the Trad SHM assay, 22.5% from the *Allium* root anatelophase assays, about 14% from tests with *Zea mays*, the remaining 11.4% include the Allium MN assay, Arabidopsis gametic mutation test and Vicia MN and sister chromatid exchange assays.

DETECTION OF GENOTOXIC CONTAMINATION IN SOILS WITH PLANT BIOASSAYS

The number of terrestrial sites which are contaminated with potentially genotoxic chemicals is unexpectedly high. In the US, more than 45 000 locations have been identified [55], in Germany more than 200 000, in Denmark, Switzerland and Austria the ranges are between 10 000 and 37 000 [56,57]. The most relevant sources of pollution are metal mining as well as smelting and processing followed by electric, gas and sanitary services, solvent recovery operation in the chemical industry [7].

The most important genotoxic chemicals deposited in soil are metals (As, Cd, Cr, Ni, Co) which account for more than 90% of overall deposits, followed by polycyclic aromatic hydrocarbons (PAHs), and halogenated organic compounds. Interestingly that genotoxic agrochemicals (herbicide and pesticides) account only for a very small genotoxic fraction (<0.1%) of the contamination [7].

It has been shown in a large number of investigations that plant bioassays are in general highly sensitive towards genotoxic effects of heavy metals [17,27,58-60]. Systematic investigations with Trad MN and root tip MN assays showed that positive results are obtained in particular with those heavy metals which are human and rodent carcinogens, for example As, Cd, Cr [61-64] while only moderate or negative effects were found with metals such as Cu and Zn [65,66] which have not been classified as human carcinogens [67,68]. The results of these comparative studies show also that the Trad MN assay is by far more sensitive than the root MN tip assay [17,27]. The fact that the effects of metals can be easily detected in PBAs is of particular importance in regard to investigations of soil since it is known that bacterial assays (Salmonella/microsome) which are frequently used for soil monitoring does not enable the detection of heavy metals [17,27,58].

On the other hand, PAHs which constitute the second most important group of soil relevant genotoxic carcinogens according to White and Claxton [7] cannot be adequately detected with PBAs due to lack of adequate metabolic activation systems [58,69,70]. Although some investigations with benzo(a)pyrene which is frequently used as a model compound, indicate that positive results may be obtained in plants [71], the high concentrations required distract from the assumption that such compounds can be detected in native environmental samples. Also aromatic amines which are occasionally found in soils do not cause DNA damage in plants. It was shown by Plewa et al. [72,73] in number of studies that these compounds are metabolized in plant cells via different pathways as in mammals and that reactive metabolites do not reach the DNA in the plants.

Nitroaromatic compounds are primarily found in munition deposit sites at relatively high concentrations [74] and it is known that some of these compounds cause cancer in rodents [75].

The sensitivity of PBAs towards these compounds has not been fully evaluated but some model compounds yielded positive results [76]; also with ground waters from munition contaminated sites in Germany clear positive results were obtained in Trad MN assays [75].

It is also well documented that halogenated organic compounds which are frequently released by industries such as tetrachlorethylene, trichlormethane, trichlorethane, trichloroethylene and vinylchloride can be detected with PBAs [28,77-80].

The results obtained with agrochemicals (i.e. pesticides and herbicides) in PBAs are reviewed in the papers of Rodriguez et al. [81] and Ma et al. [28]. It was shown that many of these compounds (for example atrazine, cyanazine, dicamba, dichlorvos, chlorpyrifos, maleic hydrazide) cause positive results in PBAs [28,30] and other genotoxicity assays [30,81]. Also with sodium azide positive results were obtained [70]. This compound is a typical example for false positive results as it is well documented that the enzyme which required converting this agent to DNA reactive metabolites is not present in mammals and humans [82].

The high sensitivity of PBAs (in particular of the Trad MN assay) towards ionizing radiation is well documented in numerous studies [26,77,83,84] and this feature renders these tests ideal for detection of radionuclides in soils (see next chapter).

EXAMPLES OBTAINED IN RECENT EXPERIMENT WITH SOILS

Detailed descriptions of the results obtained with PBAs in soils can be found in several reviews [7,59,77,81,85]. Excellent response ranges and significant differences between heavy contaminated (industrial, agricultural) and reference sites were found in experiments with *Zea mays* and in anaphase aberration and MN assays with *Allium*, in which direct root contacts were used, while Tradescantia assays with leachates were less responsive. This latter problem has been solved by development of a direct exposure protocol for the Trad MN assay [17,27] in which intact plants with roots are exposed to soils; and we showed that the latter approach is by far more sensitive for the detection of the effects of metal contaminated soils as the other methods.

Table 1 lists up the results of a number of newer investigations which were not evaluated in the aforementioned reviews. Taken together, these findings further support the assumption that PBAs are highly efficient biomonitoring tools for soil contamination.

As mentioned above, plants are also highly sensitive to ionizing radiation, therefore it is not surprising that they were successfully used to detect effects caused by radionuclides in soils and sediments [86-88]. In our own experiments, we studied recently the effects of radioactive sediments from different locations (i.e. from Kerala, India; Bahia, Brazil; Jachymov, Czech Republic). In the Trad MN we found significant effects even at low radiation doses ($\geq$2.0μGy/h) when the inflorescence were directly placed on the surface of the samples [89].

Taken together, the currently available data show that PBAs are a valuable tool to monitor genotoxins in contaminated soils. Since they are able to detect the effects of important groups of compounds (such as heavy metals) which give negative results in other biomonitoring systems, they should be integrated in test batteries for the study of soils and sediments.

Table 1.

Test systems	Soil - type of contaminations	Results / Remarks	References
Trad MN	military munition contaminated soil	+ / several positive results from sites and water contaminated by exploded military munitions	[75]
Trad MN	industrially contaminated soil (heavy metals)	+? / MN frequencies were high in some water leachates, however an insufficient number of tetrads was counted	[90]
Trad MN	natural radionuclides contaminated sediments and soils (Th, U)	+ / MN frequencies has been positive when direct exposure was used	Submitted [89]
Vicia MN, AA	industrial sites, near fuel tanks / PAH	- / no significant increase of the frequency of MN or AA in cells	[91]
Vicia MN	industrially and artificially contaminated soils	+ / correlations between results of the Comet assay and results of Vicia MN assays were found	[92]
Vicia MN	waste water irrigated soil (PAH, agrochemicals)	+ / the highest frequencies of MN were obtained from soil with highest concentration of PAH	[93]
Flow cytometry / AFPL	industrially contaminated soil (PAH)	- / the extent of damage caused by the tested PAH concentrations was not sufficient to affect plant development.	[94]
Crop plants CA	radionuclide contaminated soil (Cs, Sr)	+ / chronic low dose exposure can cause cytogenetic damage and karyotypic variability in the offspring of irradiated organisms.	[86]
Pinus silvestris / CA	radionuclide contaminated soil	+ / adaptation to radiation – radio resistance which cannot be explained by natural selection	[88]

REFERENCES

[1] L.L. Woo, K. Onel and N.A. Ellis (2007). The broken genome: genetic and pharmacologic approaches to breaking DNA. *Ann Med,* 39, 208-218.

[2] M. Mahmoudi, J. Mercer and M. Bennett (2006). DNA damage and repair in atherosclerosis. *Cardiovasc Res,* 71, 259-268.

[3] V.A. Bohr, O.P. Ottersen and T. Tonjum (2007). Genome instability and DNA repair in brain, ageing and neurological disease. *Neuroscience,* 145, 1183-1186.

[4] M. Joffe (2003). Infertility and environmental pollutants. *Br Med Bull,* 68, 47-70.

[5] J.W. Bickham, S. Sandhu, P.D. Hebert, L. Chikhi and R. Athwal (2000). Effects of chemical contaminants on genetic diversity in natural populations: implications for biomonitoring and ecotoxicology. *Mutat Res,* 463, 33-51.

[6] J. Salminen, C.A. van Gestel and J. Oksanen (2001). Pollution-induced community tolerance and functional redundancy in a decomposer food web in metal-stressed soil. *Environ Toxicol Chem,* 20, 2287-2295.

[7] P.A. White and L.D. Claxton (2004). Mutagens in contaminated soil: a review. *Mutat Res,* 567, 227-345.

[8] K. Sax (1938). Chromosome aberration induced by X-rays. *Genetics,* 23, 494-516.

[9] L.M. Steinitz (1944). The effects of lack of oxygens on meiosis in Tradescantia. *Am J Botany,* 31, 428-443.

[10] T.H. Ma, A.H. Sparrow, L.A. Schairer and A.F. Nauman (1978). Effect of 1,2-dibromoethane (DBE) on meiotic chromosomes of Tradescantia. *Mutat Res,* 58, 251-258.

[11] T.H. Ma, G.L. Cabrera, R. Chen, B.S. Gill, S.S. Sandhu, A.L. Vandenberg and M.F. Salamone (1994). Tradescantia micronucleus bioassay. *Mutat Res,* 310, 221-230.

[12] M. Fenech (2000). The in vitro micronucleus technique. *Mutat Res,* 455, 81-95.

[13] G.S. Rodrigues, T.H. Ma, D. Pimentel and L.H. Weinstein (1997). Tradescantia Bioassays as Monitoring Systems for Environmental Mutagenesis: A Review. *Critical Reviews in Plant Sciences,* 16, 325-359.

[14] S. Monarca, D. Feretti, I. Zerbini, A. Alberti, C. Zani, S. Resola, U. Gelatti and G. Nardi (2002). Soil contamination detected using bacterial and plant mutagenicity tests and chemical analyses. *Environmental Research,* 88, 64-69.

[15] S. Cotelle, J.F. Masfaraud and J.F. Férard (1999). Assessment of the genotoxicity of contaminated soil with the Allium/Vicia-micronucleus and the Tradescantia-micronucleus assays. *Mutat Res,* 426, 167-171.

[16] T. Gichner and J. Velemínsky (1999). Monitoring the genotoxicity of soil extracts from two heavily polluted sites in Prague using the Tradescantia stamen hair and micronucleus (MNC) assays. *Mutat Res,* 426, 163-166.

[17] H. Steinkellner, K. Mun-Sik, C. Helma, S. Ecker, T.H. Ma, O. Horak, M. Kundi and S. Knasmüller (1998). Genotoxic effects of heavy metals: Comparative investigation with plant bioassays. *Environ Mol Mutagen,* 31, 183-191.

[18] T.H. Ma, G.L. Cabrera, A. Cebulska-Wasilewska, R. Chen, F. Loarca, A.L. Vandenberg and M.F. Salamone (1994). Tradescantia stamen hair mutation bioassay. *Mutat Res,* 310, 211-220.

[19] M. Emmerling-Thompson and M.M. Nawrocky (1979). Genetic studies of flower color in Tradescantia. *Journal of Heredity,* 70, 115-122.

[20] M. Emmerling-Thompson and M.M. Nawrocky (1980). Genetic basis for using Tradescantia clone 4430 as an environmental monitor of mutagens. *Journal of Heredity,* 71, 261-265.

[21] O.P. Arora, V.C. Shah and R. S.R. (1969). Studies on micronuclei induced by mitomycin-C in the root cells of Vicia faba. *Experimental Cell Research,* 56, 443-448.

[22] W.F. Grant (1982). Chromosome aberration assays in Allium. A report of the U.S. Environmental Protection Agency Gene-Tox Program. *Mutat Res,* 99, 273-291.

[23] T.H. Ma (1982). Vicia cytogenetic tests for environmental mutagens. A report of the U.S. Environmental Protection Agency Gene-Tox Program. *Mutat Res,* 99, 257-271.

[24] T.H. Ma, Z. Xu, C. Xu, H. McConnell, E.V. Rabago, G.A. Arreola and H. Zhang (1995). The improved Allium/Vicia root tip micronucleus assay for clastogenicity of environmental pollutants. *Mutat Res,* 334, 185-195.

[25] O. Kovalchuk, I. Kovalchuk, A. Arkhipov, P. Telyuk, B. Hohn and L. Kovalchuk (1998). The Allium cepa chromosome aberration test reliably measures genotoxicity of soils of inhabited areas in the Ukraine contaminated by the Chernobyl accident. *Mutat Res,* 415, 47-57.

[26] A.H. Sparrow and W.R. Singleton (1953). The use of radiocobalt as a source of gamma rays and some effects of chronic irradiation on growing plants. . Amer. Nat., 87, 29-48.

[27] S. Knasmüller, E. Gottmann, H. Steinkellner, A. Fomin, C. Pickl, A. Paschke, R. Goed and M. Kundi (1998). Detection of genotoxic effects of heavy metal contaminated soils with plant bioassays. *Mutat Res,* 420, 37-48.

[28] T.H. Ma, G.L. Cabrera and E. Owens (2005). Genotoxic agents detected by plant bioassays. *Reviews on Environmental Health,* 20, 1-13.

[29] T.H. Ma (2001). Tradescantia-micronucleus bioassay for detection of carcinogens. *Folia Histochem Cytobiol, 39 Suppl* 2, 54-55.

[30] T.H. Ma, M.M. Harris, V.A. Anderson, I. Ahmed, K. Mohammad, J.L. Bare and G. Lin (1984). Tradescantia-Micronucleus (Trad-MCN) tests on 140 health-related agents. *Mutat Res,* 138, 157-167.

[31] T.H. Ma, S.S. Sandhu, Y. Peng, T.D. Chen and T.W. Kim (1992). Synergistic and antagonistic effects on genotoxicity of chemicals commonly found in hazardous waste sites. *Mutat Res,* 270, 71-77.

[32] M.S. Kong and T.H. Ma (1999). Genotoxicity of contaminated soil and shallow well water detected by plant bioassays. *Mutat Res,* 426, 221-228.

[33] K. Kunova (1999). Bioindication of phytotoxicity and genotoxicity through higher plants in the ore mine area. *Ekologia-Bratislava* 18 271-276.

[34] G.D. Sayles, C.M. Acheson, M.J. Kupferle, Y. Shan, Q. Zhou, J.R. Meier, L. Chang and R.C. Brenner (1999). Land treatment of PAH-contaminated soil: Performance measured by chemical and toxicity assays. *Environ Science Tech,* 33, 4310-4317.

[35] J. Rank and M.H. Nielsen (1993). A modified Allium test as a tool in the screening of the genotoxicity of complex mixtures. *Hereditas,* 118, 49-53.

[36] J. Rank and M.H. Nielsen (1997). Allium cepa anaphase-telophase root tip chromosome aberration assay on N-methyl-N-nitrosourea, maleic hydrazide, sodium azide, and ethyl methanesulfonate. *Mutat Res,* 390, 121-127.

[37] J. Rank and M.H. Nielsen (1998). Genotoxicity testing of wastewater sludge using the Allium cepa anaphase-telophase chromosome aberration assay. *Mutat Res,* 418, 113-119.

[38] N. Sang and G. Li (2004). Genotoxicity of municipal landfill leachate on root tips of Vicia faba. *Mutat Res,* 560, 159-165.

[39] A.J. Müller (1965). On the time period of mutation induction after treatment with N-nitroso-N-methylurea of swelling seeds of Arabidopsis. *Uber den Zeitpunkt der Mutationsauslösung nach Einwirkung von N-nitroso-N-methylharnstoff auf quellende Samen von Arabidopsis.,* 2, 426-437.

[40] T. Gichner, S.A. Badayev, S.I. Demchenko, J. Relichova, S.S. Sandhu, P.D. Usmanov, O. Usmanova and J. Veleminsky (1994). Arabidopsis assay for mutagenicity. *Mutat Res,* 310, 249-256.

[41] V.I. Kriukov, V.A. Shashkin and S.F. Sokolenko (1996). [Effects of chronic effects of lead nitrate and ionizing radiation on mutagenesis in Arabidopsis Thaliana (L.) Heynh]. *Radiats Biol Radioecol,* 36, 209-218.

[42] M.J. Plewa (1982). Specific-locus mutation assays in Zea mays. A report of the U.S. Environmental Protection Agency Gene-Tox Program. *Mutat Res,* 99, 317-337.

[43] L.J. Stadler (1928). Genetic Effects of X-Rays in Maize. *Proc Natl Acad Sci U S A,* 14, 69-75.

[44] G.S. Rodrigues, D. Pimentel and L.H. Weinstein (1998). In situ assessment of pesticide genotoxicity in an integrated pest management program: II. Maize waxy mutation assay. *Mutat Res,* 412, 245-250.

[45] W.F. Grant and E.T. Owens (2006). Zea mays assays of chemical/radiation genotoxicity for the study of environmental mutagens. *Mutat Res,* 613, 17-64.

[46] A.R. Collins (1999). The comet assay: Some considerations. *Neoplasma,* 46, 3-5.

[47] A.R. Collins (2004). The comet assay for DNA damage and repair: Principles, applications, and limitations. *Applied Biochemistry and Biotechnology,* 26, 249-261.

[48] A. Hartmann, E. Agurell, C. Beevers, S. Brendler-Schwaab, B. Burlinson, P. Clay, A. Collins, A. Smith, G. Speit, V. Thybaud and R.R. Tice (2003). Recommendations for conducting the in vivo alkaline Comet assay. *Mutagenesis,* 18, 45-51.

[49] T. Gichner, Z. Patkova, J. Szákova and K. Demnerova (2004). Cadmium induces DNA damage in tobacco roots, but no DNA damage, somatic mutations or homologous recombination in tobacco leaves. *Mutat Res,* 559, 49-57.

[50] T. Gichner, Z. Patkova, J. Szákova and K. Demnerova (2006). Toxicity and DNA damage in tobacco and potato plants growing on soil polluted with heavy metals. *Ecotoxicol Environ Saf,* 65, 420-426.

[51] T. Gichner (2003). Differential genotoxicity of ethyl methanesulphonate, N-ethyl-N-nitrosourea and maleic hydrazide in tobacco seedlings based on data of the Comet assay and two recombination assays. *Mutat Res,* 538, 171-179.

[52] T. Gichner, A. Mukherjee and J. Velemínsky (2006). DNA staining with the fluorochromes EtBr, DAPI and YOYO-1 in the comet assay with tobacco plants after treatment with ethyl methanesulphonate, hyperthermia and DNase-I. *Mutat Res,* 605, 17-21.

[53] T. Gichner, O. Ptaček, D.A. Stavreva, E.D. Wagner and M.J. Plewa (2000). A comparison of DNA repair using the comet assay in tobacco seedlings after exposure to alkylating agents or ionizing radiation. *Mutat Res,* 470, 1-9.

[54] O. Ptaček, Z. Muehlfeldova, J. Dostálek, T. Čechák and T. Gichner (2002). Monitoring DNA damage in wood small-reed (Calamagrostis epigejos) plants growing in a sediment reservoir with substrates from uranium mining. *Journal of Environmental Monitoring,* 4, 592-595.

[55] http://www.epa.gov/superfund/sites/phonefax/products.html.

[56] Ad Hoc International Working Group on Contaminated Land, Management of Contaminated Sites and Land in Central and Eastern Europe, *Danish Cooperation for Environment in Eastern Europe, Ministry of Environment and Energy,* Copenhagen, Denmark, 2000, pp. pp. 151.

[57] G. Prokop, M. Schamann and I. Edelgaard. Management of Contaminated Sites in Western Europe., *European Environment Agency,* Copenhagen, Denmark, 2000, pp. pp 171.

[58] B.J. Majer, D. Tscherko, A. Paschke, R. Wennrich, M. Kundi, E. Kandeler and S. Knasmüller (2002). Effects of heavy metal contamination of soils on micronucleus induction in Tradescantia and on microbial enzyme activities: a comparative investigation. *Mutat Res,* 515, 111-124.

[59] S.S. Sandhu and W.R. Lower (1989). In situ assessment of genotoxic hazards of environmental pollution. *Toxicol Ind Health,* 5, 73-83.

[60] H. Wang (1999). Clastogenicity of chromium contaminated soil samples evaluated by Vicia root-micronucleus assay. *Mutat Res,* 426, 147-149.

[61] S. Satarung, J.R. Baker, S. Urbenjapol, M. Maswell-Elkins, P.E.B. Reilly, J. Williams and M. Moore (2007). A global perspective on cadmium pollution and toxicity in non-occupationally exposed population. *Toxicol Lett,* 137, 65-83.

[62] L. Patrick (2003). Toxic metals and antioxidants: Part II. The role of antioxidants in arsenic and cadmium toxicity. *Altern Med Rev,* 8, 106-128.

[63] T. Gebel (1997). Arsenic and antimony: comparative approach on mechanistic toxicology. *Chem Biol Interact.,* 107, 131-144.

[64] M. Waisberg, P. Joseph, B. Hale and D. Beyersmann (2003). Molecular and cellular mechanisms of cadmium carcinogenesis. *Toxicology,* 192, 95-117.

[65] D. Lewinska, J. Palus, M. Stepnik, E. Dziubaltowska, J. Beck, K. Rydzynski , A.T. Natarajan and R. Nilsson (2007). Micronucleus frequency in peripheral blood lymphocytes and buccal mucosa cells of copper smelter workers, with special regard to arsenic exposure. *Int Arch Occup Environ Health,* 80, 371-380.

[66] Z. Hamurcu, H. Donmez, R. Saraymen and H. Demirtas (2001). Micronucleus frequencies in workers exposed to lead, zinc, and cadmium. *Biol Trace Elem Res,* 83, 97-102.

[67] M. Valko, C.J. Rhodes, J. Moncol, M. Izakovic and M. Mazur (2006). Free radicals, metals and antioxidants in oxidative stress-induced cancer. *Chem Biol Interact,* 160, 1-40.

[68] D.H. Lee, K.E. Anderson, L.J. Harnack, A.R. Folsom and D.R. Jacobs, Jr. (2004). Heme iron, zinc, alcohol consumption, and colon cancer: Iowa Women's Health Study. *J Natl Cancer Inst,* 96, 403-407.

[69] B.J. Majer, T. Grummt, M. Uhl and S. Knasmüller (2005). Use of plant bioassays for the detection of genotoxins in the aquatic environment. *Acta Hydrochim Hydrobiol,* 33, 45-55.

[70] J. Velemínsky and T. Gichner (1988). Mutagenic activities of promutagens in plants indirect evidence of their activation. *Mutat Res,* 99, 339-347.

[71] T.H. Ma (1981). Tradescantia micronucleus bioassay and pollen tube chromatid aberration test for in situ monitoring and mutagen screening. *Environ Health Persp,* 37, 85-90.

[72] M.J. Plewa The plant activation of aromatic amines into unique high molecular weight agents that induce genomic DNA damage in mammalian cells, in: J. Maluszynska and M.J. Plewa (Eds.), *Bioassays in Plant Cells for Improvement of Ecosystem and Human Health,* Wydawnictwo Universytetu Slaskiego, Katowice, 2003.

[73] M.J. Plewa, E.D. Wagner, D.A. Stavreva and T. Gichner (1996). Plant activation and its role in environmental mutagenesis and antimutagenesis. *Mutat Res,* 350, 163-171.

[74] T. Grummt, H.G. Wunderlich, A. Chakraborty, M. Kundi, B. Majer, F. Ferk, A.K. Nersesyan, W. Parzefall and S. Knasmuller (2006). Genotoxicity of nitrosulfonic acids,

nitrobenzoic acids, and nitrobenzylalcohols, pollutants commonly found in ground water near ammunition facilities. *Environ Mol Mutagen,* 47, 95-106.

[75] T. Grummt. *Nachweis von genotoxischen Wirkungen und deren Ursachen in Oberflächenwasser: Verifizierung und Spezifizierung der genotoxischen Bosivihy, German Env. Protection Agency*. 2003.

[76] P. Gong, R.G. Kuperman and G.I. Sunahara (2003). Genotoxicity of 2,4- and 2,6-dinitrotoluene as measured by the Tradescantia micronucleus (Trad-MCN) bioassay. *Mutat Res,* 538, 13-18.

[77] T.H. Ma (1982). Tradescantia cytogenetic tests (root-tip mitosis, pollen mitosis, pollen mother-cell meiosis). A report of the U.S. Environmental Protection Agency Gene-Tox Program. *Mutat Res,* 99, 293-302.

[78] S. Sandhu, T.H. Ma, Y. Peng and X.D. Zhou (1989). Clastogenicity evaluation of seven cheicals commonly found at hazardous industrial waste sites. *Mutat Res,* 224, 437-445.

[79] T.H. Ma, C. Xu, S. Liao, H. McConnell, B.S. Jeong and C.D. Won (1996). In situ monitoring with the Tradescantia bioassay on the genotoxicity of gasseous emission from a closed landfill site and an incinerator. *Mutat Res,* 359, 39-52.

[80] J. Van't Hoff and L.A. Schairer (1982). Tradescantia assay system for gaseous mutagens. A report of the U.S. Environmental Protection Agency. Gene/Tox Program. *Mutat Res,* 99, 303-315.

[81] G.S. Rodrigues, T.H. Ma, D. Pimentel and L.H. Weinstein (1997). Tradescantia Bioassays as Monitoring Systems for Environmental Mutagenesis: A Review. *Crit. Rev. Pl. Sci,* 16, 325-359.

[82] P. Arenaz, L. Hallberg, F. Mancillas, G. Gutierrez and S. Garcia (1989). Sodium azide mutagenesis in mammals: inability of mammalian cells to convert azide to a mutagenic intermediate. *Mutat Res,* 227, 63-67.

[83] D. Anderson and T.H. Ma (1982). Micronuclei induced by low-dose cobalt-60 gamma-irradiation in Tradescantia pollen mother cells. *Environ Mutagen,* 4, 348.

[84] M. Minouflet, S. Ayrault, P.M. Badot, S. Cotelle and J.F. Ferard (2005). Assessment of the genotoxicity of 137Cs radiation using Vicia-micronucleus, Tradescantia-micronucleus and Tradescantia-stamen-hair mutation bioassays. *J Env Rad,* 81, 143-153.

[85] B.S. Gill and S.S. Sandhu (1992). Application of the Tradescantia micronucleus assay for the genetic evaluation of chemical mixtures in soil and aqueous media. *Mutat Res,* 270, 65-69.

[86] S.A. Geraskin, V.G. Dikarev, Y.Y. Zyablitskaya, A.A. Oudalova, Y.V. Spirin and R.M. Alexakhin (2003). Genetic consequences of radioactive contamination by the Chernobyl fallout to agricultural crops. *J Env Rad,* 66, 155-169.

[87] C. Chenal, R. Zaka, F. Legue and M.T. Misset (2006). Effects of low-dose irradiation on two plant models: Stipa capillata and Pisum sativum. *International Journal of Low Radiation,* 3, 234-240.

[88] S.A. Geraskin, L.M. Zimina, V.G. Dikarev, N.S. Dikareva, V.L. Zimin, D.V. Vasiliyev, A.A. Oudalova, L.D. Blinova and R.M. Alexakhin (2003). Bioindication of the anthropogenic effects on micropopulations of Pinus sylvestris, L. in the vicinity of a plant for the storage and processing of radioactive waste and in the Chernobyl NPP zone. *J Env Rad,* 66, 171-180.

[89] M. Mišík, G. Krupica, K. Mičieta, M. Bichler, T.H. Ma and S. Knasmüller (2007). *Detection of genotoxic effects caused by natural radioactive sands with Tradescantia micronucleus assay.* Mutat Res.

[90] B. Lah, T. Vidic, E. Glasencnik, T. Cepeljnik, G. Gorjanc and R. Marinsek-Logar (2007). *Genotoxicity evaluation of water soil leachates by Ames test, comet assay, and preliminary Tradescantia micronucleus assay.* Environ Monit Assess.

[91] E. Sturchio, P. Boccia, B. Ficociello, C. Riccardi, S. Spicaglia, C. De Simone, S. Marconi, M. Errichetti and M. Pezzella (2004). Mutagenesis and molecular analysis to monitor the effects of chemicals released in the environment. *Fresen Environ Bull,* 13 1385-1394

[92] E. Sturchio, P. Boccia, S. Marconi, G. Bagni, S. Hernandez, M. Mascini and C. Beni (2006). Detection of DNA damage by genotoxic substances: Comparative investigation with comet assay, micronucleus test and DNA biosensor. *Fresen Environ Bull,* 15, 724-730.

[93] Y. Chen, C. Wang, Z. Wang and S. Huang (2004). Assessment of the contamination and genotoxicity of soil irrigated with wastewater. *Plant and Soil,* 261, 189-196.

[94] R. Aina, L. Palin and S. Citterio (2006). Molecular evidence for benzo[a]pyrene and naphthalene genotoxicity in Trifolium repens L. *Chemosphere,* 65, 666-673.

In: Soil Contamination Research Trends
Editor: Javier B. Dominguez, pp. 217-234
ISBN 978-1-60456-319-1

Chapter 10

Statistical Approaches to Establish Background Values of Potentially Toxic Elements

C. Micó*, L. Recatalá and J. Sánchez
Land Use Planning Department, Desertification Research Centre-CIDE
(University of Valencia, CSIC, GV), Camí de la Marjal s/n,
46470 Albal, Valencia, Spain

Abstract

The assessment of potentially toxic element (PTE) contents in soils and the establishment of background values are priority objectives within the European Union and throughout the world. This is a necessary task in order to improve or even maintain the quality of soil resources and for environmental and health protection of livestock and humans. Different methods have been used to derive background values of PTEs in soils. Statistical approaches used elsewhere include descriptive statistics, probability plots, the Bootstrap method, and the recent publication of the International Organization for Standardization (ISO/DIS 19258). Most of these methods do not consider soil properties and characteristics, despite soil components such as organic matter, clay, hydrous oxides and carbonates play a relevant role on the content and dynamics of most contaminants in soils due to their binding properties. Methods based on regression equations take into account some selected soil properties, such as organic matter and clay content, and have been applied in different works. This chapter tries to present an overview of statistical approaches, modifications of existing methods and recent new approaches that have been used to establish background values of PTEs in soils. The application of these various approaches is explored in a representative area of the Mediterranean region, and the limitations and usefulness are also outlined. Moreover, it is discussed to what extent these methods have, up to now, satisfied the expectations or have been satisfactory in particular applications. The correct choice of the most useful statistical approach is finally

* Corresponding author (email: carolina.mico@uv.es; tel: +00-34-961220540; fax +00-34-961270967)

dependent on the objectives of the background study and on the characteristics of the study area.

Keywords: Background values; Potentially toxic elements (PTEs); ISO/DIS 19258; Probability plots; Bootstrap; Regression analysis.

INTRODUCTION

Soil plays a relevant role in the regulation of potentially toxic elements (PTEs) in ecosystems since soil is the interface for most of the anthropic activities and is greatly impacted by humans. Furthermore, soil is an important long-term sink for PTEs due to the immobilization by mechanisms of adsorption and precipitation, which determine the natural abundance of such chemical elements. When soil is not affected by anthropogenic inputs, soil exhibits an average composition close to the earth's crust [1]. However, the parent material from which soils are derived is not uniform and soil-forming processes differ markedly from one climatic region to another, accounting for considerable overall variability in PTE concentrations. It is well known that human activities may raise these levels. Therefore, sources of PTEs in soils may be from natural geological materials or may occur from human activity. In terms of environmental pollution, it is necessary to have reliable data on the naturally occurring amounts of these elements in order to assess the significance of contributions from man and industry.

The normal abundance of an element in soil is commonly referred to by the geochemists as background values [2,3], which vary according to the nature of the parent material for any particular element. Background values were originally defined in exploration geochemistry in order to differentiate between normal element concentrations and anomalies that might indicate an ore occurrence [3]. In soil science, background values are usually referred to as natural concentrations of PTEs in soils uninfluenced by human activities [4]. However, this term has proven to be a theoretical rather than a practical concept applied to soils because of both natural variability and widespread anthropic input [5]. Indeed, the establishment of background values is a difficult task in ancient populated areas such as the European Mediterranean region, where pristine soils are almost impossible to find. Many terms have been used to refer to background values. Some authors [6,7,8,9,10] have used the term baseline values as an expected range of natural variability associated with the background levels specific to one area and time. Baseline values consider diffuse or non-point input of PTEs to soils and, therefore, these levels are not always true background values. Particularly, baseline values in agricultural soils would correspond to total concentrations of elements in soils not influenced by point input due to local human activities (e.g. industries), but include diffuse input due to atmospheric deposition and agricultural practices (e.g. fertilization) [6,11]. Then, the term of baseline values is sometimes used in equivalency to background values in order to measure current levels of PTE so that future change can be quantified [4]. However, the former has been rejected by some authors because it seems to imply that the natural background is constant for each element. Despite the appearance of both terms, the International Organization for Standardization (ISO) has recently published an international guideline (ISO/DIS 19258) using the term background values against baseline values, but differentiating between natural and usual background values, respectively. The natural background value is defined as the statistical characteristic of the natural concentration of a substance in soils resulting from natural geological and pedological processes and therefore excluding any addition of human origin, whereas the usual background value is the statistical characteristic of the concentration of a substance in soils resulting from both the natural pedo-geochemical content and the diffuse source input into the soil [12]. Consequently, the term of background values continues being used on a worldwide scale [4,5,12,13,14,15,16,17,18]. Without paying attention to the drawbacks of each term, which are out of discussion of this chapter, the term background values will be used with the same meaning as that of usual background concentrations defined by ISO.

The assessment of the PTE contents in soils and the establishment of background values are priority objectives within the European Union, according to the Thematic Strategy to Soil Protection [19,20], and also throughout the world [15,21]. This is a necessary task in order to improve or even maintain the quality of soil resources, and for environmental and health protection of livestock and humans. Some works have used common ranges or average concentration of PTEs established, for example, by Adriano [8] or by Kabata-Pendias [22] as an indicator of whether a soil is potentially contaminated. However, this approach is not appropriate since the native concentration in a specific soil may fall out of the listed ranges [23]. Then, the most common test to identify contamination processes by PTEs in soils consists of a site-to-background comparison [24], which involves comparing the maximum detected concentration of the site data to the background variation established in control samples with similar soil properties and characteristics, and collected as far away as possible from polluting sources [14]. The use of background values as a preliminary tool may indicate the degree of contamination and provide basic information about the contamination status of a site even with relatively few soil samples [25,26]. The comparison based on background values has a great advantage, since characterisation of contaminated samples can be quickly and cheaply done in many cases once the background values have been set up. However, this comparison has some drawbacks that threaten its reliability [26]. One refers to the soil sampling scheme and methods for analysing soil samples at a laboratory. There are no standardised and accepted procedures for soil sampling and laboratory analyses. Different laboratories use distinct soil analysis methods and, consequently, comparisons of results are difficult if not impossible. Furthermore, the reliability of approaches based on background values becomes critical when one considers the complexity of soil [26,27]. Depending on its characteristics and properties (e.g. texture, type of clay, content of organic matter, pH, porosity, permeability, etc.) each soil has a different capacity for retention and mobility of contaminants. Adsorption and solubility of PTEs in soils are mainly affected by properties such as pH [28,29,30], cation exchange capacity [31], organic matter content [32], soil mineralogy [31] or carbonate content [33]. The inherent problem associated to this approach refers to the great number of factors and their interactions involved in the mobility of PTEs in soil. It is clear that the same background values cannot be valid for all kinds of soils, since they may change from area to area within a region and among regions [4,26]. For instance, Reimann and Garrett [4] demonstrated the variability of background values using data from two different scale geochemical surveys carried out in northern Europe. In order to overcome this drawback, background values should be individualised to the particularities of soil or should be estimated considering soil properties and characteristics. Another inherent disadvantage refers to the procedure for estimating background values, since a wide range of statistical approaches has been used for such purposes.

There are many ways to establish background values, for example, using descriptive statistics or applying specific statistical approaches. In relation to the former, the international guideline ISO/DIS 19258 establishes the use of descriptive statistics as a method for estimating background values [12]. In relation to the latter, some statistical approaches such as probability plots or the Bootstrap method have been also used to this purpose. However, those methods do not consider soil properties and characteristics, which play an important role on the dynamics of PTEs in soil. Some countries (e.g. The Netherlands or Belgium) and regions (e.g. Basque Country in Spain) have expressed background values as linear equations between the total concentration of PTEs in soils and selected soil components such as organic matter and clay [27,34]. Generally, such equations are proposed for inorganic elements, as they can be natural constituents of soil and are narrowly related to those soil components.

Collectively, the existing studies on the establishment of background values for PTEs in soils are too variable since background values have been derived by different methods. This chapter presents an overview of the different statistical approaches applied to estimate background values of PTEs in soils. Statistical approaches, modifications of existing methods, and recent new approaches are briefly reviewed and discussed. These approaches include the use of descriptive statistics, the international guideline ISO/DIS 19258, probability plots, the Bootstrap method and methods based on regression equations. Finally, these various approaches have been implemented to establish background values for lead (Pb) in agricultural soils of the Alicante province (Spain), a representative area of the European Mediterranean region, in order to compare results and to explore limitations and usefulness.

Descriptive Statistics

Some works have applied simple descriptive statistics (e.g. arithmetic or geometric mean, median, upper confidence limit, nth percentile, etc.) in order to propose background values for PTEs [35], or much better the upper limit of the background variation. For instance, Kabata-Pendias and Pendias [36] suggested that mean values and ranges of PTE concentrations seem to provide acceptable background values to assess changes at local and regional level [18]. The main drawback of some descriptive statistics as the arithmetic mean is that its use is not statistically justified because soil geochemical data rarely show a normal distribution [37]. Therefore, other statistics such as geometric mean, median or nth percentile have been preferred for legislative purposes since these statistics are less sensitive to extreme scores and are preferred for skewed distributions. Gough et al. [38] reported background values for 29 PTEs in soils collected at Bull Island, South Carolina (USA), using the geometric mean (GM) and the geometric deviation (GD). Specifically, the background values were expressed as a range between GM/GD^2 and $GMxGD^2$. At regional level, De Miguel et al. [39] proposed background values based on the 99th percentile in the Madrid region (Spain).

Among methods that use descriptive statistics, the most useful method consists in establishing background values expressed as the upper limit of normal concentrations, covering the range of natural variability associated to certain elements at regional level.

According to this statistical approach, the upper limit of background values (UL) consists in adding n-times the standard deviation to the mean value for each element studied. Assuming a normal distribution, about 95% of the population is included into the interval X±2SD, whereas 99.7% is included into the interval X±3SD (Figure 1). Therefore, background values are statistically defined as a percentage of the expected range of normal concentrations. The UL of the background values [3, 41] is usually calculated considering the mean value (X) for each element and the standard deviation (SD) following the equation:

$$UL = X + 2SD \qquad (1)$$

Values within the range [X±2SD] were initially defined as the geochemical background range in exploration chemistry [3]. There have been a number of studies on the establishment of background values using this statistical approach. Ramos et al. [9] reported 0.8 mg kg^{-1} for Cd, 37 mg kg^{-1} for Ni and 90 mg kg^{-1} for Pb as the upper limit of the background values in greenhouses of the Almería province (SE Spain). Results for Cd and Ni showed a close correspondence with results achieved by Micó et al. [42] in a similar study area, such as the Alicante province (SE Spain), and using the same statistical approach. However, Pb showed three-fold higher value in greenhouses due to the presence of some contaminated soils (for example, 405 mg kg^{-1} for Pb). Vázquez et al. [43] also obtained background values for 16 PTEs in two areas of Central Spain, comparing background values in soils derived from carbonate and granite material. Results achieved by these authors for carbonate soils were similar to values obtained by Ramos et al. [9] and Micó et al. [42] in the Spanish Mediterranean region due to the narrow correspondence to calcareous material, with the exception of Cu (Fe and Ni were not analysed by Vázquez et al., [43]). Other studies reported similar results, indicating the wide applicability of this statistical approach. Chen et al. [5,44] established background concentrations in Florida (USA), whereas Crock et al. [45] calculated the upper limit of the normal range for 29 elements as the expected 95% element range concentration in surface soils of the Kenai Peninsula (Alaska). Tack et al. [6] identified outliers before establishing background values for nine elements in the Flemish region (Belgium) considering mean value plus twice the standard deviation.

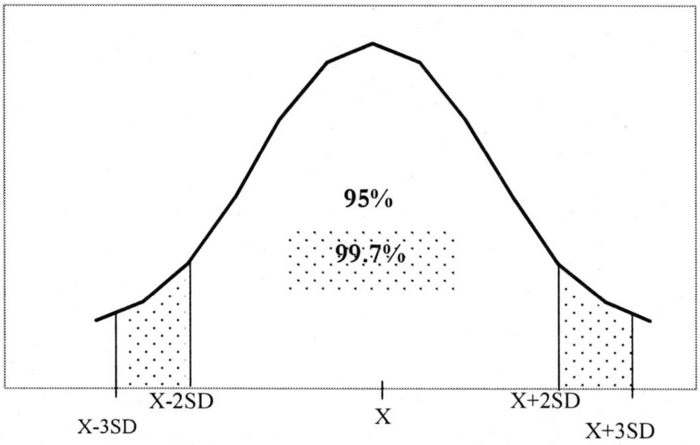

Figure 1. Graphical interpretation of intervals X±2SD and X±3SD for a normal distribution [40].

The main advantage of calculating background values based on descriptive statistics is that only measured values for a single element in question are needed to site-to-background comparison. Nevertheless, it is necessary to consider uncontaminated soils in order to establish background values by this statistical approach, for example, sampling in soils from natural areas or in soils as far away as possible from polluting sources. An overlooked disadvantage of such backgrounds is that they are based on the overall data distribution, which might be shifted towards higher values due to enrichments. In fact, the raw data sets often include normal and anomalous values especially in ancient populated areas such as the European Mediterranean region. Particularly, background values in agricultural soils are referred to total concentrations of elements in soils not influenced by point input due to local human activities [11,12], but non-point inputs could be included. In the case of diffuse pollution (e.g. atmospheric deposition, fertilization), enrichments below the threshold are not recognised and, therefore, the mean value would be overestimated in that case. Then, it is absolutely necessary to identify outlier values before establishing background values, for example using boxplots [4,18]. Outliers are values belonging to a different population because they have been originated from another process or source, for example, from contaminating distributions [46]. Nevertheless, some authors have proposed the use of [median±MAD], since the median and the median absolute deviation (MAD) are robust against extreme values [10,38,40]. Alternatively, other statistical approaches shown in the following sections are used when the raw data set includes outliers.

PROBABILITY PLOTS

Given that some areas may include high levels of PTEs associated with contaminated soils, different works suggested the need to separate normal and anomalous values. Probability plots were initially applied by geochemists in mineral exploration studies to partition data into component populations thus obtaining background and anomalous values [2,47]. This statistical method has been also applied by several authors to establish background values in soils [14,18,48,49,50].

Probability plots can be obtained using a general statistical software package or by PROBPLOT® v. 1.00 [51] provided by *The Association of Applied Geochemists* (Ontario, Canada). Probability plots are based on the cumulative frequency plotted against the measured value, where complex curves can be partitioned in two statistical populations, the uncontaminated or background population (A-population) and the potentially contaminated population (B-population), as was proposed by Tobías et al. [14] in an agro-industrial area near Barcelona (NE Spain). Both populations can be separated through the detection of the inflection point that can be identified graphically. Due to the high uncertainty associated to this process, Fleischauer and Korte [50] proposed an approach based on skewness to determine the inflection point. Given that skewness is a measure of the symmetry of a distribution, a skewness of zero or the nearest zero indicates normally distributed populations (or log-transformed populations if the distribution is lognormal). Therefore, the inflection point could be referred to as the upper limit of the background population.

Using this approach, Davies [48] established the normal Pb levels in soils of Great Britain, and Davies and Wixson [49] reported the background levels of Co, Cu, Mn, Ni, Pb and Zn in soils of Missouri. Fleischauer and Korte [50] also used probability plots to establish background values of different elements in the Kansas City metropolitan area in Missouri (USA) and to formulate cleanup standards for soil remediation. In the European Mediterranean region, Tobías et al. [14] carried out a study in an agro-industrial area of Catalonia (NE Spain) in order to establish background levels for Ba, Cd, Cr, Cu, Ni, Pb, V and Zn. Recently, Zhao et al. [18] also applied cumulative frequency distributions to derive background values for Cd, Co, Cr, Cu, Ni, Pb and Zn using a database of soil samples from England and Wales.

Probability plots allow the partitioning of different populations in a raw data set and after that the estimation of the background population, even when there is overlap of the constituent populations. In most cases, the data sets consist of two well established populations, which can be easily separated due to the presence of outliers. However, it is possible to find many different types of cases. Sometimes, inflection point cannot be identified since the raw data set as a whole represents the background population. Other times, some elements show a departure from linearity in the lower part of the background linear portion. For minor elements such as Cd, this fact could be explained in part by lower values because here one is working near the detection limit of analytical technique, whereas for other major elements such Fe or Mn departure from linearity could be due to deficiency problems [52]. On the other hand, Zhao et al. [18] also highlighted another important shortcoming for probability plots since the uncontaminated population represented by the lowest values could include soils of light textures, suggesting that their relatively low concentrations are partly due to the effect of soil type. Finally, complex curves with three or more populations might be found due to the occurrence of high variability of PTE contents. In that case, the identification of inflection point could not be an easy work and then other statistical approaches such as the Bootstrap method might be applied. Despite these disadvantages, the use of probability plots represents a more applicable approach because it permits the establishment of background values where contaminated soils may be present considering all samples analysed.

BOOTSTRAP METHOD

Other statistical approaches may estimate a certain number of unknown parameters through a random process from the raw data set, including outliers. The Bootstrap method refers to the process of randomly resampling the original data set to generate new data sets. This statistical approach assumes that the empirical cumulative distribution function is a reasonable estimate of the unknown cumulative distribution function of a population [53]. The most common uncertainty calculation is generating a confidence interval for a specific statistics. For this task, a subsample of a size less than or equal to the size of the raw data set is built from the data. Each subsample is generated with replacement so that any data point can be sampled multiple times or not sampled at all. Consequently, high levels of PTEs get a lesser significance at the end of the process. Resampling is repeated for many subsamples, typically between 500 and 1000, in order to estimate confidence intervals.

The Bootstrap method can be applied for establishing background values using a general statistical software package or by the R® statistical program v. 1.9.0. The distribution of the

[X+2SD] values based on the new data sets can be plotted in a histogram, where 25th and 75th percentiles of the distributions can be adopted as the confidence interval of statistics. The selection of a confidence level for an interval determines the probability that the confidence interval produced will contain the true parameter value. Common choices for the confidence level are 0.90, 0.95, and 0.99. These levels correspond to percentages of the area of the normal density curve. For example, a 95% confidence interval covers 95% of the normal curve and then the probability of observing a value outside of this area is less than 0.05. Once distribution of the [X+2SD] has been plotted, background values for each PTE can be estimated for example as the average of the [X+2SD] values obtained.

Despite the great applicability and usefulness of this method, the Bootstrap method has not been extensively used for background proposal. Di Stefano et al. [54] used the Bootstrap method to estimate the background, or literally the reference level, of ^{137}Cs in an uneroded site in Sicily (Italy). Chirenje et al. [55] reported the 95% upper confidence limit (UCL) of the mean for arsenic in Florida soils through Bootstrap method.

The Bootstrap method is a powerful statistical approach because it permits the establishment of background values considering all the variability of the population. Furthermore, the main advantage of the Bootstrap method is that does not require assumptions about the distribution and, for example, it works for both normal (or lognormal) and non-normal distributions. However, it is expected higher background values using the Bootstrap method than for other statistical approaches due to the consideration of the raw data sets as a whole, where outliers are included. Therefore, a probably overestimation of results might be obtained when potentially contaminated soils are included in the study area.

Regression Equations

Approaches mentioned above do not consider soil components for establishing background values. However, soil is not a simple or static medium, since soil properties and characteristics may change across relatively short distances. Moreover, some soil components such as organic matter, clay, hydrous oxides and carbonates play a relevant role in the dynamics and accumulation of PTEs in soils. Indeed, soils containing high amounts of such components tend to retain naturally higher PTE concentrations due to their binding properties.

Some works reported background values for PTEs considering some soil properties. For example, the German Federal Soil Protection and Contaminated Site Ordinance [56] defined precautionary soil values according to the texture of soil (Table 1). These values of PTE concentration are dependent on soil type and soil pH. The precautionary values for the soil type "loam/silt" are applied for strong silty sands, too. This is only valid for soils with pH>6, while precautionary values for soil type "sand" are used for soils with pH<6. Zhao et al. [18] also reported background values for PTE taking into account different soil texture, since this property is considered a relevant soil characteristic for mobility of these contaminants.

Table 1. Precautionary soil values (mg/kg dry weight) for some PTEs base on different textures of soils in Germany [56]

Metal	Soil type clay	Soil type loam/silt	Soil type sand
Cd	1.5	1	0.4
Cr	100	60	30
Cu	60	40	20
Hg	1	0.5	0.1
Ni	70	50	15
Pb	100	70	40
Zn	200	150	60

Regression equations between PTE contents and soil parameters provide a quantitative description of the relationship between the upper limit of the background values for each PTE and selected soil components, indicating the influence of some soil properties on naturally concentrations. The Environmental Program of the Netherlands 1986-1990 applied this statistical approach to establish background values of PTEs in soils under a multipurpose soil use [57] and, since then, it has been applied in many other works [6,16,34,43]. Initially, the Netherlands only took into account soil parameters such as organic matter and clay. However, given the importance of other soil components in the chemistry of contaminants in soils, hydrous oxides or carbonates have been also included in other works [43,52]. The upper limit of the background values (UL) for a specific element can be expressed as a function of different soil components following the general equation:

$$UL = X + a\,H + b\,L + c\,C + d\,D \ldots \quad (2)$$

The mean value (X) is referred to uncontaminated soils. The coefficients a, b, c and d are the slopes of the single regression equation between each element and the single soil component. If the PTE shows significant correlation with only one soil component, the coefficient fits with the slope of the single regression equation. However, if the PTE presents significant correlation with two or three soil components, the slope of the single regression equation must be divided by two or three, respectively, in order to take into account the contribution factor of each component [34]. Finally, H, L, C and D are percentages of organic matter, clay, carbonates and hydrous oxides, respectively, although other soil components could be included according to the specific characteristics of a region.

Regression equations proposed in the Netherlands and the Basque Country (Spain) are shown in Table 2. However, these regression equations do not establish a numerical value to evaluate the extent of soil pollution. Background values can be calculated for a so-called "standard soil", which presents the average properties of soils in the study area. For the Dutch conditions, this is a soil containing 25% of clay and 10% of organic matter [57], whereas the Basque Country establishes 5% of organic matter and 30% of clay for a standard soil [34] (see Table 2). Tack et al. [6] also used a multiple regression approach to calculate background values of nine elements in Flanders (Belgium) as function of clay and organic matter contents. Crommentuijn et al. [16] established the background concentrations of 18 elements in the Netherlands taking into account both soil components. Vázquez et al. [43] reported background values of 16 elements in the Madrid region (Spain) through empirical formulae that consider organic matter, clay content and Fe-oxyhydroxides.

Table 2. Regression equations and background values for a standard soil (mg kg^{-1} dry weight) proposed in The Netherlands and Basque Country [27,34]

Element	The Netherlands		Basque Country	
	Regression equation	Standard soil[a]	Regression equation	Standard soil[b]
Cd	0.4+0.007(L+3H)	0.8	0.17+0.013 L	0.8
Co	20	20	6+0.5L	20
Cr	50+2L	100	25+L	53
Cu	15+0.6(L+H)	36	10+0.5L	24
Ni	10+L	35	12+L	40
Pb	50+L+H	85	16+0.7L+2.1H	44
Zn	50+1.5(2L+H)	140	50+2L	106

[a] 25% of clay and 10% of organic matter.
[b] 30% of clay and 5% of organic matter.

The main disadvantage of this statistical approach supposes the establishment of a standard soil, since it displays the average contents of soil components (e.g. organic matter and clay) but not represent all cases. Therefore, background values estimated using regression equations are expected to be lower than background values established by other statistical approaches, since the consideration of a standard soil represents the average conditions for all soils present in a geographical area. Consequently, these values can result highly restrictive for legislative purposes since remedial actions might be proposed for sites that are not really contaminated and therefore economic and human resources would be unnecessarily spent. It would be more recommendable introducing the organic matter, clay and carbonate contents for a specific soil in order to establish specific background values. Although the statistical approach based on linear equations is undoubtedly more realistic, it does not take into account all the factors and processes that can be crucial for characterising the behaviour of PTEs under some particular conditions. For a given total content of some heavy metals, the free fraction is greater in acid than in basic soils [28,59] and therefore it would not be valid the application of the same background value for all kind of soils. Therefore, soil reaction seems to be a determinant factor for the mobility of contaminants in soil [58], although this property is not always considered in the regression equations. Nevertheless, it seems clearly difficult to include within the regression equations all the parameters related to PTE dynamics due to the complexity of the relationships between toxic elements and soil properties.

Application in a Representative Area of the Mediterranean Region: A Case of Study for Lead

The application of these various approaches was explored to establish background values for lead (Pb) in agricultural soils of the Alicante province (SE Spain), a representative area of the European Mediterranean region. Concentrations of PTEs in this area were previously analysed by Micó et al., [60], and results showed evidence of human inputs of Pb from atmospheric deposition and the use of waste water. Therefore, Pb was selected to illustrate the comparison of statistical methods in order to estimate background values, since this element represents an anthropogenic element in the study area due to the presence of potentially contaminated soils. Fifty-four agricultural plots devoted to vegetable crops in the Alicante province were randomly selected during July and October 2001. Total concentration of Pb in soil samples were analysed by the method 3051A of the United States Environmental Protection Agency [61]. Further details about the sampling procedure, distribution of sampling sites, and analytical methods are described in a full by Micó et al. [42,60].

The comparison of the background values obtained using different statistical approaches is presented in Table 3. The Pb population was heavily skewed indicating a deviation of the normal distribution. A total of seven anomalous values were identified as outliers using the Pb boxplot. The data set fitted with a normal distribution when outliers were not considered as part of the population data. The upper limit of the background variation using the X+2SD approach was 28 mg/kg, as was reported in a previous work [42].

The probability plot of the Pb raw data set (Figure 2) was obtained using PROBPLOT® v. 1.00 [51] provided by *The Association of Applied Geochemists* (Canada). The plot fitted well with a sigmoidal curve and, therefore, it was possible to identify an inflection point that separates two populations. The linearity of the lower part of curve is consistent with a hypothesis that the metal values are derived from uncontaminated soils for which the frequency distribution is normal. The inflection point of the cumulative frequency curve represents the transition from the background population (A-population) to the potentially contaminated population (B-population). The upper limit of the background variation was 29 mg/kg, according to this statistical approach (Table 3).

The distribution of the estimated [X+2SD] values for Pb was plotted in a histogram using the R® statistical program v. 1.9.0 after 999 resampling from the raw data set. The 25th and 75th percentiles computed as the upper and lower bounds of the confidence interval are shown in Figure 3. The confidence interval obtained was 33 mg/kg and 72 mg/kg, whereas the average of the [X+2SD] distribution was 53 mg/kg (Table 3).

Regression equation between Pb concentration and selected soil components was also obtained. In this case, Pb showed a significant correlation with the organic matter ($p<0.05$), whereas did not show a significant correlation with clay, carbonates or other soil properties. Since organic matter was the only component in predicting background values for Pb, the coefficient of the regression equation agrees with the slope of the single regression line. The upper limit of the background variation (UL) can be estimated following the formula 18+4H (Table 3). However, this equation does not establish a numerical value to evaluate the extent of soil pollution. According to a previous work [60], a standard soil in the study area presents an organic matter content of 2%, a clay content of 30% and a total carbonate content of 50%, since these values represent the average of these properties. In this case, only organic matter was considered in the regression equation and therefore the upper limit of the background variation was 26 mg/kg (Table 3).

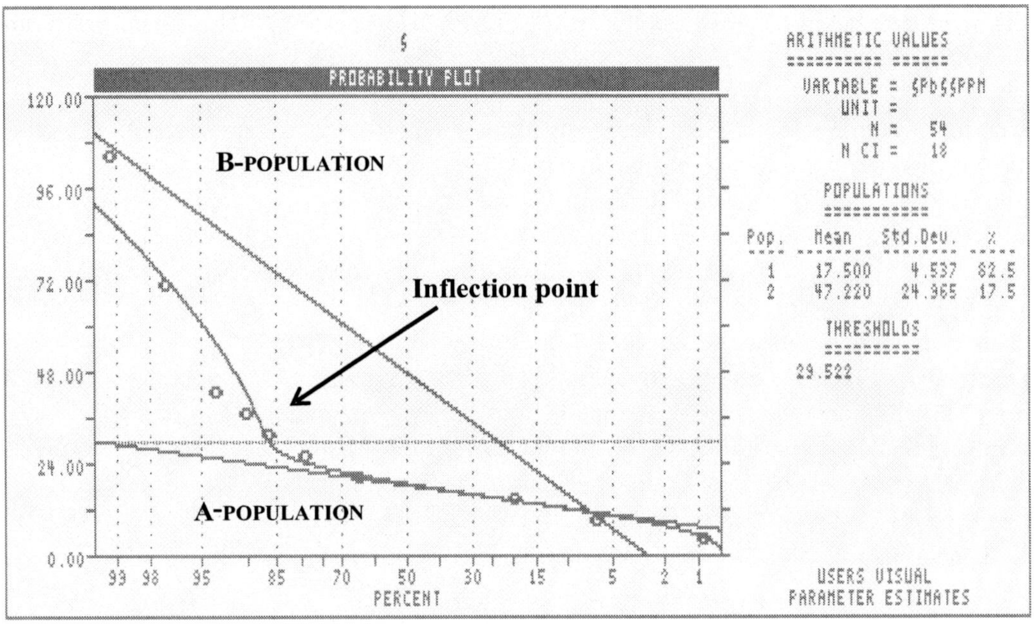

Figure 2. Probability plot for Pb (A-population is the background population and B-population is the potentially contaminated population) (unpublished data).

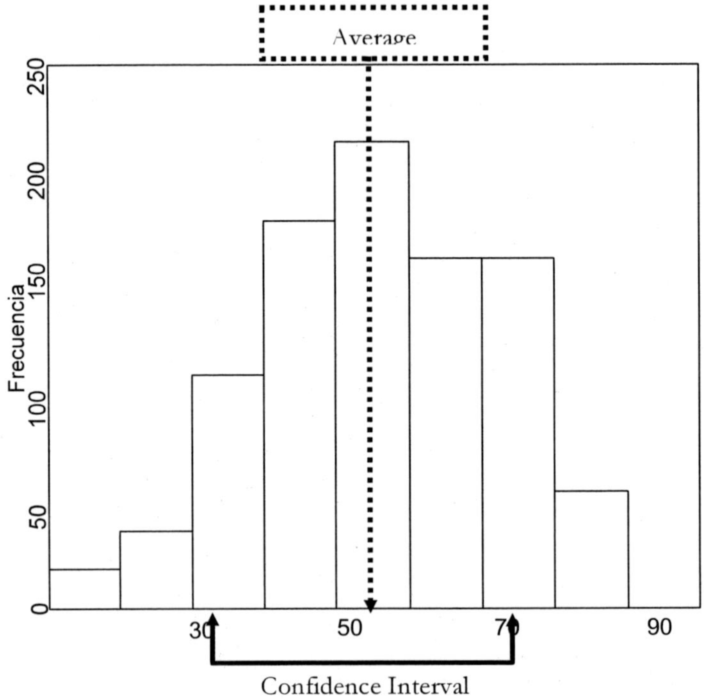

Figure 3. Histogram about distribution of [X+2SD] values for Pb in the Alicante province (unpublished data).

Table 3. Background Values for Pb in Agricultural Soils of the Alicante Province Using a Range of Statistical Approaches (Unpublished Data)

Element	X+2SD [a]	Probplot	Bootstrap	Regression equation	Standard soil [b]
Pb	28	29	53	18+4H	26

[a] Micó et al. [42].

[b] 2% of organic matter, 30% of clay, and 50% of total carbonates.

In summary, background values for Pb were quite similar using descriptive statistics, the probability plot, and the regression equation applied to a standard soil, taking into account the magnitude of those values. To apply the ISO/DIS 19258 method to soils in the study area, it is necessary to consider soils not influenced by point input due to local human activities. Therefore, a previous identification of outliers is always required to use descriptive statistics in order to not include them into the data set. In the other two cases, the raw data set was used considering the whole population variability. The regression equation produced the lowest value due to the use of a standard soil. As was commented before and for this particular case, it is more recommendable to introduce the organic matter content for a specific soil in order to estimate specific Pb background values, since the standard soil shows average properties in the study area. On the other hand, the Bootstrap method reported the highest value due to the consideration of outliers that probably determines an overestimation of the final result during the resampling process. According to the findings achieved, the selection of the statistical approach will be finally dependent on the objectives of the background study and, particularly, on the characteristics of the study area.

Conclusion

In the last few decades, the establishment of background values of PTEs in soils has become increasingly active. A number of works and countries have published background values at local or regional level. Different statistical approaches are currently under development, including descriptive statistics, probability plots, Bootstrap method, and methods based on regression equations taking into account soil properties. Alternatively, multivariate statistical analysis has been applied by some authors [18,60,62] in order to discriminate between elemental patterns of natural and anthropogenic origin that can be used to estimate lithogenic elements in order to propose background values. Additionally, many studies have explored the use of other methods for establishing background values.

For instance, Blaser et al. [63] suggested that element concentrations measured in the deeper soil horizon (e.g. the C horizon) can be taken as the local background for the top soil, after correction for the variation of a reference element such as Zr. This method has been named "the enrichment factor approach", since it is based on the ratio between PTE concentrations in the A-horizon (top soil) and the B- and/or C-horizon. This method has been criticised for failing to take into account the natural biogeochemical processes that redistribute elements in the soil profile [18]. Moreover, the top soil contains considerably more organic matter than the deeper horizon, increasing the amount of some elements due to its binding properties. This fact leads to a relative depletion of PTE concentrations in the deeper horizon [4].

Other works have used the relationship between PTEs and semi-conservative elements such as Fe, Al and Mn to predict the expected values in surface horizons [24]. For example, Hamon et al. [64] analysed soil geochemical data in Namibia, Australia and Asia and used relationships with Fe or Mn to make estimates of the background values, assuming that the soils from remote areas were uncontaminated. Sterckeman et al. [65] also used a similar approach to estimate background values in soils developed from loess deposit in northern France. Zhao et al. [18] applied regression analysis on the $\log_{10}$-transformed data of Al, Fe and seven elements, excluding data that were below the detection limits or outliers above an upper limit.

The development of a consensus method to be applied would be timely. However, it is unlikely that a single statistical approach was accepted on a worldwide scale, although the international guideline ISO/DIS 19258 can help to achieve a common procedure to be applied by soil scientist or environmental impact assessment practitioners in order to compare results among areas. Nevertheless, the experience of going through a case study in the Alicante province generally shows little differences among results achieved by different statistical approaches, although a great difference might appear when highly contaminated soils are included. It is clear that no global background levels of PTEs can be defined in soils and only regional or local estimations can be made [4,26] because of the background values are narrowly related to soil properties. Background values must be derived from a particular scenario and are not applicable outside their own area, since soils may be significantly different from the standard soil type. The selection of the statistical approach is dependent on the objectives of the background study and on the characteristics of the study area (i.e. presence of potentially contaminated soils). In any case, a detailed observation of the study area and PTE populations is required before selecting the statistical approach to be applied for establishing background values for PTEs.

REFERENCES

[1] Wedepohl, K. H. (1991). The composition of the upper earth's crust and the natural cycles of selected metals: Metals in natural raw materials; Natural resources. In E. Merian (Ed.), *Metals and their compounds in the environment* (pp. 3-17). Weinheim: VCH-Verlagsges.

[2] Lepeltier, C. (1969). A simplified statistical treatment of geochemical data by graphical representation. *Economic Geology, 64*, 538-550.

[3] Rose, A. W., Hawkes, H. E., & Webb, J. S. (1979). *Geochemistry in mineral exploration* (second edition). New York: Academic Press.

[4] Reimann, C., & Garrett, R. G. (2005). Geochemical background: Concept and reality. *Science of the Total Environment, 350*, 12-27.

[5] Chen, M., Ma, L. Q., Hoogeweg, C. G., & Harris, W. G. (2001). Arsenic background concentrations in Florida (USA) surface soils: Determination and interpretation. *Environmental Forensics, 2*, 117-126.

[6] Tack, F. M. G., Verloo, M. G., Vanmechelen, L., & Van Ranst, E. (1997). Baseline concentration levels of trace elements as a function of clay and organic carbon contents in soils in Flanders (Belgium). *Science of the Total Environment, 201*, 113-123.

[7] Salminen, R., & Gregorauskiene, V. (2000). Considerations regarding the definition of a geochemical baseline of elements in the surficial materials in areas differing in basic geology. *Applied Geochemistry, 15*, 647-653.

[8] Adriano, D. C. (2001). *Trace elements in terrestrial environments. Biogeochemistry, bioavailability and risks of metals* (second edition). New York: Springer-Verlag.

[9] Ramos, J., Gil, C., & Boluda, R. (2002). Baseline concentrations of cadmium, lead and nickel in greenhouses-western-Almeria surface soils (Spain). Cartagena, Murcia, España. In A. Faz, R. Ortiz, & A. R. Mermut (Eds.), *Sustainable use and management of soils in arid and semiarid region* (pp. 466-468). Cartagena, Murcia: Quaderna Editorial.

[10] Tume, P., Bech, J., Longan, L., Tume, L., Reverter, F., & Sepulveda, B. (2006). Trace elements in natural surface soils in Sant Climent (Catalonia, Spain). *Ecological Engineering, 27*, 145-152.

[11] Holmgren, G. G. S., Meyer, M. W., Chaney, R. L., & Daniels, R. B. (1993). Cadmium, lead, zinc, copper and nickel in agricultural soils of the United States of America. *Journal of Environmental Quality, 22*, 335-348.

[12] ISO/DIS. (2005). *ISO 19258. Soil quality: Guidance on the determination of background values*. Switzerland: International Standard Organisation (ISO).

[13] Singh, A., Singh, A. K., & Flatman, G. (1994). Estimation of background levels of contaminants. *Mathematical Geology, 26*, 361-388.

[14] Tobías, F. J., Bech, J., & Sánchez, P. (1997). Establishment of the background levels of some trace elements in soils of NE Spain with probability plots. *Science of the Total Environment, 206*, 255-265.

[15] Breckenridge, R. P., & Crockett, A. B. (1998). Determination of background concentrations of inorganics in soils and sediments at hazardous waste sites. *Environmental Monitoring and Assessment, 51*, 621-656.

[16] Crommentuijn, T., Sijm, D., De Brujin, P. J., Van den Hoop, M., Van Leeuwen, K., & Van de Plassche, E. (2004). Maximum permissible and negligible concentrations for metals and metalloids in the Netherlands, taking into account background concentrations. *Journal of Environmental Management, 60*, 121-143.

[17] Baize, D., & Sterckeman, T. (2001). Of the necessity of knowledge of the natural pedo-geochemical background content in the evaluation of the contamination of soils by trace elements. *Science of the Total Environment, 264*, 127-139.

[18] Zhao, F. J., McGrath, S. P., & Merrington, G. (2007). Estimates of ambient background concentrations of trace metals in soils for risk assessment. *Environmental Pollution, 148*, 221-229.

[19] CEC. (2006). *Communication from the Commission to the Council, the European Parliament, the European Economic and Social Committee and the Committee of the Regions: Thematic Strategy for Soil Protection*. SEC(2006)620, SEC(2006)1165. Brussels: Commission of the European Communities (CEC).

[20] Van Camp, L., Bujarrabal, B., Gentile, A. R., Jones, R. J. A., Montanarella, L., Olazabal, C., et al. (2004). *Reports of the Technical Working Groups Established*

Under the Thematic Strategy for Soil Protection. Volume IV. Contamination and Land Management (No. EUR 21319 EN/4). Luxembourg: European Communities.

[21] Tiller, K. G. (1992). Urban soil contamination in Australia. *Australian Journal of Soil Research, 30*, 937-957.

[22] Kabata-Pendias, A. (2000). *Trace elements in soils and plants* (third edition). Boca Raton, Florida: CRC Press.

[23] McLean, J. E., & Bledsoe, B. E. (1992). *Behaviour of metals in soils* (EPA/540/S-92/018). Washington DC: United States Environmental Protection Agency (US-EPA).

[24] Myers, J., & Thorbjornen, K. (2004). Identifying metals contamination in soil: a geochemical approach. *Soil and Sediment Contamination, 13*, 1-16.

[25] Moorman, T.B. (1996) Assessing organic chemical contaminants in soil. In J. W. Doran, & A. J. Jones (Eds.), *Methods for assessing soil quality* (pp. 311-326). Madison, WI: Soil Science Society of America, SSSA Special Publication No. 49, SSSA Inc.

[26] Recatalá, L., Micó, C., Sánchez, J., & Boluda, R. (2001). Approaches for characterising contaminated sites: an analysis considering the role of soil. In Y. Villacampa, C.A. Brebbia, & J.L. Usó (Eds.), *Third International Conference on Ecosystems and Sustainable Development (ECOSUD)* (pp. 347-356). Boston: Wessex Institute of Technology.

[27] Visser, W. J. F. (1995). Contaminated land policies in Europe. *Chemistry & Industry, 3*, 496-499.

[28] Jopony, M., & Young, S. D. (1994). The solid-solution equilibria of lead and cadmium in polluted soils. *European Journal of Soil Science, 45*, 59-70.

[29] Martínez, C. E., & Motto, H. L. (2000). Solubility of lead, zinc and copper added to mineral soils. *Environmental Pollution, 107*, 153-158.

[30] De Matos, A. T., Fontes, M. P. F., Da Costa, L. M., & Martínez, M. A. (2001). Mobility of heavy metals as related to soil chemical and mineralogical characteristics of Brazilian soils. *Environmental Pollution, 111*, 429-435.

[31] Zipper, C., Komarneni, S., & Baker, D. E. (1988), Specific cadmium sorption in relation to the crystal chemistry of clay minerals. *Soil Science Society of American Journal, 52*, 49-53.

[32] Elliot, H. A., Liberati, M. R., & Huang, C. P. (1986). Competitive adsorption of heavy metals by soils. *Journal of Environmental Quality, 15*, 214-219.

[33] Madrid, L., & Diaz-Barrientos, E. (1992). Influence of carbonate on the reaction of heavy metals in soils. *European Journal of Soil Science, 43*, 709-721.

[34] IHOBE. (1998). *Calidad del suelo. Valores indicativos de evaluación (VIE-A, VIE-B y VIE-C)*. Pais Vasco: Sociedad Pública de Gestión Ambiental (IHOBE), Departamento de Urbanismo, Vivienda y Medio Ambiente (in Spanish).

[35] McKeague, J. A., & Wolynetz, M. S. (1980). Background levels of minor elements in some Canadian soils. *Geoderma, 24*, 299-307.

[36] Kabata-Pendias, A., Dudka, S., Chlopecka, A., & Gawinowska, T. (1992). Background levels and environmental influences on trace metals in soils of the temperate humid zone of Europe. In D. C. Adriano (Ed.), *Biogeochemistry of Trace Metals* (pp. 61-84). Boca Raton, Florida: Lewis Publishers.

[37] Reimann, C., & Filzmoser, P. (2000). Normal and lognormal data distribution in geochemistry: death of a myth. Consequences for the statistical treatment of geochemical and environmental data. *Environmental Geology, 39*, 1001-1014.

[38] Gough, L. P., Severson, R. C., & Jackson, L. L. (1994). Baseline element concentrations in soils and plants, Bull Island, Cape Romain National Wildlife Refuge, South Carolina, USA. *Water, Air and Soil Pollution, 74*, 1-17.

[39] De Miguel, E., Callaba, A., Arranz, J. C., Cala, V., Chacón, E., Gallego, E., Alberruche, E., Alonso, C., Fdez-Canteli, P., Iribarren, I., Palacios, H., & Barettino, D. (2002). *Determinación de niveles de fondo y niveles de referencia de metales pesados y otros elementos traza en suelos de la Comunidad de Madrid.* Madrid: Instituto Geológico y Minero de España, Ministerio de Ciencia y Tecnología, Consejería de Medio Ambiente (in Spanish).

[40] Miller, J. C., & Miller, J. N. (1993). *Estadística para química analítica.* Delaware: Addison-Wesley Iberoamericana S.A (in Spanish).

[41] Reimann, C., Filzmoser, P., & Garrett, R. G. (2005). Background and threshold: critical comparison of methods of determination. *Science of the Total Environment, 346*, 1-16.

[42] Micó, C., Peris, M., Recatalá, L., & Sánchez, J. (2007). Baseline values for heavy metals in agricultural soils in an European Mediterranean region. *Science of the Total Environment, 378*, 13-17.

[43] Vázquez, I., Martín, J. A., Moreno, A. M., & González, J. (2000). Aplicación de fórmulas empíricas en el cálculo de valores de referencia de elementos traza en suelos de la Comunidad de Madrid. *Boletín Geológico y Minero, 111*, 63-78 (in Spanish).

[44] Chen, M., Ma, L. Q., & Harris, W. G. (1999). Baseline concentrations of 15 trace elements in Florida surface soils. *Journal of Environmental Quality, 28*, 1173-1181.

[45] Crock, J. G., Severson, R. C., & Gough, L. P. (1992). Determining baselines and variability of elements in plants and soils near the Kenai National Wildlife Refuge, Alaska. *Water, Air and Soil Pollution, 63*, 253-271.

[46] Barnett, V., & Lewis, T. (1994). *Outliers in statistical data* (third edition). New York: Wiley & Sons.

[47] Sinclair, A. J. (1974). Selection of threshold values in geochemical data using probability graphs. *Journal of Geochemical Exploration, 3*, 129-149.

[48] Davies, B. E. (1983). A graphical estimation of the normal lead content of some British soils. *Geoderma, 29*, 67-75.

[49] Davies, B. E., & Wixson, B. G. (1985). Trace elements in surface soils from the mineralized area of Madison County, Missouri, USA. *Journal of Soil Science, 36*, 551-570.

[50] Fleischhauer, H. L., & Korte, N. (1990). Formulation of cleanup standards for trace elements with probability plots. *Environmental Management, 14*, 95-105.

[51] Stanley, C. R. (1987). *Instruction manual for PROBPLOT: An interactive computer program to fit mixtures of normal (or lognormal) distributions with maximum likelihood optimization procedures* (Version 1.00). Ontario: The Association of Applied Geochemists.

[52] Micó, C. (2005). *Estudio de metales pesados en suelos con cultivos hortícolas de la provincia de Alicante* (Doctoral Thesis). Valencia: Universitat de València (in Spanish).

[53] Efron, B., & Tibshirani, R. J. (1993). *An introduction to the bootstrap*. New York: Chapman & Hall.

[54] Di Stefano, C., Ferro, V. & Porto, P. (2000) Applying the bootstrap technique for studying soil redistribution by caesium-137 measurements at basin scale. *Hydrological Sciences Journal, 45*, 171-783.

[55] Chirenje, T., Ma, L. Q., Szulczewski, M., Littell, R., Portier, K. M., & Zillioux, E. (2003). Arsenic distribution in Florida urban soils. *Journal of Environmental Quality, 32*, 109-119.

[56] Umweltdaten. (1999). *Federal Soil Protection and Contaminated Sites Ordinance (BBodSchV)*. Berlin: Ministry of Environment (Umweltdaten).

[57] Vegter, J. J. (1995). Soil protection in The Netherlands. In W. Salomons, U. Förstner, & P. Mader (Eds.), *Heavy Metals: Problems and Solutions* (pp. 79-100). Berlin: Springer-Verlag.

[58] Ritchie, G. S. P., & Sposito, G. (1995). Speciation in soils. In A. M. Ure, & C. M. Davidson (Eds.), *Chemical speciation in the environment* (pp. 201-233). Glasgow: Blackie Academic & Professional.

[59] McBride, M., Sauvé, S., & Hendershot, W. (1997). Solubility control of Cu, Zn, Cd and Pb in contaminated soils. *Journal of Soil Science, 48*, 337-346.

[60] Micó, C., Recatalá, L., Peris, M., & Sánchez, J. (2006). Assessing heavy metal sources in agricultural soils of an European Mediterranean area by multivariate analysis. *Chemosphere, 65*, 863-872.

[61] USEPA. (1998). *Method 3051A. Microwave assisted acid digestion of sediments, sludges, soils and oils* (Revision 1). Washington, DC: United States Environmental Protection Agency (US-EPA).

[62] Facchinelli, A., Sacchi, E., & Mallen, L. (2001). Multivariate statistical and GIS-based approach to identify heavy metal sources in soils. *Environmental Pollution, 114*, 313-324.

[63] Blaser, P., Zimmermann, S., Luster, J., & Shotyk, W. (2000). Critical examination of trace element enrichment and depletions in soils: As, Cr, Cu, Ni, Pb and Zn in Swiss forest soils. *Science of the Total Environment, 249*, 257-280.

[64] Hamon, R. E., McLaughlin, M. J., Gilkes, R. J., Rate, A. W., Zarcinas, B., Robertson, A., Cozens, G., Radford, N., & Bettenay, L. (2004). Geochemical indices allow estimation of heavy metal background concentrations in soils. *Global Biogeochemical Cycles, 18*, GB1014.

[65] Sterckeman, T., Douay, F., Baize, D., Fourrier, H., Proix, N., Schvartz, C., & Carignan, J. (2006). Trace element distributions in soils developed in loess deposits from northern France. *European Journal of Soil Science, 57*, 392-410.

INDEX

A

abiotic, x, 68, 120, 138
absorption spectroscopy, 175
access, 100
accounting, 218
acid, 54, 79, 103, 104, 111, 112, 114, 120, 124, 125, 129, 130, 133, 140, 159, 160, 168, 169, 170, 172, 175, 226, 234
acidic, ix, 109, 111, 112, 114, 115
acidity, 173
Acinetobacter, x, 143, 145, 146, 149, 150, 151, 152, 157, 158, 159, 160
acrylonitrile, 102
Actinobacteria, viii, ix, 73, 74, 79, 80, 81, 92, 93, 95, 96, 97, 98, 99, 100, 101, 103, 106, 149
actinomycetes, 141
activation, 207, 213
adaptation, viii, 35, 51, 78, 97, 128, 131, 139, 209
additives, 6
adjustment, 24
adsorption, 29, 38, 78, 111, 167, 168, 174, 175, 178, 218, 232
adults, 29
aerobic, viii, xi, 73, 157, 177, 178, 201
aerobic bacteria, xi, 177
agar, 133, 134
age, xii, 203, 204
agent, 131, 170, 208
aggregates, xi, 81, 178, 180, 189, 191
aging, 14, 19, 36, 38, 40, 45, 47, 48, 121, 209
aging process, 14, 19, 36, 38
agricultural crop, 214
agriculture, 49, 164, 172
agrochemicals, 207, 208, 209
Alaska, 221, 233
alcohol consumption, 213
alcohol(s), 49, 120, 189, 213
alfalfa, 66
algae, 3, 27, 29, 30, 32, 34, 37, 38, 42, 44, 178, 194, 197
Algeria, 174
algorithm, 48, 85, 87, 90, 91, 98
alkaline, 110, 111, 212
alkane hydroxylase gene (alkB), x, 144, 152
alkane(s), x, 143, 144, 145, 146, 149, 150, 151, 152, 153, 157, 158, 159, 160
allele, 205
Allium cepa, xii, 203, 205, 211
allosteric, 129
alluvial, 82, 96
alternative(s), vii, 1, 2, 11, 14, 15, 18, 19, 24, 31
aluminium, 167
Aluminum, 162
amendments, 114, 115, 164, 173, 176
amine(s), xii, 188, 189, 203, 207, 213
amino, xi, 129, 130, 160, 177, 180, 185, 188, 189
amino acid(s), 129, 130, 160, 189
amino groups, 188
ammonia, xi, 106, 133, 177, 180, 185, 197
ammonium, 122, 123
anaerobic, xi, 177, 178
aneuploidy, 204
animal waste, 184
animals, 3, 8, 28, 54, 55, 56, 71, 162
anion(s), 66, 67, 78, 141
ANOVA, 3, 32, 34, 36, 38
antagonism, 24
anthracene, 148, 149
anthropogenic, 13, 74, 75, 76, 96, 110, 113, 163, 164, 214, 218, 227, 229
antibiotic(s), 79, 160
antimicrobial, 41
antimony, 213
antioxidants, 120, 213

appendix, 180
aquatic systems, 14, 178
Arabidopsis thaliana, xii, 203, 206
Archaebacteria, 79
arithmetic, 220
arkoses, viii, 51, 65
aromatic hydrocarbons, xi, xii, 41, 49, 144, 145, 146, 154, 203, 207
aromatics, 145
arrest, 69
arsenic, 48, 99, 105, 140, 141, 162, 165, 213, 224
arthropods, 28
ash, 63
Asia, 230
Aspergillus niger, ix, 119, 123, 133, 140, 141
assessment, vii, xii, 1, 2, 3, 4, 5, 6, 8, 9, 10, 11, 12, 13, 14, 15, 16, 17, 18, 19, 20, 21, 23, 24, 25, 26, 27, 28, 29, 30, 31, 32, 34, 35, 36, 38, 39, 40, 41, 42, 43, 44, 45, 47, 48, 49, 50, 157, 163, 173, 212, 213, 217, 219, 230, 231
assessment tools, vii, 1
assumptions, 11, 15, 16, 17, 224
ASTM, vii, 1, 3, 27, 29, 30, 41, 47
atherosclerosis, 204, 209
atmosphere, xi, 84, 161, 164, 188
atmospheric deposition, 163, 164, 171, 218, 222, 227
attachment, xi, 9, 177
attention, 52, 100, 163, 168, 219
atypical, 129
Australia, 45, 71, 230, 232
Austria, 50, 203, 207
autotrophic, viii, 52, 62
availability, 10, 12, 13, 14, 26, 30, 38, 47, 77, 100, 104, 112, 133, 162, 163, 166, 167, 168, 170, 171, 173, 193
avoidance, 28, 80
awareness, 76

B

Bacillus, 122, 131, 140, 146, 181
Bacillus subtilis, 131, 140
bacteria, ix, xi, 44, 74, 79, 80, 99, 100, 102, 103, 107, 121, 130, 131, 133, 139, 141, 144, 146, 148, 150, 152, 153, 155, 156, 157, 158, 159, 177, 178, 179, 180, 181, 184, 188, 191, 197, 198
bacterial, viii, ix, x, xii, 30, 44, 73, 74, 75, 79, 80, 81, 84, 85, 86, 88, 92, 93, 95, 96, 97, 99, 100, 101, 102, 103, 104, 105, 106, 119, 121, 125, 130, 137, 138, 140, 144, 145, 151, 152, 153, 154, 155, 156, 157, 159, 160, 203, 204, 207, 210
bacterial cells, 130, 159
bacterial strains, 121
bacterium, 30, 128, 130, 138, 139
banks, viii, 52, 62, 67, 182
barley, 49
barriers, 53
base pair, 137
battery(ies), xii, 30, 31, 32, 164, 204, 208
behavior, viii, 4, 5, 8, 27, 28, 31, 51, 56, 62, 66, 68, 69, 111, 129, 199
behavioral effects, 27
Belgium, 220, 221, 225, 231
benefits, 60, 179, 197
benthic soil, xi, 177, 178, 180
benzene, 49, 148, 149
benzo(a)pyrene, 207
bias, 97
binding, xii, 80, 112, 168, 217, 224, 230
bioaccumulation, 8, 9, 13, 19, 29, 57
bioassay(s), vii, xii, 2, 24, 25, 32, 33, 39, 41, 44, 45, 48, 49, 56, 57, 66, 68, 70, 203, 204, 207, 210, 211, 213, 214
bioavailability, 9, 13, 14, 15, 17, 18, 19, 24, 26, 36, 38, 40, 44, 45, 47, 49, 53, 77, 105, 107, 110, 111, 112, 133, 166, 167, 171, 231
bioconcentration, 9, 19, 25, 28
biodegradation, x, xi, 57, 102, 135, 137, 138, 141, 144, 157
biodiversity, 18, 23, 42, 65, 66, 76, 78, 100
biogas, xi, 177, 180, 181, 182, 185, 187, 189, 198
bioindicators, 112
biological activity, 169
biological processes, 52, 56
biological systems, 56
biological toxicity, 74
bioluminescence, 30
biomarker(s), 9, 22, 152
biomass, viii, x, 4, 28, 34, 35, 52, 61, 75, 77, 78, 79, 80, 81, 99, 101, 102, 103, 104, 106, 132, 133, 134, 144, 145, 151, 152, 153, 154, 155, 156, 159, 160, 169, 174
biomonitoring, 208, 210
bioremediation, x, 48, 49, 136, 141, 143, 144, 145, 146, 151, 152, 153, 154, 155, 156, 158, 178
biosorption, 80, 106
biotechnology, 79, 138
biotic, x, 45, 49, 68, 120, 138
biotic factor, 68
birds, 17, 18, 43
blood, 213
Bohr, 209
bonds, 136
boom system, xi, 177, 180, 193, 195, 196, 197, 198, 199
bootstrap, 234

Bootstrap method, xii, 217, 220, 223, 224, 229
borderline, 168
boron, 162
bounds, 227
bovine, 123, 127
brain, 209
Brazil, 208
breakdown, 133, 136, 188
Britain, 222
British Columbia, 8, 200
Brussels, 42, 43, 231
bubbles, xi, 177, 180, 181, 182, 185, 187, 188, 189, 191, 198
Bubbles, 192
buccal mucosa, 213
buffer, 84, 85, 104
Burkholderia, 145
burning, 63, 64

C

Ca^{2+}, 110, 136
cadmium, 81, 99, 104, 105, 124, 134, 141, 162, 173, 174, 175, 176, 213, 231, 232
calcium, 77, 133, 167, 168
calcium carbonate, 167, 168
cambisols, viii, 51, 54
Canada, 6, 16, 44, 48, 177, 180, 187, 222, 227
cancer, xii, 203, 204, 207, 213
candida, 28, 43, 44, 45
candidates, 54
carbon, xi, 4, 29, 35, 36, 37, 38, 39, 46, 76, 77, 78, 84, 92, 96, 99, 100, 101, 103, 105, 121, 137, 146, 158, 177, 185, 192, 231
carbon dioxide, xi, 78, 177, 185
carbonates, xii, 54, 110, 167, 168, 170, 171, 217, 224, 225, 227, 229
carcinogenesis, 213
carcinogens, xii, 12, 41, 203, 207, 211
carrier, 84
case study, 157, 230
catalase, ix, 119, 120, 121, 122, 123, 124, 125, 126, 127, 128, 129, 130, 131, 133, 134, 138, 139, 140
catalase-1, ix, 119, 122, 128, 129
catalase-2, ix, 119, 122, 128
catalytic properties, 120
cation, ix, 104, 109, 112, 141, 168, 219
cattle, 54, 63, 71
cell, viii, ix, xii, 4, 29, 30, 34, 41, 51, 80, 106, 112, 119, 122, 127, 128, 131, 136, 146, 203, 205, 214
cell division, 205
cell line, 34, 41
cell surface, 80, 106
cereal crop, 64
cereals, 54, 62, 63
certainty, 41
CH3COOH, 170
channels, 55, 69
chaos, 140
chelating agents, 14, 77, 111
chemical, vii, viii, xi, 1, 2, 3, 6, 8, 10, 12, 13, 14, 15, 21, 24, 25, 27, 28, 29, 31, 32, 33, 35, 36, 39, 40, 43, 47, 48, 52, 56, 57, 61, 62, 65, 75, 77, 112, 120, 121, 133, 136, 138, 161, 162, 163, 164, 167, 171, 175, 176, 178, 179, 180, 185, 189, 193, 194, 197, 198, 207, 210, 211, 212, 214, 218, 232
chemical industry, 207
chemical interaction, 40
chemical properties, 138, 162, 175
Chernobyl, 211, 214
Chernobyl accident, 211
China, 103
chloride, 47, 66, 68, 133, 167, 170, 195
chloroform, 151
chlorophenols, 15
chlorophyll, 206
chromatid, 207, 213
chromatography, 145, 158
chromium, 104, 162, 165, 175, 194, 195, 196, 213
chromosome(s), xii, 203, 204, 205, 206, 210, 211
classes, 32
classification, 24, 32, 33, 39, 41, 47, 79, 106, 129
clay, viii, xii, 9, 21, 52, 54, 77, 92, 106, 110, 154, 158, 167, 171, 217, 219, 220, 224, 225, 226, 227, 229, 231, 232
cleaning, 180
cleavage, 128, 136
clone(s), 75, 80, 105, 210, 204
cloning, 139, 160
clustering, 87
clusters, 87, 97, 100
CO2, 111
coal, 123, 124, 126, 127, 131, 132, 133, 134, 135, 137
coal dust, 123, 124, 126, 127, 131, 132, 133, 134, 135, 137
cobalt, 214
cocoon, 4
coding, 105, 130
coffee, 113
coke, 45
collateral, 59
colon cancer, 213
colonization, 22, 64, 65
Comamonas terrigena, ix, 119, 121, 134, 138, 139, 140

Comamonas testosteroni, ix, 119, 121, 139
combined effect, 15
combustion, 110, 120
commercial, 7, 66, 121, 164
community, viii, ix, 21, 22, 25, 33, 40, 45, 52, 53, 57, 59, 61, 67, 68, 69, 70, 72, 73, 74, 75, 77, 78, 80, 81, 82, 85, 86, 87, 89, 92, 93, 94, 95, 96, 97, 98, 99, 100, 101, 102, 103, 104, 105, 106, 107, 131, 157, 210
competition, 168
complement, 24, 36, 40, 70
complexity, 3, 8, 14, 16, 19, 40, 53, 57, 64, 69, 74, 75, 219, 226
components, xii, 52, 55, 62, 69, 77, 112, 135, 145, 164, 170, 171, 204, 217, 220, 224, 225, 226, 227
composition, 27, 45, 54, 61, 62, 77, 79, 92, 99, 102, 103, 105, 107, 136, 159, 168, 218, 230
compost, 114, 164, 166, 168, 174, 176
composting, 168
compounds, 9, 15, 33, 62, 113, 115, 120, 122, 129, 133, 137, 138, 164, 167, 188, 204, 206, 207, 208, 230
concentration, vii, x, xii, 1, 2, 5, 6, 8, 9, 10, 11, 12, 13, 14, 15, 17, 19, 23, 25, 29, 31, 32, 33, 35, 36, 37, 38, 39, 40, 49, 77, 86, 92, 97, 99, 100, 101, 119, 122, 124, 128, 130, 131, 133, 135, 136, 137, 145, 166, 168, 169, 178, 180, 185, 191, 193, 194, 195, 196, 197, 203, 209, 219, 220, 221, 224, 227, 231
conceptual model, 7, 8, 33
conditioning, 189
conductivity, 67
confidence, 220, 223, 224, 227
confidence interval, 223, 224, 227
confidence intervals, 32, 223
consensus, 230
conservation, 54
construction, 139, 144, 154, 155, 179
consumers, 29, 30, 54
consumption, xii, 8, 18, 53, 61, 203, 213
contaminant(s), vii, xi, xii, 1, 6, 8, 13, 15, 17, 18, 19, 23, 24, 25, 27, 30, 31, 32, 35, 36, 37, 38, 40, 44, 57, 70, 115, 135, 136, 137, 161, 163, 195, 210, 217, 219, 224, 225, 226, 231, 232
contaminated soils, ix, x, xii, 3, 6, 12, 14, 26, 30, 31, 32, 33, 34, 35, 43, 44, 47, 48, 49, 71, 74, 78, 79, 81, 86, 99, 101, 106, 109, 111, 115, 139, 143, 144, 145, 152, 157, 158, 174, 203, 208, 209, 211, 221, 222, 223, 224, 227, 230, 234
contamination, vii, ix, x, xi, xii, 1, 5, 8, 18, 19, 22, 24, 26, 33, 40, 41, 44, 45, 59, 62, 63, 67, 69, 73, 74, 75, 77, 81, 84, 92, 94, 97, 99, 101, 102, 103, 104, 105, 106, 109, 110, 115, 143, 144, 145, 146, 153, 154, 155, 161, 169, 174, 185, 204, 207, 208, 210, 213, 214, 215, 219, 231, 232
continuous data, 32
control, ix, 3, 5, 25, 26, 31, 32, 33, 34, 35, 36, 38, 55, 81, 84, 86, 97, 98, 99, 110, 112, 113, 119, 124, 127, 128, 129, 131, 132, 133, 134, 135, 145, 152, 153, 155, 167, 171, 173, 179, 201, 204, 219, 234
cooling, 170
Copenhagen, 49, 173, 212
copper, 14, 45, 48, 60, 61, 80, 81, 107, 162, 173, 174, 194, 213, 231, 232
correlation(s), 14, 76, 97, 99, 112, 113, 168, 209, 225, 227
costs, 27, 54, 76
Council of Ministers, 7
coverage, 27, 81, 92, 95, 96, 100
covering, 2, 5, 8, 15, 21, 26, 30, 40, 53, 169, 220
critical value, 8
criticism, 47
crop production, 171
crops, 53, 54, 64, 74, 115, 171, 173, 175, 214, 227
crude oil, ix, 119, 121, 131, 135, 145, 148, 149, 157, 158
cultivation, 63, 64, 123, 124, 135, 152
culture, ix, 30, 80, 119, 121, 122, 123, 124, 126, 127, 128, 131, 132, 133, 134, 135, 137, 141, 151, 154
cumulative distribution function, 223
cumulative frequency, 222, 223, 227
cyanide, 102, 181, 201
Cyanobacteria, 4
cycles, 23, 29, 55, 78, 84, 85, 110, 230
cycling, 28, 74, 75
cyclohexyl, 150
cytochrome, 138, 158
cytometry, 209
cytosine, 79
cytotoxicity, 34
Czech Republic, 109, 208

D

data collection, 52
data distribution, 222, 233
data set, 10, 11, 21, 93, 222, 223, 224, 227, 229
database, 9, 130, 140, 223
death, 233
decay, 136
decision making, 12, 16
decision-making process, 41
decisions, 2
decomposition, xi, 27, 28, 48, 120, 168, 177, 178, 180, 181, 185, 188, 191
defense, 130

deficiency, 12, 223
definition, 9, 15, 74, 161, 184, 231
degradation, x, 8, 29, 54, 69, 78, 102, 119, 120, 121, 128, 135, 136, 144, 145, 148, 149, 150, 152, 153, 157, 158, 173, 189, 200
degradation pathway, x, 144, 149, 150, 158
degradation process, 29, 54, 69, 189
dehydrogenases, 29, 34
denaturing gradient gel electrophoresis (DGGE), viii, 73, 84, 85, 86, 87, 88, 90, 91, 92, 93, 94, 97, 98, 99, 104, 105, 106, 160
denitrification, 76
Denmark, 7, 49, 173, 207, 212
density, viii, 4, 18, 52, 105, 136, 162, 167, 224
deposition, 82, 102, 163, 164, 171, 173, 204, 218, 222, 227
deposits, 96, 207, 234
desorption, 157
destruction, 178
detachment, 34
detection, x, xii, 32, 102, 104, 122, 144, 145, 153, 158, 166, 195, 196, 203, 204, 207, 208, 211, 213, 222, 223, 230
detergents, 185, 189
deterministic, 10, 11, 20, 21, 87, 95, 98
developed countries, 6
deviation, 13, 65, 220, 221, 222, 227
diet, 17, 18, 166
differentiation, 32, 82
diffraction, 112
diffusion, 111
digestion, 114, 169, 170, 234
direct (eco)toxicity assessment (DTA), vii, 2, 24
direct measure, 19, 23
discharges, xi, 177
discourse, viii, 51, 53
displacement, 171
distillation, 145
distilled water, 83
distribution, 10, 11, 12, 13, 17, 21, 23, 24, 28, 29, 43, 71, 140, 155, 159, 163, 169, 220, 221, 222, 223, 224, 227, 228, 233, 234
distribution function, 11, 17, 223
diuron, 115
diversity, viii, ix, 7, 29, 52, 54, 66, 68, 73, 74, 75, 76, 78, 79, 81, 86, 87, 92, 93, 97, 99, 100, 101, 102, 103, 104, 105, 106, 107, 121, 130, 136, 137, 138, 140, 160, 210
division, 79, 205, 206
DNA, viii, x, xii, 73, 79, 84, 85, 86, 88, 93, 129, 137, 144, 145, 151, 152, 159, 160, 203, 204, 205, 206, 207, 208, 209, 212, 213, 215
DNA damage, 205, 207, 209, 212, 213, 215
DNA polymerase, 84, 85
DNA repair, 209, 212
DNase, 212
dominance, 54
dominant allele, 205
draft, 48
drainage, 8, 176
drinking water, 8, 197, 198, 199
dry matter, 166, 171
dumping, 64
duration, 10, 78, 174
dyes, 159
dynamic systems, 53

E

E. coli, 122
earth, 163, 173, 218, 230
earthworm, 19, 28, 35, 36, 38, 39, 45
earthworms, 27, 28, 34, 35, 44
Eastern Europe, 212
ecological restoration, 53, 56, 59, 69
ecology, 57, 79, 104, 105
ecosystem(s), vii, viii, xii, 1, 2, 3, 7, 8, 9, 11, 12, 15, 20, 21, 22, 23, 27, 28, 29, 30, 31, 33, 41, 42, 44, 48, 51, 52, 53, 54, 55, 56, 59, 60, 62, 63, 65, 66, 67, 69, 72, 75, 76, 77, 97, 103, 105, 107, 138, 144, 203, 204, 218
ecotoxicological, vii, 1, 10, 15, 18, 21, 25, 27, 31, 33, 43, 44, 50, 57, 59, 63, 69
ecotoxicological tests, 27
ecotoxicology, 3, 40, 43, 47, 56, 210
effluents, 200
eggs, 29
electric charge, 189
electric field, 206
electron, 61, 120, 128
electron microscopy, 61
electrophoresis, viii, xii, 73, 85, 86, 105, 106, 122, 123, 127, 160, 203, 206
elongation, 49
email, 217
embryo, 5, 29, 46
emission, 30, 44, 214
encoding, 103, 138
endocrine, 9, 22
endonuclease, 137
energy, x, 23, 43, 75, 76, 143, 144, 146, 158
England, 82, 91, 137, 223
environment, vii, ix, x, xii, 6, 13, 19, 21, 23, 26, 28, 39, 40, 41, 45, 46, 52, 55, 62, 70, 74, 75, 77, 79, 111, 119, 120, 130, 133, 135, 138, 140, 149, 157, 167, 169, 174, 175, 200, 203, 213, 215, 230, 234

environmental change, 12, 163
environmental conditions, 33, 134, 168
environmental degradation, 78
environmental effects, 59
environmental factors, 14, 68, 72
environmental impact, 12, 43, 52, 53, 54, 62, 65, 69, 103, 230
environmental impact analysis (EIA), 54
environmental influences, 232
environmental policy, xi, 161
Environmental Protection Agency (EPA), 7, 23, 27, 31, 44, 49, 61, 145, 162, 173, 174, 210, 212, 214, 227, 232, 234
enzymatic activity, 124
enzyme(s), ix, 21, 28, 29, 42, 56, 75, 106, 119, 120, 124, 128, 129, 131, 133,136, 138, 145, 208, 213
equating, 100
equipment, x, 143
erosion, viii, 26, 27, 40, 52, 55, 62, 69, 76, 107, 114, 115
Escherichia coli, 121, 139
ester(s), 103, 136, 139, 168
ester bonds, 136
estimating, 21, 220
Eukaryota, 79
eukaryotes, 121
Europe, 6, 8, 10, 42, 43, 47, 48, 110, 164, 173, 175, 212, 219, 232
European, xi, xii, 6, 7, 10, 12, 16, 18, 20, 21, 42, 43, 48, 71, 101, 103, 106, 161, 164, 165, 172, 173, 174, 175, 176, 212, 217, 218, 219, 220, 222, 223, 226, 231, 232, 233, 234
European Commission, 6, 10, 16, 20, 43, 48, 174
European Environment Agency, xi, 161, 173, 212
European Parliament, 231
European Soil Bureau, xi
European Union (EU), xii, 6, 14, 15, 18, 43, 50, 53, 61, 113, 217, 219
eutrophication, 69, 178
evaporation, 84, 164, 167, 178
evapotranspiration, 164, 167
evidence, 41, 80, 100, 115, 180, 213, 215, 227
evolution, 136, 139
experimental condition, 31, 72
experimental design, 31, 56
exposure, 2, 6, 7, 8, 10, 11, 12, 13, 14, 15, 16, 17, 18, 19, 20, 22, 23, 25, 26, 28, 29, 33, 39, 40, 56, 57, 74, 78, 98, 105, 133, 163, 169, 204, 205, 208, 209, 212, 213
extended X-ray absorption fine structure (EXAFS), 112
external environment, 52
extinction, viii, 52
extracellular matrix, 80
extraction, xii, 13, 14, 15, 96, 102, 112, 151, 159, 169, 170, 171, 173, 174, 203, 204
extraction process, 170
extrapolation, 2, 10, 15, 19, 20, 23, 43, 46

F

facies, 54
false negative, 32, 38
false positive, 208
family, x, 15, 121, 143, 146, 169
farm(s), 71, 179
fatty acid(s), 80, 103, 104, 189
fauna, xi, 27, 177, 184
feed additives, 6
fencing, 63
ferrite, 110
fertility, 27, 68, 74, 75, 114, 204
fertilization, 171, 218, 222
fertilizers, 69, 110, 164, 197
field trials, 23
film, 185, 189
filters, 178, 194
filtration, 194
fine materials, viii, 52, 68
fingerprinting, 104
first dimension, 95
fish, 4, 22, 29, 30, 34, 37, 38, 56, 193
flame, 158, 175
floating, 184, 193
flocculation, 178, 192
flood, 96
flooding, 26, 40, 82
flora, ix, xi, 65, 74, 78, 79, 81, 95, 96, 100, 101, 151, 177, 184
flotation, 104, 181, 188, 189
fluctuations, 22
fluorescence, 4, 159
fluoride, 165
fluorimeter, 29
fluvial, 102
food, 9, 11, 14, 18, 19, 28, 29, 48, 54, 55, 74, 174, 204, 210
food production, 204
forest ecosystem, 103
formamide, 85
fossil, 110, 120
fossil fuels, 110, 120
fractionation, 112
fragmentation, 205
France, 46, 47, 48, 113, 114, 230, 234
free radical, 178

frequency distribution, 13, 223, 227
freshwater, 28, 30, 44, 56
Freundlich isotherm, 110
fuel, 145, 209
fumigation, 151
functional aspects, 33
functional changes, 97
fungal, 78, 101, 102, 106, 123, 133, 138
fungicide, ix, 109
fungicides, ix, 42, 109, 110, 113, 115
fungus(i), xi, 27, 28, 79, 99, 113, 121, 133, 134, 139, 140, 141, 177, 178, 180, 184, 188

G

gamma rays, 211
gas chromatograph, 145, 158
gasoline, 145
gel, viii, xii, 73, 85, 86, 88, 92, 93, 99, 103, 105, 106, 122, 123, 127, 137, 158, 160, 203, 206
generation, xi, 177, 185
gene(s), ix, x, xii, 74, 75, 84, 86, 93, 101, 102, 103, 105, 119, 128, 130, 133, 137, 138, 139, 140, 144, 146, 147, 152, 158, 203, 205
genetic diversity, 105, 210
genetics, 105
Geneva, 44
genome, 209
geochemical, 47, 53, 219, 220, 221, 230, 231, 232, 233
geology, vii, 231
Germany, x, 9, 43, 143, 144, 207, 225
germination, 69, 101
glass, 84
global warming, 53
glucose, 122, 140
glucosidases, 29
Gordonia, x, 143, 144, 146, 149, 150, 153, 158
grains, 206
Gram-negative, 158
Gram-positive, 158
granites, viii, 51
graph, 23
grass, 59, 61, 64, 71, 95, 96, 114
grasses, 62, 64, 82
grassland(s), viii, ix, 53, 54, 55, 60, 61, 62, 70, 71, 73, 74, 76, 79, 81, 82, 92, 95, 96, 99, 100, 101, 102, 103, 104
gravimetric analysis, 145
gravity, 161
grazing, 55, 63, 72
Great Britain, 222
Greece, xi, 161
greenhouse, 204
ground water, viii, xii, 8, 19, 24, 37, 40, 48, 52, 55, 62, 110, 114, 115, 164, 178, 203, 204, 207, 214
groups, 6, 10, 11, 20, 21, 22, 30, 35, 37, 38, 39, 40, 76, 114, 136, 168, 188, 198, 208
growth, ix, 2, 3, 4, 5, 9, 26, 27, 28, 29, 30, 32, 34, 35, 36, 37, 38, 39, 42, 46, 56, 68, 71, 99, 101, 109, 115, 119, 121, 122, 123, 124, 125, 127, 128, 131, 132, 133, 134, 135, 137, 140, 149, 173, 194
growth rate, 28, 29, 37, 38
growth temperature, 140
guanine, 79
guidance, 6, 9, 46
guidelines, vii, 1, 6, 21, 144

H

habitat, 18, 81, 96, 197
hair cells, 205
harmonization, 15, 42
harvesting, 61
hazards, 213
health, vii, viii, xii, 1, 7, 8, 12, 52, 53, 54, 71, 73, 75, 144, 157, 169, 211, 217, 219
heat, 76, 139
heating, 170
heavy oil, 144
heavy particle, 194, 195
height, 28, 62, 64, 66
Helicobacter pylori, 122, 139
hematite, 110
heme, ix, 119, 120, 122, 129, 138
herbicide(s), xii, 197, 203, 207, 208
herbivores, viii, 52, 61
heterogeneity, 75, 77, 102
heterotrophic, 105
hexachlorocyclohexanes, 34
hexane, 149, 150
highways, 163
histogram, 224, 227
housing, 54, 64, 179
human actions, 69
human activity, 53, 218
human exposure, 29, 74, 169
human input, 227
human resources, 226
humic substances, 167, 168, 174
hybridization, 129, 130
hydrocarbons, vii, x, xi, 34, 104, 143, 144, 145, 146, 148, 149, 152, 153, 155, 157, 200
hydrochloric acid, 169
hydrogen, 120, 122, 124, 133, 140, 185
hydrogen peroxide, 120, 122, 124, 133, 140

hydrology, vii
hydrolysis, 178
hydrophobic, 191
hydrophobicity, 201
hydrous oxides, xii
hyperthermia, 212
hypothesis, 53, 76, 90, 96, 193, 196, 227

I

Iberian Peninsula, viii, 51, 53, 54, 61, 65
identification, vii, 1, 6, 15, 18, 25, 41, 137, 223, 229
immobilization, 4, 29, 34, 106, 218
impact assessment, 230
implementation, 15, 25, 47, 48
in situ, x, xii, 25, 55, 75, 100, 120, 138, 203, 213
in vitro, 29, 41, 44, 210
in vivo, 212
incidence, 11, 174
inclusion, 15, 24, 25
incubation, 83, 89
India, 208
indication, 34, 48, 99
indicators, viii, 42, 56, 73, 75, 102, 103
indices, 72, 87, 97, 234
indigenous, 25, 31, 74, 176
induction, 122, 124, 131, 140, 211, 213
industrial chemicals, 6, 50
industrial wastes, vii
industrialization, vii
industry, 9, 110, 122, 207, 218
inertia, 93, 95
infertility, xii, 203
ingestion, 14, 27
inhibition, 2, 3, 4, 5, 25, 30, 34, 35, 36, 38, 39, 44, 46, 47
inhibitors, 134
inhibitory, 44, 168
inhibitory effect, 44
injury, viii, 52
inoculation, 145, 153
insight, 62
instability, 209
insurance, 76
integration, 15
integrity, 9, 42
intensity, vii, 115, 193
interaction(s), 22, 24, 31, 36, 40, 52, 53, 68, 74, 77, 166, 167, 168, 219
interface, 40, 167, 218
International Organization for Standardization (ISO), vii, xii, 1, 3, 27, 28, 30, 44, 217, 218, 220, 229, 230, 231
internet, 130
interpretation, 97, 221, 230
interval, 30, 221, 223, 224, 227
intervention, 12
intoxication, 62
invertebrates, 3, 4, 8, 17, 19, 21, 22, 28, 30, 33, 49
iodine, 206
ionization, 158
ionizing radiation, 205, 208, 212
ions, 66, 80, 104, 124, 138, 167
iron, 164, 167, 171, 174, 178, 194, 213
irradiation, 211, 214
isolation, 139
isopods, 27, 28, 43, 49
isozyme(s), ix, x, 119, 120, 122, 123, 124, 127, 128, 130, 138
Italy, xi, 161, 170, 224

J

Japan, x, 143, 144
judgment, 11

K

K^+, 136
kerosene, 145
kidney(s), 71, 72
killing, 130, 140

L

laboratory method, viii, 52
lakes, 197
land, 7, 8, 9, 11, 15, 30, 41, 45, 50, 54, 69, 72, 74, 163, 164, 172, 178, 232
land use, 7, 8, 9, 11, 15, 30, 69, 163
landfills, vii, viii, 42, 51, 52, 53, 54, 59, 60, 61, 62, 63, 64, 65, 66, 67, 68, 69, 71, 164
landscapes, 53, 62
language, viii, 51, 53
laws, x, 143, 144
leachate(s), viii, 24, 26, 27, 29, 30, 31, 32, 33, 34, 37, 38, 39, 42, 51, 56, 59, 61, 62, 65, 70, 71, 204, 208, 209, 211, 215
leaching, vii, 8, 18, 24, 26, 40, 47, 55, 171, 172
lead, vii, viii, ix, 8, 15, 36, 44, 73, 78, 80, 81, 82, 86, 92, 96, 97, 103, 104, 109, 110, 112, 115, 131, 133, 138, 145, 151, 152, 161, 162, 167, 173, 174, 176, 194, 197, 198, 204, 212, 213, 220, 226, 231, 232, 233
Legionella pneumophila, 121, 139

legislation, 10, 39, 44, 164
legumes, 66, 80, 106
leisure, 54, 96, 101
lesions, 206
life cycle, 2, 29
lifetime, 2
ligand(s), 14, 42, 45, 49, 168, 174
likelihood, 6, 16, 17, 40, 233
limestones, viii, 51
limitation, 18, 23, 25, 29, 32, 194
links, 52, 53
literature, 36, 54, 55, 56, 62, 74, 77, 78, 79, 110, 113
liver, 41, 123, 127
livestock, xii, 8, 55, 71, 74, 217, 219
localization, 26
locus, 206, 212
London, 105, 116, 199, 200, 201
low risk, 25
low temperatures, 157
LSD, 34, 36
lubricating oil, 145
luminescence, 32
luvisols, viii, 51, 54
Lycopersicon esculentum, 112
lymphocytes, 213

M

macroalgae, 178
magnesium, 133
maize, 66, 102
mammalian cells, 204, 213, 214
management, 41, 47, 50, 69, 171, 179, 200, 212, 231
manganese, 162, 167, 171, 174
man-made, vii, xi, 177, 180, 184, 185, 188, 189, 191, 197
manufacturer, 84, 85
manufacturing, 184
manure, 41, 43, 164
marsh(es), viii, 51, 62
masking, 97
matrix, 34, 40, 77, 79, 80, 111, 112
measurement, x, 8, 9, 19, 30, 34, 144, 159
measures, 23, 28, 33, 45, 69, 75, 84, 211
meat peptone, ix, 119, 121, 122
media, vii, 1, 47, 135, 136, 146, 163, 214
median, 2, 220, 222
medicine, 79, 138
Mediterranean, viii, xi, xii, 51, 54, 55, 69, 71, 161, 164, 167, 168, 169, 171, 172, 174, 175, 176, 217, 218, 220, 221, 222, 223, 226, 233, 234
Mediterranean countries, 172
meiosis, 210, 214
menadione, 140
mercury, 80, 105, 131, 161, 162, 165, 176
metabolic changes, viii, 52
metabolic pathways, 145, 150
metabolism, 12, 29, 30, 78, 140, 157, 159
metabolites, 79, 150, 207, 208
metal content, 36, 59, 60, 61, 74, 83, 101, 169, 172, 175
metal ions, 80, 138
metal salts, 14
metalloids, 13, 162, 231
metallophytic, viii, ix, 73, 74, 81, 82, 92, 95, 96, 99, 100, 101, 104
metals, viii, xi, 9, 12, 13, 14, 21, 34, 36, 38, 40, 41, 42, 47, 52, 53, 55, 59, 61, 72, 73, 74, 75, 77, 78, 80, 81, 82, 84, 86, 97, 98, 99, 100, 101, 110, 111, 112, 113, 123, 124, 133, 134, 161, 162, 163, 164, 166, 167, 168, 169, 171, 172, 173, 174, 175, 191, 207, 213, 230, 231, 232
methane, xi, 177, 180, 185
Mg^{2+}, 110, 136
microbes, 200
microbial, viii, ix, x, 9, 21, 28, 33, 35, 41, 47, 73, 74, 75, 76, 77, 78, 80, 81, 85, 97, 99, 100, 101, 102, 103, 104, 105, 106, 107, 115, 119, 121, 124, 130, 131, 133, 135, 136, 138, 141, 145, 157, 158, 159, 165, 168, 174, 201, 213
microbial activity, 103
microbial cells, 133
microbial communities, viii, 21, 35, 73, 74, 76, 77, 78, 80, 99, 101, 102, 103, 104, 105, 106, 107, 133
microbial community, viii, 21, 33, 73, 74, 75, 77, 78, 81, 99, 100, 101, 102, 103, 104, 105, 107, 131
microcosms, viii, 3, 22, 31, 33, 34, 40, 48, 52, 57, 59, 61, 68, 70, 73, 83, 84, 86, 87, 88, 89, 97, 98
microenvironments, 81
microhabitats, 104
micromycetes, ix, 119
micronucleus, xii, 203, 210, 211, 213, 214, 215
micronutrients, 133
microorganism(s), ix, x, xi, 3, 4, 8, 14, 21, 22, 28, 29, 30, 34, 35, 36, 39, 46, 49, 77, 100, 103, 115, 119, 120, 129, 133, 135, 137, 138, 143, 144, 145, 149, 151, 152, 154, 155, 157, 158, 169, 174, 177, 178, 180, 181, 184, 188, 191, 198
microscope, 29, 112
microscopy, 61
microwave, 170
migration, 86, 169, 206
military, 209
mine soil, 169, 174
mineralization, 21, 29, 35, 36, 37, 38, 39, 106, 115, 139

mineralized, 233
minerals, 77, 106, 110, 164, 232
mines, viii, 51, 59, 124, 133
mining, viii, xi, 54, 73, 74, 81, 82, 86, 96, 104, 120, 122, 164, 169, 173, 177, 179, 197, 207, 212
Missouri, 223, 233
mites, 27
mitosis, 214
mitotic, 205, 206
mixing, x, 84, 144, 145, 154, 155, 189
mobility, 18, 34, 40, 78, 107, 110, 112, 113, 114, 164, 166, 169, 172, 173, 174, 219, 224, 226
modeling, vii, 14, 192
models, 8, 14, 17, 18, 19, 21, 26, 33, 40, 43, 214
moisture, 84
moisture content, 84
molecular oxygen, 120
molecular weight, 122, 129, 213
molecules, 120, 168, 188
mollusks, 27
molybdenum, 165
monograph, 52
monomer, 122
Monte Carlo, 17, 23
morphology, 135
mortality, viii, 2, 9, 28, 32, 35, 36, 39, 45, 52, 131
Moscow, 200
mother cell, 214
movement, viii, 18, 26, 40, 52, 167, 171
mRNA, ix, 74, 102
mucosa, 213
multiple regression, 225
multivariate, 93, 175, 229, 234
mutagen, 213
mutagenesis, 212, 213, 214
mutant, 131, 139, 140
mutation(s), xii, 203, 205, 206, 207, 210, 211, 212, 214
mycelium, 124
mycobacteria, 157
Mycobacterium, 121, 138, 139, 145, 157

N

Na^+, 110, 136
Namibia, 230
naphthalene, 135, 136, 148, 149, 215
narcotics, 30, 44
National Research Council (NRC), 6, 46
native plant, 59
native species, 66
NATO, 70, 72
natural environment, 168, 178
natural selection, 209
nematode(s), 66, 72
Netherlands, x, 5, 6, 7, 10, 12, 24, 47, 49, 70, 143, 144, 176, 220, 225, 226, 231, 234
network, viii, 52, 55
neurological disease, 209
neurological disorder, 204
New England, 137
New York, 40, 48, 49, 115, 116, 141, 157, 173, 174, 199, 200, 201, 230, 231, 233, 234
New Zealand, 45
nickel, 47, 80, 101, 103, 162, 167, 173, 174, 231
nitrate, x, 4, 119, 122, 123, 124, 126, 127, 133, 167, 170, 212
nitric acid, 169, 170
nitrification, 21, 31, 76
nitrogen, xi, 29, 34, 46, 92, 103, 122, 124, 133, 135, 157, 175, 177, 178, 185, 197
N-mineralization, 21
non-linear, 53
non-renewable, xi, 161
normal curve, 224
normal distribution, 220, 221, 224, 227
North America, 29
nucleation, 198
nuclei, 205
nucleic acid, 159
nucleotides, 79
null hypothesis, 90, 96
nutrient cycling, 74
nutrient flow, viii, 52, 54
nutrients, x, 23, 26, 27, 75, 122, 135, 136, 144, 145, 154, 155, 162, 197
nutrition, 163, 175

O

observations, 18, 65, 66, 76, 97, 206
octane, 149
oil, vii, ix, 1, 14, 18, 26, 27, 32, 33, 36, 38, 39, 42, 48, 49, 53, 56, 80, 83, 84, 119, 121, 131, 135, 144, 145, 146, 148, 149, 151, 157, 158, 161, 167, 173, 185, 197
oils, ix, xi, 66, 71, 81, 99, 109, 111, 112, 115, 144, 148, 149, 158, 161, 169, 171, 172, 176, 189, 234
optical density, 4, 136
optimization, 158, 159, 233
ores, 31, 82, 164, 171
organic C, ix, 74, 92, 93, 94, 95, 99, 100
organic chemicals, 189
organic compounds, 9, 137, 207, 208
organic matter, viii, ix, xii, 9, 18, 20, 26, 27, 28, 52, 54, 60, 101, 109, 110, 112, 113, 114, 164, 166,

167, 168, 170, 171, 172, 173, 174, 176, 178, 181, 184, 185, 188, 197, 217, 219, 220, 224, 225, 226, 227, 229
organism, xi, 2, 8, 13, 14, 27, 29, 33, 37, 38, 44, 56, 130, 161
organization(s), viii, 3, 22, 51, 52, 57
Organization for Economic Cooperation and Development (OECD), vii, 1, 11, 27, 28, 29, 30, 32, 34, 37, 46, 47
orientation, 62, 68
osmotic, 133
outliers, 221, 222, 223, 224, 227, 229, 230
oxidants, 121
oxidation, ix, x, 119, 120, 138, 144, 149, 150, 159, 178
oxidative agents, 125
oxidative damage, 120
oxidative stress, x, 119, 120, 122, 128, 130, 131, 138, 140, 213
oxide(s), ix, xii, 45, 109, 110, 112, 113, 167, 170, 171, 217, 224, 225
oxygen, ix, xi, 119, 120, 177, 185
oxyhydroxides, 226

P

pain, 24
parameter, 18, 19, 39, 198, 224
Paris, 46, 47, 48, 199
particles, xi, 27, 28, 29, 177, 178, 179, 180, 188, 189, 191, 193, 194, 195, 196, 197, 198, 199, 200
partition, 87, 222
passive, 111
pasture(s), viii, 51, 54, 55, 61, 63, 64, 66, 102, 113, 115
pathogenic, 179
pathways, 6, 7, 145, 150, 159, 164, 207
Pb, viii, ix, 35, 51, 59, 60, 61, 65, 73, 74, 83, 84, 86, 87, 88, 90, 91, 92, 97, 98, 99, 100, 101, 105, 107, 113, 162, 163, 164, 165, 166, 167, 168, 169, 172, 174, 220, 221, 222, 225, 226, 227, 228, 229, 234
peat, 176
peracetic acid (PAA), 124, 125
percentile, 10, 11, 12, 13, 33, 220
percolation, vii
performance, 20
peripheral blood, 213
permeability, 167, 219
peroxidation, x, 119
peroxide, 120, 122, 124, 131, 133, 140
perturbation, 53, 66, 69, 79
pest management, 212
pesticide, 45, 113, 115, 212
pesticides, vii, xii, 6, 16, 18, 42, 43, 110, 112, 113, 115, 197, 203, 207, 208
petroleum, x, 143, 144, 145, 153, 157
petroleum products, 158
pH, 9, 26, 35, 54, 65, 67, 78, 81, 110, 111, 112, 113, 114, 129, 133, 134, 138, 141, 165, 166, 167, 168, 170, 171, 172, 173, 174, 219, 224
pH values, 111, 133, 167
pharmaceutical(s), 6, 79
phenol, 29, 121, 128, 137, 139, 141
phenotype, 140
phosphatases, 29, 34
phosphate(s), 78, 135, 136, 167, 170, 194, 195, 196, 197, 199
phosphorus, 133, 164, 178, 194, 197
photoelectron spectroscopy, 112
physical properties, viii, xi, 52, 161, 171
physicochemical, 16, 18, 26, 34, 35, 38, 40
physicochemical properties, 26, 175
physics, 62
phytoplankton, 178, 188
phytoremediation, 59, 64
phytostabilizing, viii, 52, 61
phytotoxicity, 28, 30, 48, 110, 112, 115, 211
pigs, 72
Pisum sativum, 214
planning, 56, 65
plant bioassays (PBA), xii, 203
plants, viii, ix, xi, xii, 3, 4, 5, 8, 19, 21, 22, 27, 28, 29, 30, 33, 35, 36, 41, 46, 47, 49, 52, 55, 56, 59, 66, 68, 69, 71, 74, 82, 99, 109, 112, 113, 115, 144, 162, 163, 164, 166, 167, 169, 171, 174, 177, 178, 188, 191, 197, 198, 203, 204, 205, 206, 207, 208, 209, 211, 212, 213, 232, 233
plasmid, 80, 139
pollen, xii, 203, 205, 206, 213, 214
pollen tube, 213
pollutants, viii, ix, 2, 5, 15, 19, 24, 25, 26, 34, 36, 38, 39, 40, 44, 51, 52, 54, 56, 61, 69, 119, 120, 133, 137, 145, 162, 163, 167, 172, 175, 178, 179, 180, 181, 182, 193, 195, 196, 197, 198, 199, 209, 211, 214
pollution, viii, x, xi, 2, 8, 17, 22, 25, 26, 30, 33, 40, 44, 51, 52, 53, 54, 57, 62, 69, 73, 75, 76, 77, 106, 110, 114, 115, 119, 123, 130, 133, 134, 135, 137, 144, 152, 161, 163, 164, 166, 169, 171, 177, 178, 179, 182, 184, 185, 188, 189, 191, 197, 198, 200, 204, 207, 213, 218, 222, 225, 227
polyacrylamide, 85, 122, 123, 127
polycyclic aromatic hydrocarbons (PAHs), xii, 13, 41, 43, 49, 138, 157, 158, 145, 158, 203, 207
polymerase, 84, 85, 105, 159

polymerase chain reaction (PCR), viii, 73, 75, 84, 85, 86, 97, 102, 104, 105, 129, 130, 137, 141, 152, 159
polyploidy, 206
polysaccharides, 80
poor, 23, 54, 168, 171, 191
population, viii, ix, 9, 22, 23, 25, 28, 29, 35, 43, 52, 57, 59, 74, 75, 81, 87, 95, 100, 169, 197, 213, 221, 222, 223, 224, 227, 228, 229
population density, viii, 52
population growth, 28
porosity, 81, 219
Portugal, 71
positive correlation, 76
potassium, 133
potato, 212
power, 182
precipitation, 80, 110, 167, 218
prediction, 15, 77, 112
pressure, x, 8, 61, 98, 110, 119, 130, 188
prevention, 179
priming, 157
probability, xii, 6, 12, 22, 23, 33, 75, 198, 217, 220, 223, 224, 227, 229, 231, 233
probability distribution, 12, 23
producers, 27, 30
production, 4, 34, 68, 78, 110, 121, 122, 123, 124, 127, 133, 171, 204, 205
productivity, 53
program, 47, 53, 70, 212, 223, 227, 233
prokaryotes, 74, 105, 121, 128
prokaryotic, 102
promote, 164
protective role, 140
protein(s), 34, 80, 122, 128, 129, 131, 139, 188
protocol(s), 2, 3, 6, 10, 16, 20, 23, 29, 30, 41, 54, 56, 60, 61, 81, 102, 204, 206, 208
Pseudomonas, 130, 135, 136, 140, 145, 146, 149, 158, 160, 181
purification, x, xi, 139, 143, 159, 177, 178, 179, 180, 181, 182, 188, 189, 191, 192, 197, 198, 200
pyrene, 207, 215
pyruvic, xi, 177, 185

Q

qualifications, 76
quality assurance, 47, 56
quinones, 99

R

race, 174
radiation, 205, 208, 209, 212, 214
radio, 18, 209
radioactive contamination, 214
radioactive waste, 214
radionuclides, xii, 203, 208, 209
radius, 162, 168
rain, 64
rainfall, 33
rainwater, 31
range, ix, x, 9, 10, 12, 13, 22, 30, 32, 47, 74, 77, 78, 79, 80, 95, 96, 109, 136, 137, 144, 145, 160, 163, 193, 194, 198, 218, 220, 221
raw materials, 230
reactive oxygen, ix, 119, 120
reactive oxygen species (ROS), ix, 119, 120
reagents, 170, 189, 193
reality, 16, 21, 53, 62, 230
receptors, vii, 1, 6, 7, 8, 15, 16, 17, 24, 40
recognition, 137
recombination, 212
recovery, 22, 76, 102, 159, 207
recycling, 66
redistribution, 112, 234
redox, 167
reduction, 11, 19, 20, 78, 84, 99, 115, 120, 149, 178, 197
redundancy, 27, 28, 210
reference system, 54
refining, 16, 19
regional, 7, 220, 229, 230
regosols, viii, 51, 54
regression, xii, 3, 26, 31, 217, 220, 225, 226, 227, 229, 230
regression analysis, 230
regression equation, xii, 217, 220, 225, 226, 227, 229
regression line, 227
regulation(s) , 2, 5, 6, 8, 12, 16, 20, 24, 33, 120, 144, 160, 166, 218
regulatory bodies, 5
regulatory framework, 32
rehabilitation, 171
relationship(s), 2, 8, 9, 10, 31, 40, 54, 60, 62,66, 75, 76, 86, 87, 93, 95, 100, 159, 163, 174, 225, 226, 230
relevance, 2, 7, 9, 17, 22, 27, 28, 30, 40
reliability, 9, 16, 18, 219
remediation, vii, 3, 11, 24, 25, 26, 32, 33, 45, 49, 51, 53, 55, 72, 81, 104, 115, 169, 175, 223
remediation procedures, 115
repair, 209, 212

reparation, 198
repression, 158
reproduction, 2, 3, 4, 9, 22, 26, 28, 44, 46, 115
residential, 7
residues, 27, 56, 68, 112, 163, 164, 171
resilience, 75, 76, 103, 104
resin, 104, 145
resins, 145, 146, 158
resistance, x, 31, 41, 75, 76, 78, 80, 105, 106, 120, 122, 130, 131, 136, 140, 209
resolution, ix, 106, 119, 122, 127, 128
resource availability, 100
resources, xii, 47, 56, 76, 217, 219, 226, 230
respiration, 4, 28, 29, 31, 34, 78, 79, 104, 151
respiratory, 75, 99, 121
respiratory rate, 75
retention, 62, 110, 111, 113, 114, 167, 171, 219
reverse transcriptase, 85
rhizosphere, 80, 99, 102, 103, 106, 112, 160
Rhodococcus, x, 143, 144, 145, 146, 149, 150, 152, 153, 154, 157, 158, 159, 160
ribonucleic acid, 79
risk, vii, viii, 1, 3, 6, 7, 8, 9, 10, 11, 12, 13, 14, 15, 16, 19, 20, 22, 23, 24, 25, 26, 27, 28, 29, 32, 33, 34, 36, 37, 39, 40, 42, 43, 48, 49, 50, 52, 54, 63, 113, 114, 115, 157, 163, 173, 197, 231
risk assessment, 3, 6, 8, 9, 11, 12, 13, 14, 15, 16, 20, 22, 23, 24, 26, 28, 29, 39, 40, 42, 43, 48, 50, 157, 163, 173, 231
River Po, 200
rivers, 62, 176, 178, 179, 181, 184, 197, 198, 200
RNA, 79, 85, 92
rodents, 207
rolling, 85
room temperature, 170
root elongation, 49
Royal Society, 102
runoff, 55, 59, 65, 66, 114, 115, 164, 173, 184
rust, 163

S

safety, 47, 194
salination, 71
salinity, 66, 171, 176
salmon, 48
Salmonella, 30, 41, 130, 133, 140, 207
salt(s), 13, 14, 45, 62, 79, 168, 170, 189
sample, 2, 11, 24, 25, 26, 31, 32, 33, 39, 40, 59, 75, 82, 83, 85, 86, 95, 97, 129, 135, 158, 159, 170, 193, 194
sampling, viii, 52, 54, 61, 80, 83, 92, 103, 219, 222, 227
sanitation, 200
saturated hydrocarbons, 146
scarcity, 164, 172
schema, 20
science, 53, 218
scores, 220
sea-level rise, 164
search, 56, 130
sediment(s), vii, xii, 1, 33, 41, 43, 44, 45, 47, 56, 69, 80, 83, 97, 123, 124, 126, 131, 132, 133, 134, 135, 137, 159, 173, 203, 208, 209, 212, 231, 234
sedimentation, 78, 178
seed(s), viii, 52, 53, 61, 64, 67, 68, 69, 211
seedlings, 4, 34, 205, 212
selecting, 26, 27, 230
selenium, 165
sensitivity, 2, 9, 10, 11, 12, 20, 21, 25, 26, 27, 30, 32, 33, 35, 38, 40, 41, 75, 77, 128, 131, 141, 207, 208
separation, 103, 104, 170, 188
sequencing, 130, 146
series, 32, 37
settlements, 164
severity, 11
sewage, xi, 41, 48, 74, 76, 80, 103, 106, 107, 110, 163, 164, 165, 168, 171, 172, 173, 174, 175, 176, 177, 179, 184, 197
shaping, 100
sheep, 54, 71, 82
Shigella, 159
shoot, 4
shrimp, 140
shrubs, 54
silicate, 112
silver, 59
similarity, 86, 87, 89, 90, 95, 96, 97, 100, 129
sister chromatid exchange, 207
sites, vii, viii, ix, xi, 5, 9, 16, 25, 28, 34, 44, 45, 52, 53, 54, 56, 59, 61, 62, 66, 69, 71, 72, 74, 77, 82, 93, 95, 96, 101, 144, 155, 167, 168, 172, 177, 178, 180, 181, 184, 189, 191, 193, 195, 196, 197, 198, 207, 208, 209, 210, 211, 212, 214, 226, 227, 231, 232
skewness, 222
skills, vii
Slovakia, 203
sludge, ix, 41, 48, 74, 76, 80, 103, 106, 107, 110, 119, 121, 131, 157, 163, 164, 165, 168, 171, 172, 173, 174, 175, 176, 178, 191, 192, 200, 201, 211
smelting, 74, 120, 173, 207
smoke, 179
sodium, 47, 77, 133, 208, 211
software, 83, 84, 85, 86, 222, 223
soil erosion, 114

soil organic C, 99, 100
soil pollution, vii, viii, 8, 17, 22, 30, 40, 51, 52, 57, 169, 171, 225, 227
solid phase, 167
solid waste, viii, 51, 53
solubility, 8, 110, 133, 141, 167, 168, 219
solvent(s), vii, 26, 207
somatic mutations, 212
sorption, ix, 29, 109, 110, 115, 167, 168, 173, 176, 232
South Carolina, 220, 233
Spain, xi, 1, 7, 8, 13, 16, 24, 49, 51, 54, 62, 63, 69, 71, 72, 113, 161, 164, 165, 169, 171, 172, 173, 175, 176, 217, 220, 221, 222, 223, 225, 226, 231
specialization, 65
speciation, 14, 47, 107, 111, 234
species, ix, 2, 3, 9, 10, 11, 12, 13, 14, 15, 18, 20, 21, 22, 27, 28, 29, 30, 31, 32, 35, 36, 40, 53, 55, 56, 59, 61, 62, 64, 65, 66, 67, 68, 69, 70, 71, 72, 82, 92, 95, 99, 105, 107, 109, 115, 119, 120, 121, 123, 130, 131, 137, 141, 160, 169, 178, 204
species richness, 66
specific gravity, 161
specific tax, 79
spectroscopy, 175
spectrum, 21, 129, 179, 180
speed, 64
spore, ix, 119, 122, 123, 124, 127
St. Louis, 200
stability, viii, ix, xii, 53, 73, 74, 75, 76, 81, 96, 97, 101, 103, 106, 112, 116, 138, 167, 176, 203, 204
stabilization, 59, 115, 191
stages, 4, 5, 29, 46, 107, 180
standard deviation, 13, 65, 221
standards, vii, xi, 1, 2, 6, 11, 12, 13, 15, 48, 49, 56, 161, 223, 233
stars, 130
starvation, 112, 122, 133, 159
statistical analysis, 17, 38, 56, 86, 229
statistics, xii, 217, 220, 222, 223, 224, 229
steel, 174
sterile, 84
storage, vii, 159, 214
stormwater, 173, 174, 179
strain, ix, 28, 30, 119, 121, 122, 123, 124, 126, 131, 132, 133, 134, 138, 139, 149, 153, 160
strategies, 14, 22, 65, 66, 115
stratification, 81
stratified sampling, 61
Streptomyces, 79, 80, 101, 105
stress, ix, x, 30, 73, 74, 76, 78, 81, 84, 86, 97, 99, 101, 105, 119, 120, 122, 125, 127, 128, 130, 131, 133, 138, 140, 213
stressors, x, 119, 124
striatum, 61
structural changes, 105
subgroups, 21
substitution, 167
substrates, viii, 51, 54, 63, 65, 78, 105, 120, 122, 149, 212
subsurface, vii, 113, 164, 167
sulfate, 113
sulphate, 136, 167
Superfund, 144
superoxide dismutase, 120, 131, 138
supply, 197, 199
suppressor, 140
surface area, 167
surface tension, 185
surface water, vii, viii, xi, 8, 51, 54, 55, 62, 69, 176, 178, 180, 186, 187, 189, 191, 193, 195, 196, 197, 198
surfactant(s), xi, 49, 121, 136, 177, 178, 180, 185, 187, 189, 191, 197
surrogates, 9
surveillance, 69
survival, 4, 5, 28, 35, 36, 38, 39, 106, 131, 134, 140
survival rate, 131
sustainability, 40, 76, 112, 115
sustainable development, 53, 173
Sweden, 8, 49
swelling, 211
Switzerland, 7, 207, 231
symbols, 148
symmetry, 222
symptoms, 115
synthesis, 27
systems, viii, x, 3, 14, 22, 29, 31, 32, 33, 45, 48, 51, 52, 53, 54, 56, 61, 62, 65, 68, 69, 76, 80, 143, 144, 160, 169, 178, 193, 197, 199, 201, 206, 207, 208, 209

T

tanks, vii, 191, 209
targets, 5
taxonomic, 10, 11, 20, 21, 22, 28, 29, 30, 33, 35, 37, 38, 39, 40, 79
taxonomy, 28, 137
technology, 174
telophase, 206, 211
temperature, 34, 45, 138, 140, 170
temperature gradient, 34
temporal, 14, 15, 18, 52, 53, 62, 72
tension, 185
test data, 27

test procedure, 34
test statistic, 35
tetrad, 204
theory, 76, 78, 105
threats, 110
threshold(s), vii, 1, 3, 5, 10, 12, 20, 21, 23, 24, 25, 32, 33, 39, 40, 222, 233
time, vii, x, 2, 3, 4, 5, 14, 17, 18, 19, 20, 22, 26, 28, 31, 38, 45, 56, 57, 69, 78, 86, 87, 97, 103, 104, 136, 143, 144, 149, 159, 162, 163, 164, 165, 176, 180, 206, 211, 218
timing, 80
tissue, 61
tobacco, 212
Tokyo, 115, 157, 199
topsoil, viii, 52, 54, 61, 82, 83, 102
total petroleum hydrocarbons (TPH), x, xi, 35, 144, 145, 153, 154
toxic effect, viii, 2, 13, 25, 27, 35, 38, 39, 56, 73, 105, 131, 134, 201
toxic metals, 72
toxicity, vii, viii, xi, 1, 2, 3, 4, 5, 6, 9, 10, 11, 12, 14, 15, 17, 20, 21, 22, 23, 24, 25, 26, 27, 28, 29, 30, 31, 32, 33, 34, 35, 36, 38, 39, 40, 41, 42, 43, 44, 45, 46, 47, 48, 49, 50, 51, 52, 56, 57, 70, 74, 77, 78, 102, 103, 104, 130, 131, 138, 140, 161, 162, 163, 168, 169, 211, 213
toxicology, 2, 47, 213
toxins, 197, 204
trace elements, 55, 173, 174, 231, 233
tracking, 18
trade, xi, 161
trajectory, 105
transcription, ix, 74, 102, 176
transcription factors, 176
transference, 6
transformation, 3, 4, 29, 31, 34, 37, 46, 93
transition, 227
translocation, 112
transport, viii, 8, 33, 47, 52, 59, 80, 164
trees, 64
trend, 87
trial, 23
trichloroethylene, 208
triggers, 15
turbulence, xi, 177, 179, 180, 185, 187, 189, 197
turbulent, xi, 177, 180, 181, 184, 185, 189, 193, 198
Turkey, 167, 176
turnover, 100

U

UK, viii, 24, 45, 73, 84, 96, 165, 174
Ukraine, 211
uncertainty, 10, 11, 12, 16, 17, 19, 20, 21, 53, 75, 222, 223
UNESCO, 53, 71
uniform, 218
United Nations, xi, 161
United Nations Environment Program (UNEP), xi, 161
United States, x, 13, 44, 143, 144, 145, 227, 231, 232, 234
uranium, 178, 212
urban centers, 198
urea, 85
USEPA, vii, 1, 3, 8, 42, 49, 165, 166, 170, 234

V

validation, 11, 44
validity, 10, 56
values, xi, xii, 5, 7, 8, 9, 10, 11, 12, 13, 15, 16, 17, 18, 20, 21, 24, 26, 32, 33, 35, 36, 37, 39, 42, 47, 61, 65, 67, 86, 89, 93, 95, 96, 98, 100, 111, 113, 128, 133, 134, 161, 167, 171, 172, 175, 176, 217, 218, 219, 220, 221, 222, 223, 224, 225, 226, 227, 228, 229, 230, 231, 233
vapor, 8
variability, 3, 12, 15, 16, 17, 18, 21, 32, 38, 40, 77, 79, 86, 209, 218, 219, 220, 223, 224, 229, 233
variable(s), viii, 3, 4, 5, 14, 17, 19, 23, 26, 30, 49, 52, 54, 57, 65, 79, 93, 100, 157, 163, 171, 176, 220
variance, 3, 32, 34, 38
variation, 10, 12, 14, 17, 39, 65, 66, 219, 220, 227, 229
vegetation, viii, 27, 52, 54, 55, 64, 65, 66, 72, 81, 82, 83, 91, 95, 96, 99, 115, 169, 173, 174
vertebrates, 8, 15, 16, 18, 19, 20, 29
vessels, 170
viable-but-non-culturable (VBNC), 151, 152, 159
Vicia faba, xii, 203, 205, 210, 211
vineyard, ix, 109, 110, 111, 113, 114, 115, 176
vinylchloride, 208
visualization, 93
vitriculture, ix, 109
volatilization, 29

W

walking, 184
Washington, 42, 46, 47, 49, 232, 234
waste disposal, 53, 62, 64, 69
wastewater(s), ix, 32, 47, 119, 121, 106, 131, 163, 165, 172, 174, 191, 198, 199, 209, 211, 215, 227

wastewater treatment, 106, 131, 163, 172, 191, 198
water quality, 14, 48, 180
water supplies, vii
wavelengths, 194
weakness, 25
web, 210
Western Europe, 212
wetlands, 69
wheat, 66, 112
wildlife, 11, 17, 18, 27, 204
wind, 115
winter, 197
wood, 212
workers, 213
worms, 3

X

xenobiotics, 115, 121
X-ray, 112, 210
X-ray diffraction (XRD), 112
X-ray photoelectron spectroscopy (XPS), 112

Y

yeast, 139, 153
yield, 173

Z

Zea mays, xii, 203, 206, 207, 208, 212
zinc (Zn), viii, ix, 14, 35, 43, 44, 45, 48, 51, 59, 60, 61, 65, 66, 67, 68, 70, 71, 72, 73, 74, 77, 78, 79, 80, 81, 82, 83, 84, 86, 87, 88, 90, 91, 92, 97, 98, 99, 100, 101, 103, 104, 105, 107, 112, 113, 134, 162, 163, 164, 165, 166, 167, 168, 169, 171, 172, 173, 174, 176, 194, 196, 207, 213, 223, 225, 226, 231, 232, 234
zinc oxide, 45
zooplankton, 188